AF775595

Springer Series in Operations Research and Financial Engineering

Series Editors

Thomas V. Mikosch, Køebenhavns Universitet, Copenhagen, Denmark

Sidney I. Resnick, Cornell University, Ithaca, NY, USA

Bert Zwart, Centrum Wiskunde & Informatica, Amsterdam, The Netherlands

Ton Dieker, IEOR, Columbia University, New York, USA

George Lan, H M Stewart School of ISyE, Georgia Institute of Technology, Atlanta, GA, USA

Editorial Board Members

Torben G. Andersen, Northwestern University, Evanston, USA

Dmitriy Drusvyatskiy, University of Washington, Seattle, USA

Avishai Mandelbaum, Technion - Israel Institute of Technology, Haifa, Israel

Jack Muckstadt, Cornell University, Ithaca, USA

Per Mykland, University of Chicago, Chicago, USA

Philip E. Protter, Columbia University, New York, USA

Claudia Sagastizabal, IMPA – Instituto Nacional de Matemáti, Rio de Janeiro, Brazil

David B. Shmoys, Cornell University, Ithaca, USA

David Glavind Skovmand, Køebenhavns Universitet, Copenhagen, Denmark

Josef Teichmann, ETH Zürich, Zürich, Switzerland

The Springer Series in Operations Research and Financial Engineering publishes monographs and textbooks on important topics in theory and practice of Operations Research, Management Science, and Financial Engineering. The Series is distinguished by high standards in content and exposition, and special attention to timely or emerging practice in industry, business, and government. Subject areas include:

Linear, integer and non-linear programming including applications; dynamic programming and stochastic control; interior point methods; multi-objective optimization; Supply chain management, including inventory control, logistics, planning and scheduling; Game theory Risk management and risk analysis, including actuarial science and insurance mathematics; Queuing models, point processes, extreme value theory, and heavy-tailed phenomena; Networked systems, including telecommunication, transportation, and many others; Quantitative finance: portfolio modeling, options, and derivative securities; Revenue management and quantitative marketing Innovative statistical applications such as detection and inference in very large and/or high dimensional data streams; Computational economics

Guy Latouche

Markov-Modulated Brownian Motion

The Matrix-Analytic Perspective

Springer

Guy Latouche
Faculté des Sciences
Université libre de Bruxelles
Bruxelles, Belgium

ISSN 1431-8598 ISSN 2197-1773 (electronic)
Springer Series in Operations Research and Financial Engineering
ISBN 978-3-031-98575-1 ISBN 978-3-031-98576-8 (eBook)
https://doi.org/10.1007/978-3-031-98576-8

Mathematics Subject Classification: 60K37, 15A18, 60J22, 60J65, 65C40, 15A24, 60G17, 60J55

© The Editor(s) (if applicable) and The Author(s), under exclusive license to Springer Nature Switzerland AG 2026

This work is subject to copyright. All rights are solely and exclusively licensed by the Publisher, whether the whole or part of the material is concerned, specifically the rights of translation, reprinting, reuse of illustrations, recitation, broadcasting, reproduction on microfilms or in any other physical way, and transmission or information storage and retrieval, electronic adaptation, computer software, or by similar or dissimilar methodology now known or hereafter developed.
The use of general descriptive names, registered names, trademarks, service marks, etc. in this publication does not imply, even in the absence of a specific statement, that such names are exempt from the relevant protective laws and regulations and therefore free for general use.
The publisher, the authors and the editors are safe to assume that the advice and information in this book are believed to be true and accurate at the date of publication. Neither the publisher nor the authors or the editors give a warranty, expressed or implied, with respect to the material contained herein or for any errors or omissions that may have been made. The publisher remains neutral with regard to jurisdictional claims in published maps and institutional affiliations.

This Springer imprint is published by the registered company Springer Nature Switzerland AG
The registered company address is: Gewerbestrasse 11, 6330 Cham, Switzerland

If disposing of this product, please recycle the paper.

À Raphaël, Thomas, Noah et Léa

Preface

My presentation of Markov-modulated Brownian motion is much influenced by my long participation in the development of Matrix-analytic methods (MAMs for short), a branch of algorithmic methods in probability.

Like many things in life, these methods do not have a formal definition, but are easily recognised, and I identify two essential features. Firstly, practitioners of the art heavily rely, in the theoretical approach to stochastic models, on the analysis of their trajectories over time, and new understanding arises from the constant interplay between algebraic manipulations and the physical interpretation of the underlying expressions. Secondly, a constant goal is that solutions should lend themselves to provably efficient and stable numerical algorithms, so that we may associate qualitative information about the behaviour of a system with quantitative evaluation of its performance.

The initial impetus was driven by Marcel Neuts in the 1970s and focused on queueing models, the M/M/1 queue in a random environment serving as the iconic example of a random walk in a strip of $\mathbb{N}^2$. Neuts [116] helped draw the attention of various scientific communities, from applied probabilists to telecommunication engineers and to numerical analysts. Since these beginnings, a considerable amount of research has been conducted, and the scope of Matrix-analytic methods has been much expanded. The emphasis on computability has remained a fixed point, so that new and significant results have been obtained, even where there had been a pre-existent analytical corpus.

In the 1980s, linear fluid systems became popular as models in telecommunication networks design. Also named *Fluid queues* (or FQ), they are used to represent the content of a buffer which evolves linearly in time, at rates that change with the state of a Markovian environmental process. A succession of three papers, Rogers [133], Asmussen [13], and most notably Ramaswami [130] brought these processes in the realm of Matrix-analytic methods. Ramaswami showed how tools and techniques developed for Markov-modulated queueing systems could be recycled and used in the analysis of Fluid queues.

Meanwhile, Markov-modulated models started being used in a variety of application areas, such as economics, insurance risk, financial market stress, and environmental problems, besides the traditional queues, inventories and data communication systems. Over the past 40 years or so, Markov-modulated Brownian motion and régime-switching Lévy processes have been analysed with help of different techniques; my object in this book is to show that the Matrix-analytic approach may be adapted in a straightforward manner to these stochastic models, with just a few adaptations to the peculiar properties of Brownian motion.

More Precisely

I present in Fig. 1 a sample trajectory of the standard Brownian motion. Different sample paths have very different general aspects, but they all share one property: they wiggle up and down like crazy. This chaotic behaviour is hidden behind a rather tame-looking definition:

Definition 1. *The process $\mathcal{B} = \{B(t) : t \in \mathbb{R}^+\}$ is a standard Brownian motion, also known as a Wiener process, if*

1. *it has continuous sample paths a.s.,*
2. *increments over disjoint intervals are independent,*
3. *the difference $B(t+h) - B(t)$ is normally distributed, with mean 0 and variance h, independently of t,*
4. *and $B(0) = 0$ a.s.*

More generally, $\{X(t) : t \in \mathbb{R}^+\}$ is a Brownian motion with parameters μ and σ if it has the form

$$X(t) = X(0) + \mu t + \sigma B(t)$$

or

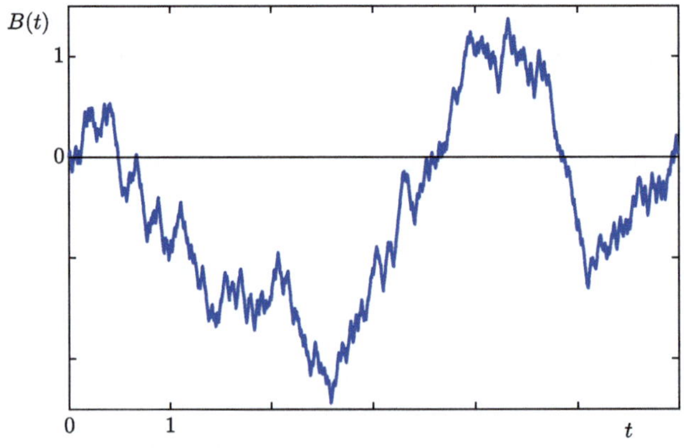

Fig. 1 A sample path of the standard Brownian motion

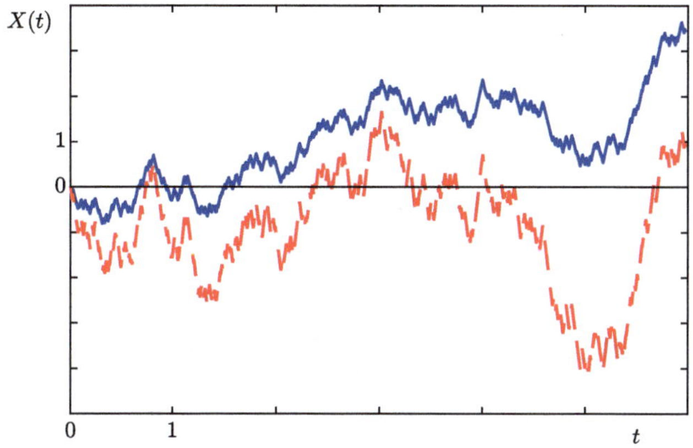

Fig. 2 Sample paths of two Brownian motions with drift $\mu = 1$. The volatility is $\sigma = 1$ for the trajectory in blue, it is $\sigma = 2$ for the one in red

$$X(t) = X(0) + \mu t + \int_0^t \sigma \, \mathrm{d}B(s),$$

where $\{B(t) : t \in \mathbb{R}^+\}$ is a standard Brownian motion independent of $X(0)$. The parameter μ is called the drift, σ is the volatility, and $X(t+h) - X(t)$ has the $\mathcal{N}(\mu h, \sigma^2 h)$ distribution—we use the shorthand $\{X(t) : t \in \mathbb{R}^+\}$ is BM(μ, σ). I show in Fig. 2 a sample trajectory for two BMs with the same drift $\mu = 1$ but different volatility: σ is equal to 1 for the process drawn in blue, and to 2 for the process drawn in red with a dashed line.

The main purpose of Markov-modulated Brownian motion (MMBM in short) is to allow for the parameters of a Brownian motion to evolve in time, in a manner sufficiently simple to allow for detailed analysis. In short, one defines a two-dimensional process $\{X(t), \varphi(t) : t \in \mathbb{R}^+\}$ where $\{\varphi(t) : t \in \mathbb{R}^+\}$ is a continous-time Markov chain on a finite state space, and $\{X(t) : t \in \mathbb{R}^+\}$ behaves like a Brownian motion with parameters that depend on $\varphi(t)$. Formally, one writes

$$X(t) = X(0) + \int_0^t \mu_{\varphi(s)} \, \mathrm{d}s + \int_0^t \sigma_{\varphi(s)} \, \mathrm{d}B(s),$$

with $\{\varphi(t) : t \in \mathbb{R}^+\}$, $X(0)$ and $\mathcal{B}$ being mutually independent. For historical reasons, $\varphi(t)$ is called the *phase* at time t and $X(t)$ is called the *level*.

Figure 3 corresponds to a two-phase MMBM. In Phase 1, the parameters are $(\mu_1, \sigma_1) = (-1, 1)$ and the drift is negative, they are $(\mu_2, \sigma_2) = (1, 1)$ in Phase 2, and the drift is positive. The phase process is such that $\varphi(t)$ is equal to 1 during two thirds of the time, so that the average drift is negative.

Despite the fact that the drifts μ_1 and μ_2 are significantly different, the trajectory of $X(t)$ in Fig. 3 is dominated by the disorderly nature of the Brownian motion, and it is all the more important that one should be able to proceed beyond qualitative features, and to numerically determine performance measures for such systems.

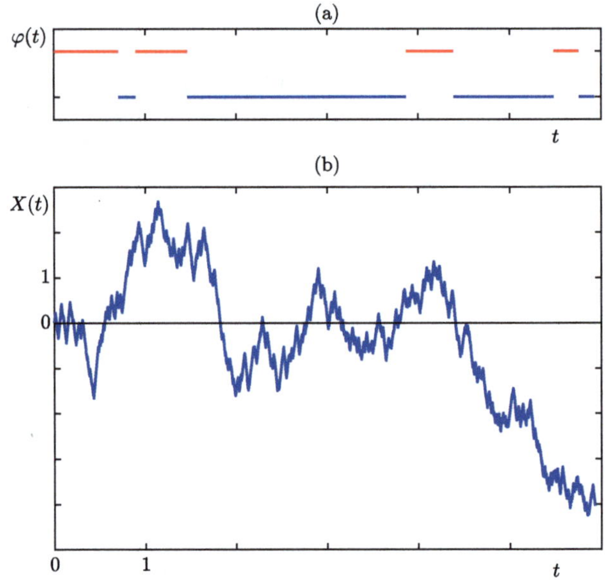

Fig. 3 Sample path of a two-phase Markov-modulated Brownian motion, with parameters $\mu_1 = -1$, $\sigma_1 = 1$, $\mu_2 = 1$ and $\sigma_2 = 1$

Road Map

Chapter 1 is there to set the stage. On the whole, Matrix-analytic methods have been mostly applied to processes for which the level takes integer values (Quasi-birth-and-death processes, M/G/1- and GI/M/1-type Markov chains, ...), or real values with piecewise linear trajectories (Fluid flows, Fluid queues). Readers familiar with that literature will find here some basic facts about Brownian motion, focusing on properties specific to MMBMs. The reader already familiar with Brownian motion will have the opportunity of getting acquainted with the notation that I shall follow.

In this preliminary chapter, I establish a connection with other processes, with Fluid queues in particular. Fluid queues are a family of MMBMs with $\sigma_i = 0$ for all i and have been extensively studied since the 1970s. I briefly mention in Sect. 1.11 that MMBMs with positive volatility are limits of rapidly oscillating Fluid queues, as the oscillation rate goes to infinity. This interpretation of MMBMs allows one to directly call upon properties long established in Fluid queues theory, and to extend them to MMBMs, as Giang T. Nguyen and I showed in the mid-2010s.

I analyse in Chap. 2 the epochs when the level is for the first time equal to $x \neq 0$. The sequence of values of the phase at such times form the *ladder process*, *ascending* for $x > 0$, *descending* for $x < 0$. The ladder processes are Markov processes on the same state space as the phase, and their generators are keys to the analysis of MMBMs, leaving their mark in all chapters. I show that the generators are characterised as solutions of matrix-quadratic equations that have a particularly simple form in two cases: for purely Brownian MMBMs, with $\sigma_i > 0$ for all phase i, and for Fluid flows with $\sigma_i = 0$ for all i. My dealing with the general case in Sect. 2.5 has two purposes: (a) it makes it clear that one may follow the same approach as in the two special cases, and (b) the messy appearance of the final equation is justification for my decision in all subsequent chapters to assume that $\sigma_i > 0$ for all i.

The question at the beginning of Chap. 3 is the following: given three levels, the initial level at time 0, one higher and one lower, does the MMBM reach the higher level or the lower level first, and what is the phase at the time when the process escapes from the interval? It turns out that the answer is expressed as the solution of a linear system, non-singular if the stationary overall drift of the level is strictly positive or strictly negative, but singular if the stationary drift is equal to 0. This is an ubiquitous feature of MMBMs (as well as QBD Markov chains, or FF processes) and I show in Sect. 3.2 how one may fully determine the solution in that case. At the end of the chapter, I determine the distribution of the phase at the time of escape from the upper boundary, for an MMBM which is not allowed to cross the lower boundary.

In Chap. 4, I deal with *local time*, a feature of primary importance in the analysis of MMBMs. I begin with establishing basic properties that connect the local time with the ladder processes analysed in Chap. 2; like in Chap. 3, the analysis turns out to be more involved when the stationary drift is equal to 0. In the early developments of Matrix-analytic methods, place of pride was given to the matrix-geometric stationary distribution of regulated QBDs and GI/M/1-type Markov chains, as witnesses the title of Neuts [116]. Over the years, however, the focus shifted to the expectation of time spent away from level 0 between successive visits to level 0. This expected time plays an important role here; its definition, however, has to be revised to take into account sample path properties of the Brownian motion.

Next, the attention turns to MMBMs with one regulated boundary at level zero, or two boundaries, at 0 and b. These would be models of queueing systems with an infinite buffer, or a finite buffer of finite capacity b. Their stationary distribution, when it exists, is determined in Chap. 5 and, using a Markov-regenerative approach, I show that the stationary density has a matrix-exponential form in the infinite buffer case, and is a mixture of two matrix-exponentials in the finite buffer case. That analysis is easily extended to MMBMs with reactive boundaries, that is, processes which are allowed to undergo a change of behaviour upon hitting a boundary.

Chapter 6 is devoted to a brief presentation of computational procedures to solve the matrix-quadratic equations obtained in Chap. 2. The algorithms are iterative in nature, and the simplest one is by functional iteration, with linear convergence if the stationary drift is different from 0. By this time, one should not be surprised that things happen if the stationary drift is equal to 0 and, indeed, convergence is sublinear in that case. A nice feature of functional iteration, and one to which I am particularly partial, is that the successive approximations come with a simple physical interpretation in terms of a constrained behaviour of the MMBM under study. The second algorithm that I discuss is related to Newton's iteration and, as one might expect, it has quadratic convergence—that is, if the stationary drift is different from 0.

The two algorithms are very easily implemented and are quite efficient in general. Functional iteration has been used for all the illustrations in this book, and most of the time has been spent not in getting numbers, but in thinking up the illustrations.

I conclude with brief mentions of possible extensions of my work, of various application areas, and of Markov-modulated Lévy processes.

Each chapter ends with a few bibliographic notes and comments, where credit and additional references are given, to put the chapter material in proper perspective with the published literature. I give a list of the main notation used in the book, and an index to complete this volume.

Prerequisites are advanced calculus and linear algebra, at the level usually taught in undergraduate curricula, and a course in stochastic processes at the level of Asmussen [14] or Resnick [132], I point the reader to specific references when additional details are required. I provide in a few places, under the heading *Parenthesis*, material that is useful in my opinion, even though it is somewhat tangential to the main topic. Such material might have been collected in appendices, but I prefer to give it in the main body of the text, close to where it is mentioned for the first time, so as to keep the rhythm of the presentation. Parentheses are indicated in the Table of Contents.

Acknowledgements

I would not have written this book had it not been for several colleagues with whom I have interacted over the years. Among them, Marcel F. Neuts has played a major role. He offered me a post-doctoral position 50 years ago, at a time when his development of Matrix-analytic methods was in full swing. Marcel made me realise how much fun it is to look for the physical meaning of one's equations.

I would be remiss indeed, if I did not mention Vaidyanathan Ramaswami, he of the fertile imagination, who has built efficient models to tackle engineering problems, who writes Carnatic music compositions, and who has written beautiful applied probability papers. Our collaboration has spanned several years and has taken many forms, including the writing of a popular book. Ramaswami is the guy who gave me the initial motivation to investigate Markov-modulated Brownian motions.

It is a pleasure to recognise the influence of Dario Bini and Beatrice Meini, for I have found much stimulation in their company and I am always amazed by the depth of their bag of tricks. Many have been the occasions when we have progressed by relay, switching from an argument from linear algebra to one from stochastic analysis and back again.

Giang T. Nguyen, who held for a couple of years a post-doctoral position at the Université libre de Bruxelles, has accompanied my first steps in applying Matrix-analytic thinking to Brownian processes. Our collaboration was very enjoyable. Among much nice work, we investigated together the fast-switching Fluid flows discussed in the first chapter of this book. In this context, let me mention Oscar

Peralta, who explained to me the argument that Giang and him had set up to prove strong convergence.

Matthieu Simon, a doctoral student of mine, worked on Markov-modulated Lévy processes under the co-supervision of Griselda Deelstra, and the two of them helped me better understand how one deals with Brownian processes. I also got a better feel for the classical approach to MMBMs through a few animated discussions with Jevgenijs Ivanovs. I am grateful to Marie-Ange Remiche, who read my first chapters, and gave me several insightful comments.

I owe a huge debt of gratitude to Peter Taylor, friend and co-author of long standing. We had started a new investigation not long before Covid-19 raised its ugly head, and over the next few years I have experienced my share of difficult episodes. Peter has always been there, seventeen thousand kilometres away but close by really, thanks to the magic of the Internet. During that time, week in, week out, we have discussed our joint work, and told each other of our new projects, but also just plain chatted about life, the universe, everything. Peter's presence has been invaluable.

Finally, I shall address the most personal thanks to my family, for the support that they offered generously and which meant so much.

I thank anonymous readers of the first draft of this book, for their kind comments and their constructive suggestions. I am fortunate to have been accompanied in this journey by Donna Chernyk, Senior Editor, and I gladly acknowledge the financial support that I received over the years from the Fédération Wallonie-Bruxelles (FSR/ARC), the Belgian Fonds de la recherche scientifique (FRS-FNRS), the Australian Research Council (ARC), and the Istituto Nazionale di Alta Matematica (INdAM).

Bruxelles, Belgium Guy Latouche

Competing Interests The author has no competing interests to declare that are relevant to the content of this manuscript.

Contents

Acronyms

a.s.	almost surely, with probability 1
BM	Brownian motion
EM	Estimation-maximisation
FF	Fluid flow
FQ	Fluid queue
GTH	Grassmann, Taksar, and Heyman
iff	if and only if
LF	Linear fluid
LST	Laplace-Stieltjes transform
MAM	Matrix-analytic method
MAP	Markov additive process
MMBM	Markov-modulated Brownian motion
PH	Phase-type
QBD	Quasi-birth-and-death

Chapter 1
Preliminaries

◇ Brownian motion ◇ Running maximum ◇ Regulated Brownian motion ◇ Skorokhod regulation ◇ Sample path properties ◇ Lévy processes ◇ Characteristic exponent ◇ Markov modulation ◇ Killing mechanism ◇ Moments for MMBMs ◇ Fluid flow approximation ◇

We give in this chapter a brief description of Brownian motion and some associated processes: the running maximum and the running minimum, and regulators. We also discuss two of its extensions: Lévy processes and Markov-modulated Brownian motion (MMBM)—the topic of our book. We present a few basic properties, so as to render intuitively plausible, and wholly reasonable, the approach that we adopt throughout the book, that is, to connect physical properties of the various processes with the algebraic structures which we have to manipulate.

This chapter is also the place where we define some of our notation. For instance, the expression $\{X\}$ is shorthand for the stochastic process $\{X(t) : t \in \mathbb{R}^+\}$, it allows us to avoid the systematic repetition of the parameter set $\mathbb{R}^+$. The standard Brownian motion may be represented equally with the shorthand $\{B\}$ as with the expression $\mathcal{B}$ used in Definition 1.

1.1 Brownian Motion

Obviously, one would first and foremost wish to make sure of the existence of Brownian motion, this is verified with Theorem 1.1. Define C to be the space of *continuous functions* on $[0, \infty)$, equipped with the metric

$$d(x, y) = \| x - y \| = \sup_t |x(t) - y(t)|,$$

© The Author(s), under exclusive license to Springer Nature Switzerland AG 2026
G. Latouche, *Markov-Modulated Brownian Motion*, Springer Series in Operations Research and Financial Engineering, https://doi.org/10.1007/978-3-031-98576-8_1

where x and y are two functions in C. Next, define $\mathcal{A}$ as the Borel σ-algebra on C. Finally, define $\{X(t) : t \in \mathbb{R}\}$ on C as $X(t;\omega) = \omega(t)$ for all functions ω in C, this is called the coordinate process on C.

Theorem 1.1. *There exists a unique probability measure W on $(C, \mathcal{A})$ such that the coordinate process on $(C, \mathcal{A}, W)$ is a standard Brownian motion.* □

This is Wiener's theorem. For a proof, see Billingsley [33, Section 8] in the case of continuous functions on $[0, 1]$; to extend it over $\mathbb{R}^+$, one builds countably many independent copies and pastes them together, as shown in Schilling and Partzsch [136, Corollary 3.4].

Although the measure W is unique, that does not mean that the standard Brownian motion is unique. As a simple example, take $\mathcal{B}$ to be a standard Brownian motion and define $\widetilde{B}(t) = -B(t)$ for all t. It is easy to verify that $\widetilde{\mathcal{B}}$ satisfies the criteria enumerated in Definition 1 in the foreword.

Once the existential question is settled, we may define new, interesting, processes. The most immediate is the general Brownian motion, with drift and volatility parameters μ and σ, and arbitrary initial value $X(0)$; it takes the form

$$X(t) = X(0) + \mu t + \sigma B(t), \tag{1.1}$$

where $\mathcal{B}$ is a standard Brownian motion independent of $X(0)$, and $\sigma > 0$; we shall write that $\{X\}$ is BM(μ, σ).

We gave in the foreword the usual definition of Brownian motion. It turns out that there is an even more economical characterisation, see Breiman [43, Chapter 12.1]:

Theorem 1.2. *If $\{X\}$ is a continuous process with stationary independent increments, then $\{X\}$ is a Brownian motion.* □

This is a very nice property, it shows that Brownian motion may be characterised in purely qualitative terms, and it reinforces the uniqueness of Brownian motion among stochastic processes. This is reminiscent of the Poisson process on the real line, which is often defined as having stationary increments, independent over disjoint intervals, with the increment over $(t, t+h)$ having a Poisson distribution with parameter λh for some λ; there is also a purely qualitative definition, where the distributional assumption is replaced with the requirement that trajectories are step functions with unit increment, see Çinlar [49, Chapter 4.1].

Life would be unbearable without some form of strong Markov property, and we may use the following: see Harrison [80, Section I.4].

Theorem 1.3. *Assume that $\{X\}$ is a BM(μ, σ) and assume that T is a stopping time. If $T < \infty$ a.s., then $\{X^*\}$ defined by $X^*(t) = X(T + t) - X(T)$, $t \geq 0$, is a BM(μ, σ) independent of $\{X(s) : 0 \leq s \leq T\}$.* □

The usual Markovian property comes as a special case since any constant is a stopping time.

1.2 Running Maximum, Running Minimum

The running maximum $\{M_X\}$ of a random process $\{X\}$ and its twin the running minimum $\{m_X\}$ are defined as

$$M_X(t) = \sup\{X(s) : 0 \le s \le t\}$$

and

$$m_X(t) = \inf\{X(s) : 0 \le s \le t\},$$

for $t \ge 0$. Their distribution is given below for any fixed t. When t is a random time with an exponential distribution, their distribution is given in Lemma 2.6, a property that will be used repeatedly in this book.

Lemma 1.4. *The maximum $M_B(t)$ of a standard Brownian motion $\mathcal{B}$, the absolute value $|m_B(t)|$ of its minimum, and the absolute value $|B(t)|$ have the same distribution for any given t.*

Proof. The joint distribution of $B(t)$ and $M_B(t)$ is

$$P[M_B(t) \le y, B(t) \le x] = P[B(t) \le x] - P[B(t) \le x - 2y] \tag{1.2}$$

for $0 \le x \le y$. To see this, define τ to be the epoch of the first visit to level y and define the process $\{B'(s) : s \ge 0\}$ with $B'(s) = B(\tau + s) - y$. We have

$$P[M_B(t) > y, B(t) \le x] = P[\tau < t, B'(t - \tau) \le x - y]. \tag{1.3}$$

By the strong Markov property, $\{B'\}$ is independent of $\{B(s) : 0 \le s \le \tau\}$, and it is a standard Brownian motion. Therefore, $B'(s)$ and $-B'(s)$ have the same distribution, and (1.3) may be written as

$$\begin{aligned} P[M_B(t) > y, B(t) \le x] &= P[\tau < t, B'(t - \tau) \ge y - x] \\ &= P[\tau < t, B(t) \ge 2y - x] \\ &= P[B(t) \ge 2y - x] \end{aligned}$$

since $B(t) \ge 2y - x$ implies that $\tau < t$. We transform the last equation as

$$P[M_B(t) > y, B(t) \le x] = P[B(t) \le x - 2y], \tag{1.4}$$

which proves (1.2). Take $x = y$ to obtain

$$P[M_B(t) \le y] = P[|B(t)| \le y] \tag{1.5}$$

for $y \ge 0$.

Obviously, $|m_B(t)| = M_{-B}(t)$ and it results from (1.5) that

$$|m_B(t)| \overset{d}{=} |B(t)| \overset{d}{=} M_B(t), \tag{1.6}$$

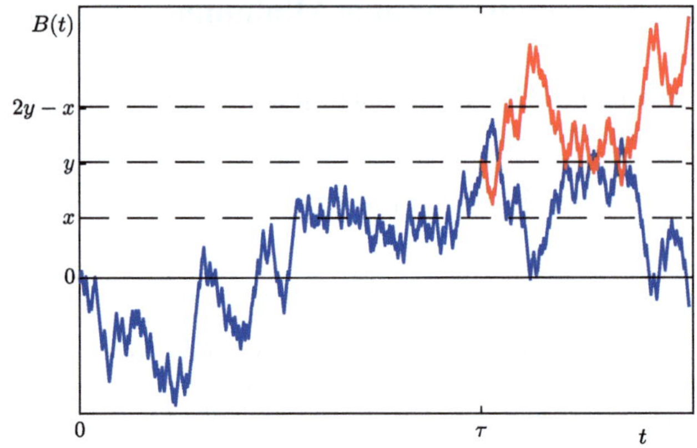

Fig. 1.1 Reflection of the Brownian motion after the first passage to level y; the blue line is a path of $\{B\}$, the path that finishes in red after time τ is a path of $\{B^*\}$

where $\stackrel{\mathrm{d}}{=}$ means "has the same distribution as." □

The argument from (1.3) to (1.4) critically depends on μ being equal to 0. The property itself is related to the reflection principle, which states that for every trajectory that hits level $y > 0$ in $[0, t]$ and is below $x < y$ at time t, there is another trajectory, equally probable, which is above $2y - x$ at time t. As an illustration, see Fig. 1.1; the line that terminates in red is a trajectory for the standard Brownian motion $\{B^*\}$, where

$$\begin{aligned} B^*(t) &= B(t) && \text{for } 0 \le t \le \tau \\ &= 2y - B(t) && \text{for } t > \tau. \end{aligned}$$

Although the absolute value of the running minimum, the absolute value of the Brownian motion, and the running maximum have all the same distribution for any fixed t, the *processes* themselves are different of course, as $M_B(\cdot)$ and $|m_B(\cdot)|$ are non-decreasing, while $|B(\cdot)|$ goes up and down.

For a BM(μ, σ) process $\{X\}$ with $\mu \neq 0$, things are not as neat, and we have, assuming that $X(0) = 0$,

$$P[M_X(t) \le y] = P[X(t) \le y] - e^{2\mu y/\sigma^2} P[X(t) \le -y], \tag{1.7}$$

for $y \ge 0$, a proof being given in Harrison [80, Chapter 1.8]. As $|m_X(t)| = M_{-X}(t)$ and $\{-X\}$ is BM$(-\mu, \sigma)$, we may write

$$\begin{aligned} P[|m_X(t)| \le y] &= P[X(t) \ge -y] - e^{-2\mu y/\sigma^2} P[X(t) \ge y] \\ &= 1 - e^{-2\mu y/\sigma^2} P[M_X(t) > y]. \end{aligned} \tag{1.8}$$

Remark 1.5. The joint distribution of the standard Brownian motion $\{\mathcal{B}\}$, its running maximum $\{M_B\}$, and its running minimum $\{m_B\}$ is given in Schilling and Partzsch [136, Section 6.5] under the name *Lévy's triple law* as

$$\begin{aligned}
&\mathrm{P}[m_B(t) > a, M_B(t) < b, B(t) \in \mathrm{d}x] \\
&\quad = \frac{\mathrm{d}x}{\sqrt{2\pi t}} \sum_{n=-\infty}^{\infty} \left(\exp\left(-\frac{(x+2n(b-a))^2}{2t}\right) - \exp\left(-\frac{(x-2a-2n(b-a))^2}{2t}\right)\right)
\end{aligned}$$

for any t and $a < 0 < b$. The proof is by repeated use of the reflection principle.

1.3 Regulated Brownian Motion

In some applications from queueing theory or production systems analysis, one needs models that behave like a Brownian motion, except that they are constrained to take positive values only. To that end, we may transform a Brownian motion $\{X\}$ into a *regulated Brownian motion* $\{Z(t) : t \in \mathbb{R}^+\}$ defined by

$$Z(t) = X(t) + S(t), \qquad \text{for } t \geq 0, \tag{1.9}$$

where $S(t) = -m_X(t)$ is called the regulator. It is clear that $Z(t) \geq 0$ for all t, and we show in this section how much it resembles the Brownian motion as it takes strictly positive values. An illustration is given in Fig. 1.2: In Figure (a) above, we show in blue a sample path of the standard Brownian motion, and its running minimum in red. The regulated process is shown in Fig. 1.2b.

A first property is that $Z(t)$ has the same distribution for any given t as the running maximum $M_X(t)$. Indeed,

$$\begin{aligned}
Z(t) &= X(t) + S(t) \\
&= \sup\{X(t) - X(s) : 0 \leq s \leq t\} \\
&= \sup\{X(t) - X(t-s) : 0 \leq s \leq t\} \\
&= \sup\{X'(s) : 0 \leq s \leq t\} \\
&= M_{X'}(t),
\end{aligned}$$

where $X'(s) = X(t) - X(t-s)$. The process $\{X'\}$ is a Brownian motion with the same parameters as $\{X\}$, so that $Z(t)$ has the same distribution as $M_X(t)$. In consequence, we see from (1.7) that

$$\mathrm{P}[Z(t) \leq y] = \mathrm{P}[X(t) \leq y] - e^{2\mu y/\sigma^2}\mathrm{P}[X(t) \leq -y] \tag{1.10}$$

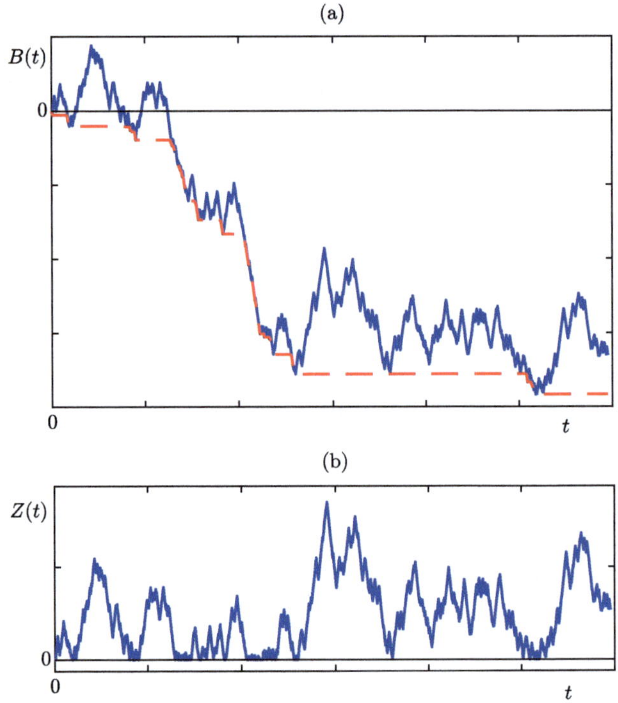

Fig. 1.2 (**a**) Shows a sample path of the standard BM, along with the running minimum m_B (the dashed line). (**b**) Shows the regulated sample path $Z(t) = B(t) - m_B(t)$

for $y > 0$, and

$$\lim_{t\to\infty} \mathrm{P}[Z(t) \le y] = 1 - e^{2\mu y/\sigma^2} \quad \text{if } \mu < 0 \tag{1.11}$$
$$= 0 \quad \text{if } \mu > 0.$$

The Brownian motion itself does not have an asymptotic distribution because it is null recurrent if $\mu = 0$ while it is transient otherwise, drifting to $+\infty$ if $\mu > 0$ and to $-\infty$ if $\mu < 0$. By contrast, the regulated Brownian motion is positive recurrent and has a limiting distribution when $\mu < 0$; a physical justification is that whenever $Z(t)$ is strictly positive, it acts like a Brownian motion with negative drift and returns to 0, a behaviour that is repeated over and over again.

For a standard Brownian motion $\{B\}$, we may complete as follows the set of equalities (1.6):

$$Z_B(t) \stackrel{\mathrm{d}}{=} S(t) \stackrel{\mathrm{d}}{=} |B(t)| \stackrel{\mathrm{d}}{=} M_B(t) \tag{1.12}$$

for any given t. In view of the fact that $Z_B(t)$ is the sum of $B(t)$ and its regulator $S(t)$ itself, the first equality comes as a mild surprise, but such are the peculiarities of the Brownian motion.

Actually, the link between $\{Z_B\}$ and the process $\{|B|\}$ of absolute values is stronger than it appears above, as the *processes* themselves have the same distribution as shown in Freedman [66, Theorem 1.42]. That may be the reason why $\{Z\}$ is often called a *reflected* Brownian motion in the literature, even for general Brownian motion, when $Z(t)$ and $|X(t)|$ do not have the same distribution. We shall, in this book, follow Harrison [80] and systematically use the term *regulated.*

Parentheses, such as the one that comes now, are about useful technical material which is not directly related to Markov-modulated Brownian motions. They may be thought of as glorified footnotes or dwarf appendices.

Parenthesis 1.6. The Skorokhod Regulator

The pair (Z, S) in (1.9) is the solution to the Skorokhod problem, meaning it is the unique pair of functions such that for a given function X

- $Z(t) = X(t) + S(t)$ for all $t \geq 0$,
- $Z(\cdot)$ is nonnegative and $S(\cdot)$ is non-decreasing, with $S(0) = 0$,
- $S(\cdot)$ increases only when $Z(t) = 0$,

see Asmussen [14, Section IX.2], Kruk et al. [97]. The function S is called the Skorokhod regulator, and for Brownian motion, it is the absolute value $|m_X|$ of the running minimum.

This is a very popular mechanism to regulate Brownian motion and related processes. Nevertheless, it is not the only one, and it is not always adequate as a model for real systems. As an example of what we have in mind, consider the well-known M/G/1 queue (Resnick [132, Chapter 3.7]): Customers arrive at epochs determined by a Poisson process and service durations are i.i.d. random variables, with an arbitrary distribution. When the queue becomes empty, the service mechanism becomes idle and it resumes with a new service upon the arrival of a first customer to the empty queue. This description may be termed the *operational view*.

Alternately, one might start from a description whereby services are performed one after another without interruption, even when the queue is empty, potentially resulting in a negative queue. If we apply the Skorokhod regulator, a customer that arrives to an empty queue receives the remainder of the service being performed.

If the service durations are exponentially distributed, then the Skorokhod and the operational views lead to identically distributed queueing systems, otherwise, the two are different. The Skorokhod view, therefore, is not always relevant, and one should be able on occasion to use other regulation mechanisms. We return to this in Chap. 5.

$\triangle$

1.4 Sample Path Properties

The trajectories of Brownian motion have some peculiar properties, to say the least, beyond what one might guess from the samples shown in the foreword.

As a first property, we might mention that sample paths are dense in C, the set of continuous functions, and may take any shape. This means that for any continuous function f with $f(0) = 0$, and for every $\varepsilon > 0$,

$$\mathrm{P}[|B(t) - f(t)| \leq \varepsilon \text{ for all } t \text{ in } [0, 1]] > 0,$$

see Freedman [66, Theorem 1.38]. Although paths may be arbitrarily close to smooth continuous functions, the fact remains that, with probability 1, the trajectories of Brownian motion are differentiable nowhere, a consequence being that representations like Fig. 1 are very crude approximations. A related property is that the set of local maxima of $B(\cdot; \omega)$ is a.s. dense: For all t_0 and all $\varepsilon > 0$, there exists t with $|t - t_0| < \varepsilon$ for which $B(t; \omega)$ is a local maximum. Furthermore, each local maximum is strict.

Now, take some arbitrary y in $\mathbb{R}$ and define $\mathcal{L}_y(\omega)$ to be the set of epochs when the standard Brownian motion visits level y:

$$\mathcal{L}_y(\omega) = \{t : B(t; \omega) = y\}.$$

One finds that $\mathcal{L}_y(\omega)$ is made up of a denumerably infinite number of separate subsets, symbolised by wiggly lines on the graph below.

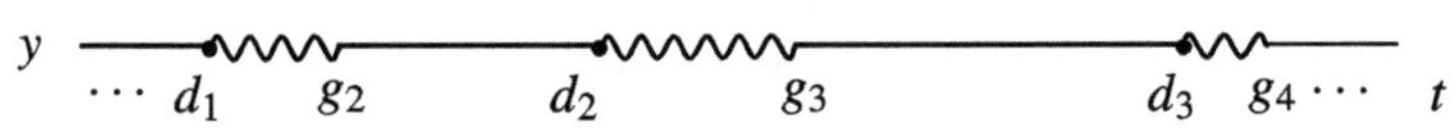

The subsets are characterised as follows.

- The open intervals (g_i, d_i) do not contain any point of $\mathcal{L}_y(\omega)$, they are called *excursion intervals*, they have a strictly positive length, and they are maximal among the open intervals on which $B(t; \omega) \neq y$.
- Each closed interval $[d_i, g_{i+1}]$ contains an uncountable number of points of $\mathcal{L}_y(\omega)$, with Lebesgue measure 0. In between these points, the trajectory oscillates above and below y.
- The points d_i are epochs of visit to y that are isolated from the left: There is, to the left of each, an open interval of strictly positive length during which $B(t; \omega) \neq y$.
- The same points are, however, limit points from the right, and the Brownian motion visits y infinitely often in any interval of the form $(d_i, d_i + \varepsilon)$ with $\varepsilon > 0$, so that every point in $\mathcal{L}_y(\omega)$ is an accumulation point.
- Assuming that $B(s) = y$,

  $$\min\{B(s + t) : 0 \leq t \leq \varepsilon\} < y < \max\{B(s + t) : 0 \leq t \leq \varepsilon\},$$

  for all $\varepsilon > 0$, with probability 1.

See Bertoin [32, Page 105] and Freedman [66, Page 38] for further details.

Finally, the processes $\{Z\}$ and $\{|B|\}$ have the same distribution, as mentioned in Sect. 1.3, and one concludes that the set

$$T_0(\omega) = \{t : Z(t; \omega) = 0\}$$

has the same properties as $\mathcal{L}_0(\omega)$. Since the points of increase of the regulator are the points where $Z = 0$, we find that T_0 is also the set

$$T_0(\omega) = \{t : S(t;\omega) > S(s;\omega) \text{ for all } s < t\}.$$

For general Brownian motions with $\mu \neq 0$, the process eventually drifts to $+\infty$ or $-\infty$, never to return to y, and so the number of excursion intervals is finite a.s., but the remaining features still hold true, and they will have a significant impact on the analysis that we perform.

1.5 Occupation Measure and Local Time

The *occupation measure* is the time spent by the Brownian motion in some set during some interval of time, it is defined as

$$\nu(t, A, \omega) = \int_0^t \mathbb{1}\{X(s,\omega) \in A\}\,\mathrm{d}s, \tag{1.13}$$

for $t \geq 0$ and A in the Borel σ-algebra $\mathcal{R}$ on $\mathbb{R}$. For any fixed time t and trajectory ω, $\nu(t,\cdot,\omega)$ is a measure on $(\mathbb{R},\mathcal{R})$ of total mass t.

It so happens that the occupation measure of the Brownian motion has a density, that is, a function ℓ: $\mathbb{R} \times [0,\infty) \times \Omega \to \mathbb{R}$ such that

$$\nu(t, A, \omega) = \int_A \ell(x,t,\omega)\,\mathrm{d}x,$$

for $t \geq 0$, $A \in \mathcal{R}$. This density is named the *local time* of the process $\{X\}$ at level x and time t, it is jointly continuous in x and t for almost all ω, and it is given by the limit

$$\ell(x,t,\omega) = \lim_{\varepsilon \downarrow 0} \frac{1}{2\varepsilon} \int_0^t \mathbb{1}\{|X(s,\omega) - x| < \varepsilon\}\,\mathrm{d}s \tag{1.14}$$

with probability 1: We start from the time spent during $[0,t]$ in a band of width 2ε around x, normalise it with the width, and take the limit as ε goes to 0.

From this definition, $\ell(x,\cdot,\omega)$ is a continuous function that increases only at the time points where $X(t,\omega) = x$. Furthermore, for any nonnegative measurable function f, and for almost all ω, we have the identity

$$\int_0^t f(X(s,\omega))\,\mathrm{d}s = \int_{-\infty}^{\infty} f(x)\ell(x,t,\omega)\,\mathrm{d}x : \tag{1.15}$$

The *time* integral on the left is replaced with the *space* integral on the right and the local time acts as the exchange factor. Equation (1.15) is called the *occupation density formula* and can be generalised to the form

$$\int_0^t \mathrm{E}_x[f(X(s)]\,\mathrm{d}s = \int_{-\infty}^{\infty} f(y)\mathrm{E}_x[\ell(y,t)]\,\mathrm{d}y. \tag{1.16}$$

Remark 1.7. It is something of a misnomer to call ℓ a local *time*, as it is a density and its physical dimension is not just that of time. The terminology is well established, however, and we do use it throughout this book.

1.6 Lévy Processes

To generalise Brownian motion, we may start from the characterisation given in Theorem 1.2 and weaken some of the constraints; Lévy processes are obtained by allowing trajectories to be discontinuous.

Definition 1.8. *A Lévy process* $\{X(t) : t \in \mathbb{R}_+\}$ *has stationary independent increments. That is, for any given* t *and* h, $X(t+h) - X(t)$ *is independent of* $\{X(s) : 0 \le s \le t\}$ *and its distribution depends on* h *only; in particular,* $X(0) = 0$ *a.s. The sample trajectories are a.s. right continuous on* $[0, \infty)$ *and have left limits on* $(0, \infty)$.

Beyond the obvious Brownian motion, simple examples of Lévy processes are the Poisson process and the compound Poisson process: Take a Poisson process $\{N(t) : t \ge 0\}$, and a sequence $\{\xi_n : n \ge 1\}$ of i.i.d. random variables with common distribution function $F(\cdot)$, it is easy to verify that the process $\{X\}$ defined by

$$X(t) = \sum_{1 \le n \le N(t)} \xi_n$$

satisfies the criteria of Definition 1.8.

Lévy processes are popular in finance (Papapantoleon [125]) precisely because they allow for jumps at epochs that are not preordained. Furthermore, they allow for distributions with heavier tails than the Gaussian distribution of Brownian motion. As an example, we give in Fig. 1.3a a sample path of the Cauchy process. The difference with Fig. 1 is striking: Here, the trajectory is made up of very many very small jumps, with an occasional bigger one. The distribution of $X(t)$, $t \ge 0$, is Cauchy, with density.

$$f_t(x) = \frac{1}{\pi} \frac{\gamma t}{\gamma^2 t^2 + x^2},$$

for $x \in \mathbb{R}$ and $\gamma > 0$, γt is the half-length of the interquartile range (Kyprianou [98, Section 1.2.6]). We display in Fig. 1.3b the Cauchy$(0, \gamma)$ and the $\mathcal{N}(0, 1)$ densities, with γ chosen such that the two distributions have the same interquartile interval. The Cauchy density has a much heavier tail, which is to be expected as its variance is infinite.

There is a one-to-one correspondence between Lévy processes and *infinitely divisible distributions*. Taking $n \ge 1$, we have by definition of the Lévy process

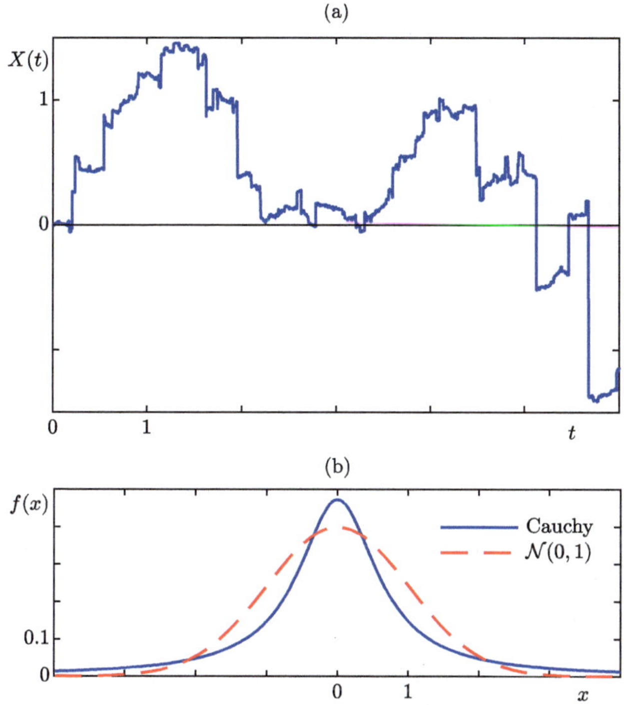

Fig. 1.3 (**a**) Above is a sample path of a Cauchy process, we show in (**b**) the Cauchy density at time $t = 1$ (blue line) and the $\mathcal{N}(0, 1)$ density (dashed red line). The parameter is $\gamma = 0.67448$

$$\begin{aligned} X(t) &= X(t/n) + [X(2t/n) - X(t/n)] + \cdots + [X(t) - X((n-1)t/n)] \\ &\stackrel{\mathrm{d}}{=} X_1(t/n) + X_2(t/n) + \cdots + X_n(t/n), \end{aligned} \tag{1.17}$$

where the $X_i(t/n)$s are i.i.d. copies of $X(t/n)$. As (1.17) holds for any $n \geq 1$, we conclude that if $\{X\}$ is a Lévy process, then for any t, $X(t)$ has an infinitely divisible distribution. Theorem 1.9 gives a signature property of such distributions, (1.19) is known as the Lévy-Khinchine formula.

Theorem 1.9. *The distribution of a random variable X is infinitely divisible if and only if there exist $a \in \mathbb{R}$, $\sigma \geq 0$, and a measure ν on $\mathbb{R}$,with $\nu(0) = 0$ and*

$$\int_{\mathbb{R}} \min(1, x^2)\, \nu(\mathrm{d}x) < \infty, \tag{1.18}$$

such that $\mathrm{E}[e^{isX}] = e^{-\psi(s)}$, *where*

$$\psi(s) = ias + \frac{1}{2}\sigma^2 s^2 + \int_{\mathbb{R}} (1 - e^{isx} + isx\, \mathbb{1}[|x| < 1])\, \nu(\mathrm{d}x), \tag{1.19}$$

for $s \in \mathbb{R}$. In (1.19), $\mathbb{1}[\cdot]$ is the indicator function. The measure ν is called the Lévy measure. □

Now, Theorem 1.10, known as the Lévy-Ito decomposition theorem, allows us to associate a Lévy process with any infinitely divisible probability distribution (Bertoin [32, Theorem 1], Kyprianou [98, Theorem 1.6]). Before we state the theorem, however, we need to introduce the *characteristic exponent* of a Lévy process. This is a function $\psi(s)$, $s \in \mathbb{R}$, with $\psi(0) = 0$, such that

$$\mathrm{E}[e^{isX(t)}] = e^{-t\psi(s)}, \tag{1.20}$$

for all $t \geq 0$. Its existence is easily proved for a positive integer n as

$$\begin{aligned} \mathrm{E}[e^{isX(n)}] &= (\mathrm{E}[e^{isX(1)}])^n && \text{by (1.17)} \\ &= e^{-n\psi(s)}, \end{aligned}$$

where $e^{-\psi(s)} = \mathrm{E}[e^{isX(1)}]$ is the characteristic function of $X(1)$. The extension to arbitrary $t > 0$ in $\mathbb{R}$ is given in [98, Page 4]. As $X(1)$ is infinitely divisible, the function $\psi(\cdot)$ is available through the Lévy-Khinchine formula (1.19).

Theorem 1.10. *Take Ω to be the set of functions right continuous on $[0, \infty)$ with left limits on $(0, \infty)$, and take $\mathcal{A}$ to be its Borel σ-field.*

For any characteristic exponent belonging to an infinitely divisible distribution, there exists a unique measure L on $(\Omega, \mathcal{A})$ such that the coordinate process is Lévy with the same characteristic exponent. □

Three examples are:

- The BM(μ, σ), for which $X(1)$ has the normal $\mathcal{N}(\mu, \sigma^2)$ distribution, with

$$\psi(s) = -is\mu + \frac{1}{2}\sigma^2 s^2, \tag{1.21}$$

  the Lévy measure ν is identically 0.
- The compound Poisson process, where the jumps are i.i.d. with distribution F, and $\{N\}$ is Poisson with parameter λ. In this case,

$$\psi(s) = \lambda \int_{\mathbb{R}} (1 - e^{isx}) F(\mathrm{d}x),$$

  and the parameters for the Lévy-Khinchine representation (1.19) are

$$a = -\lambda \int_{-1<x<1} x F(\mathrm{d}x), \qquad \sigma = 0, \qquad \nu(\mathrm{d}x) = \lambda F(\mathrm{d}x).$$

- The Cauchy(0,γ) process, for which

$$\psi(s) = \gamma |s|$$

and the Lévy-Khinchine parameters are

$$a = \sigma = 0, \qquad \nu(\mathrm{d}x) = \frac{\gamma}{\pi x^2}\,\mathrm{d}x.$$

A nice feature of (1.19) is that it gives us a description of the physical properties of a Lévy process. To begin with, assume that the measure ν is finite, that is, assume that $\int_{\mathbb{R}} \nu(\mathrm{d}x) < \infty$. In that case, the Lévy process results from the superposition of a Brownian motion and an independent compound Poisson process. To see this, we rewrite (1.19) as

$$\psi(s) = is(a + \int_{|x|<1} x\nu(\mathrm{d}x)) + \frac{1}{2}\sigma^2 s^2 + \lambda \int_{\mathbb{R}} (1 - e^{isx})\, F(\mathrm{d}x),$$

where $\lambda = \int_{\mathbb{R}} \nu(\mathrm{d}x)$ and $F(\cdot) = \nu(\cdot)/\lambda$. The sum of the first two terms above is the characteristic exponent of a Brownian motion, while the third is the characteristic exponent of a compound Poisson process.

If, however, the Lévy measure is infinite, then the process results from the superposition of a Brownian motion and a denumerably infinite number of high-frequency compound Poisson processes with tiny jumps. Such is the case of the Cauchy process: Its Lévy measure γx^{-2} is infinite, and we do see on Fig. 1.3 that its trajectory is made of very many small jumps, with a few big ones.

We give here a high-level explanation for this claim, precise details are found in [98, Chapter 2]. We know by (1.18) that

$$\int_{|x|\geq 1} \nu(\mathrm{d}x) + \int_{|x|<1} x^2\, \nu(\mathrm{d}x) < \infty.$$

When $\int_{\mathbb{R}} \nu(\mathrm{d}x) = \infty$, therefore, the frequency $\int_{|x|<1} \nu(\mathrm{d}x)$ of jumps with absolute size less than 1 is infinite. Since $|x| = 1$ is an arbitrary cut point, it means that the frequency of arbitrarily small non-zero jumps is infinite. Now, define the disjoint intervals

$$A_0 = [1, \infty), \qquad A_n = [2^{-n}, 2^{-(n-1)}), \quad n \geq 1,$$

with length decreasing to 0 as n increases to ∞. The integral in (1.19) may be written as the series

$$S = \lambda_0 \int_{|x|\in A_0} (1 - e^{isx})\, F_0(\mathrm{d}x) + \sum_{n\geq 1} \lambda_n \int_{|x|\in A_n} (1 - e^{isx} + isx)\, F_n(\mathrm{d}x), \qquad (1.22)$$

where $\lambda_n = \nu(A_n) < \infty$ and $F_n(\mathrm{d}x) = \nu(\mathrm{d}x)/\lambda_n$. By the same argument that we gave above, the terms in the series (1.22) are the characteristic exponents of independent compound Poisson processes with intensity $\nu(A_n)$, jump size in the interval A_n, and an additional drift component.

1.7 Markov Modulation

In Sect. 1.6, we allow trajectories to be discontinuous. Here, we hold on to the continuity assumption, but we weaken the requirement that increments should be independent over disjoint intervals of time. A popular mechanism is to allow a Markov process to select the parameters of a Brownian motion: We take a continuous-time Markov chain on a finite state space, during interval of time spent in state i, say, the new process behaves like a Brownian motion with parameters that depend on i. A short technical definition follows.

Definition 1.11. *Denote by $\{\varphi(t) : t \in \mathbb{R}^+\}$ an irreducible continuous-time Markov chain on the state space $\mathcal{E} = \{1, \ldots, m\}$, m finite, with generator Q. Define the jump epochs $\{T_n : n \in \mathbb{N}\}$ as follows: $T_0 = 0$,*

$$T_n = \inf\{t > T_{n-1} : \varphi(t) \neq \varphi(T_{n-1})\}, \qquad \text{for } n \geq 1.$$

Define the jump chain $\{\varphi_n; n \in \mathbb{N}\}$, with $\varphi_n = \varphi(T_n)$.

Denote by $\{X_i(t) : t \in \mathbb{R}^+\}$, i in $\mathcal{E}$, a set of Brownian motions with parameters μ_i and σ_i, respectively. Assume that $\{\varphi\}$, $\{X_1\}, \ldots, \{X_m\}$ are independent of each other, and independent of $X(0)$.

The process $\{X\}$ is defined through

$$X(t) = X(T_n) + (X_{\varphi_n}(t) - X_{\varphi_n}(T_n)), \qquad \text{for } T_n \leq t < T_{n+1}. \tag{1.23}$$

The two-dimensional process $\{X(t), \varphi(t), t \in \mathbb{R}^+\}$ is called a Markov-modulated Brownian motion, abbreviated as MMBM. The component X is called the level and φ is called the phase.

In this book, the hard-wired parameters of the MMBM are the vectors $\boldsymbol{\mu} = \begin{bmatrix}\mu_1 \cdots \mu_m\end{bmatrix}^{\mathrm{T}}$, $\boldsymbol{\sigma} = \begin{bmatrix}\sigma_1 \cdots \sigma_m\end{bmatrix}^{\mathrm{T}}$ and the generator Q. The initial value of $X(0)$ and the distribution of $\varphi(0)$ are chosen according to circumstances.

One also finds in the literature the expression that we gave on Page ix:

$$X(t) = X(0) + \int_0^t \mu_{\varphi(s)}\, \mathrm{d}s + \int_0^t \sigma_{\varphi(s)}\, \mathrm{d}B(s), \tag{1.24}$$

where $\{B\}$ is a standard Brownian motion, independent of $\{\varphi\}$ and $X(0)$. Compared to (1.23) where we need m independent Brownian motions, (1.24) may seem to be more economical, but we like Definition 1.11, inspired from Asmussen and Pihlsgård [17, Section 5], because it makes the next statement dazzlingly obvious.

Theorem 1.12. *Assume that $\{X, \varphi\}$ is an MMBM with parameters Q, $\boldsymbol{\mu}$, and $\boldsymbol{\sigma}$, and assume that T is a stopping time.*

If $T < \infty$ a.s., then $\{X_T(t), \varphi_T(t) : t \in \mathbb{R}^+\}$ defined by

$$X_T(t) = X(T + t) - X(T), \qquad \varphi_T(t) = \varphi(T + t), \qquad t \geq 0$$

is an MMBM with the same parameters, conditionally independent of $\{X(s), \varphi(s), 0 \le s \le T\}$ *given* $\varphi(T)$.

Moreover, if $\varphi(T) = i$, *then* $\{X_T, \varphi_T\}$ *has the same law as* $\{X, \varphi\}$ *with* $X(0) = 0$ *and* $\varphi(0) = i$. □

We refer to this as the strong Markov property, for a formal argument, see [98, Chapter 3].

Now, for any given t, $X(t)$ is the sum of m random variables $\widetilde{X}_i$, $1 \le i \le m$ normally distributed with parameters $\mu_i S_i(t)$ and $\sigma_i^2 S_i(t)$, where

$$S_i(t) = \int_0^t \mathbb{1}[\varphi(s) = i]\,\mathrm{d}s$$

is the total time spent in phase i during the interval $(0, t)$. This is not a very convenient representation of the distribution of $X(t)$ and so we turn to a transform, as we did for Lévy processes. Here, however, we need to take the Markovian environment into account, and we define a *matrix* Laplace transform $L(t; s)$ with entries

$$L_{ij}(t; s) = \mathrm{E}[e^{sX(t)}\mathbb{1}[\varphi(t) = j] | X(0) = 0, \varphi(0) = i],$$

for i, j in $\mathcal{E}$, s in $\mathbb{R}$, $t \ge 0$.

We condition on the phase $\varphi(t)$ and easily show that $L(t + h; s) = L(t; s)L(h; s)$ with $L(0; s)$ equal to the identity matrix I, and we conclude that there exists a matrix function $\mathcal{F}(s)$ such that

$$L_{ij}(t; s) = (e^{t\mathcal{F}(s)})_{ij} \qquad \text{for } i, j \text{ in } \mathcal{E}. \tag{1.25}$$

It is customary to write $\mathrm{E}_{x,i}[\cdot]$ for the expectation of a random variable, conditionally on $X(0) = x$, $\varphi(0) = i$, and $\mathrm{E}_{x,\boldsymbol{\alpha}}[\cdot]$ if $\varphi(0)$ has the distribution $\boldsymbol{\alpha}$. In addition, we use the shorthand $\mathrm{E}[\,\cdot\,; \varphi(t)]$ for the matrix of order m with elements

$$(\mathrm{E}[\,\cdot\,; \varphi(t)])_{ij} = \mathrm{E}[\,\cdot\,\mathbb{1}[\varphi(t) = j] | X(0) = 0, \varphi(0) = i], \tag{1.26}$$

for $i, j \in \mathcal{E}$ and, in case it is necessary to specify the value of $X(0)$, we write $\mathrm{E}_{x,\varphi(0)}[\,\cdot\,; \varphi(t)]$ for the matrix with entries

$$(\mathrm{E}_{x,\varphi(0)}[\,\cdot\,; \varphi(t)])_{ij} = \mathrm{E}[\,\cdot\,\mathbb{1}[\varphi(t) = j] | X(0) = x, \varphi(0) = i]. \tag{1.27}$$

Similarly, we use $\mathrm{P}_{x,\alpha}[\cdot]$, $\mathrm{P}[\,\cdot\,; \varphi(t)]$, and $\mathrm{P}_{x,\varphi(0)}[\,\cdot\,; \varphi(t)]$ for probabilities.

With this notation, (1.25) will be written as

$$L(t; s) = e^{t\mathcal{F}(s)} = \mathrm{E}[e^{sX(t)}; \varphi(t)], \qquad s \text{ in } \mathbb{R}, t \ge 0. \tag{1.28}$$

By analogy with (1.20), $\mathcal{F}(s)$ is called the *matrix Laplace exponent* of the Markov-modulated Brownian motion.

Lemma 1.13. *The matrix Laplace exponent of an MMBM with parameters* $\boldsymbol{\mu}$, $\boldsymbol{\sigma}$, *and* Q *is given by*

$$\mathcal{F}(s) = s\Delta_\mu + \frac{s^2}{2}\Delta_\sigma^2 + Q, \tag{1.29}$$

where $\Delta_v = \mathrm{diag}(\boldsymbol{v})$ *for any vector* $\boldsymbol{v}$.

Proof. To determine $\mathrm{E}_{0,i}[e^{sX(t)}\mathbb{1}[\varphi(t) = j]]$, we condition on the epoch T of the first jump and on the value of $\varphi(T)$. If $T > t$, then $\varphi(t) = i$ and $X(t) = X_i(t)$ has the $\mathcal{N}(\mu_i t, \sigma_i^2 t)$ distribution.

If $T = u < t$ and $\varphi(u) = k$, then $X(t) = X_i(u) + X^*(t-u)$ and $\varphi(t) = \varphi^*(t-u)$ by the strong Markov property, where $\{X^*, \varphi^*\}$ is an MMBM with the same parameters $\boldsymbol{\mu}$, $\boldsymbol{\sigma}$ and Q, and $\varphi^*(0) = k$, independent of $X_i(u)$.

Thus,

$$\begin{aligned}(e^{t\mathcal{F}(s)})_{ij} &= e^{t(s\mu_i + \sigma_i^2 s^2/2)} e^{-\lambda_i t} \delta_{ij} \\ &+ \sum_{k \neq i} \int_0^t e^{u(s\mu_i + \sigma_i^2 s^2/2)} \lambda_i e^{-\lambda_i u} \widetilde{P}_{ik} (e^{(t-u)\mathcal{F}(s)})_{kj} \, \mathrm{d}u,\end{aligned}$$

where $\lambda_i = |Q_{ii}|$ is the parameter of the exponential sojourn time distribution in phase i, $\delta_{ij} = \mathbb{1}[i = j]$, and

$$\widetilde{P} = I + \Delta_\lambda^{-1} Q \tag{1.30}$$

is the transition matrix of the jump chain for the phase process.

In matrix form, this is written as

$$e^{t\mathcal{F}(s)} = e^{t(s\Delta_\mu + s^2/2\Delta_\sigma^2)} e^{-\Delta_\lambda t} + \int_0^t e^{u(s\Delta_\mu + s^2/2\Delta_\sigma^2)} e^{-\Delta_\lambda u} (\Delta_\lambda + Q)\, e^{(t-u)\mathcal{F}(s)} \, \mathrm{d}u.$$

We take the derivative on both sides with respect to t, evaluated at $t = 0$, and this completes the proof. □

1.8 A Killing Mechanism

It is sometimes useful to implement some device to stop the evolution of an MMBM. This is often called a *killing mechanism* and its simplest version goes as follows. In addition to the process $\{\varphi\}$ with generator Q, we consider a random variable e_α exponentially distributed with parameter α, independent of everything else, such that the process $\{X, \varphi\}$ no longer evolves after time e_α.

One way of giving it a formal representation is to write that $X(t)$ is given by (1.24) for $t < e_\alpha$ and that $(X(t), \varphi(t)) = \dagger$ for $t \geq e_\alpha$, where $\dagger$ is a cemetery state of no specific property, beyond the fact that it is not in $\mathcal{E}$.

It is easily verified that the killed process has a matrix Laplace exponent $\mathcal{F}_\alpha(s)$ defined by

$$e^{t\mathcal{F}_\alpha(s)} = \mathrm{E}[e^{sX(t)}\mathbb{1}[t < e_\alpha]; \varphi(t)]$$

and given as

$$\mathcal{F}_\alpha(s) = s\Delta_\mu + \frac{s^2}{2}\Delta_\sigma^2 + (Q - \alpha I).$$

In short, it suffices to replace the generator Q with the non-conservative, or defective, generator $Q_\alpha = Q - \alpha I$.

We use this mechanism in a few places (including the proofs of Theorems 3.14, 4.9, and 4.14) for which arguments become simpler with the addition of an artificial killing process.

1.9 Moments for MMBMs

With the Laplace exponent given by (1.29), we have the Laplace transform $\boldsymbol{\alpha}^{\mathrm{T}} e^{t\mathcal{F}(s)}\mathbf{1}$ of $X(t)$, where $\boldsymbol{\alpha}$ is the distribution of $\varphi(0)$, and we may obtain the distribution of $X(t)$ by numerical inversion, as we illustrate on two examples in Figs. 1.4 and 1.5. We may also determine various moments of $X(t)$ by taking the derivative with respect to s, evaluated at $s = 0$.

The expected value $\mathrm{E}[X(t)]$ is obtained in Lemma 1.15 by direct argument. This lemma, together with Lemma 1.17, gives us the opportunity of introducing the *group inverse* of singular matrices. We will meet this generalised inverse on several occasions and we give its main characteristics now.

Parenthesis 1.14. Group Inverse

Assume that M is a singular matrix. Its group inverse, if it exists, is the unique matrix $M^\#$ such that

$$MM^\#M = M, \qquad M^\#MM^\# = M^\#, \qquad \text{and} \qquad MM^\# = M^\#M. \tag{1.31}$$

In the special case when M has a simple eigenvalue equal to 0, we denote as $\boldsymbol{v}$ and $\boldsymbol{u}$ the corresponding right and left eigenvectors, respectively. The solution of the system $MX = B$, $\boldsymbol{u}^{\mathrm{T}}B = \mathbf{0}$, is

$$X = M^\#B + \boldsymbol{v} \cdot \boldsymbol{y}^{\mathrm{T}}, \tag{1.32}$$

where $\boldsymbol{y}$ is an arbitrary vector. Similarly, the solution of the system $XM = B$, $B\boldsymbol{v} = \mathbf{0}$, is

$$X = BM^\# + \boldsymbol{z} \cdot \boldsymbol{u}^{\mathrm{T}}, \tag{1.33}$$

where $\boldsymbol{z}$ is arbitrary.

In particular, the generator Q of an irreducible continuous-time Markov chain on a finite state space $\mathcal{E}$ does have a group inverse and $Q^\#$ is the unique solution of each of the two systems

$$QX = I - \mathbf{1} \cdot \boldsymbol{\pi}^{\mathrm{T}} \qquad \text{and} \qquad XQ = I - \mathbf{1} \cdot \boldsymbol{\pi}^{\mathrm{T}}, \tag{1.34}$$

where $\boldsymbol{\pi}$ is the stationary probability vector of Q: $\boldsymbol{\pi}^{\mathrm{T}}Q = \mathbf{0}$, $\boldsymbol{\pi}^{\mathrm{T}}\mathbf{1} = 1$.

The matrix $Q^\#$, like Q, has one eigenvalue equal to 0, and the two matrices have the same associated left and right eigenvectors:

$$\boldsymbol{\pi}^{\mathrm{T}} Q^\# = \boldsymbol{\pi}^{\mathrm{T}} Q = \mathbf{0} \qquad \text{and} \qquad Q^\# \mathbf{1} = Q\mathbf{1} = \mathbf{0}; \tag{1.35}$$

the non-zero eigenvalues of $Q^\#$ are the reciprocals of the non-zero eigenvalues of Q.

In addition, there is a physical property associated with $Q^\#$:

$$Q^\# = \int_0^\infty (\mathbf{1} \cdot \boldsymbol{\pi}^{\mathrm{T}} - e^{Qs})\, \mathrm{d}s \tag{1.36}$$

and we may write that

$$(Q^\#)_{ij} = \lim_{t\to\infty} (\mathrm{E}_\pi[N_j(t)] - \mathrm{E}_i[N_j(t)]), \qquad i, j \in \mathcal{E},$$

where $N_j(t)$ is the total time spent in state j during the interval of time $(0, t)$. The integral in (1.36) converges whenever the state space is finite, even though the integral of e^{Qt} by itself is infinite.

The matrix $\mathcal{D} = -Q^\#$ is also called the *deviation matrix* of the Markov chain, for which a good reference is Coolen-Schrijner and van Doorn [52], and the matrix $Q^\#$ is related to $(Q + \mathbf{1} \cdot \boldsymbol{\pi}^{\mathrm{T}})^{-1}$, known as of the *fundamental matrix* of the Markov chain, via the relation

$$(Q + \mathbf{1} \cdot \boldsymbol{\pi}^{\mathrm{T}})^{-1} = Q^\# + \mathbf{1} \cdot \boldsymbol{\pi}^{\mathrm{T}}. \tag{1.37}$$

In the case of an irreducible *discrete-time* Markov chain with transition matrix P, the matrix $P - I$ has the same algebraic property as Q above. If the Markov chain is aperiodic, then (1.36) is replaced with

$$(I - P)^\# = \sum_{\nu \ge 0} (P^\nu - \mathbf{1} \cdot \boldsymbol{\pi}^{\mathrm{T}}), \tag{1.38}$$

$$= (I - P + \mathbf{1} \cdot \boldsymbol{\pi}^{\mathrm{T}})^{-1} - \mathbf{1} \cdot \boldsymbol{\pi}^{\mathrm{T}}. \tag{1.39}$$

For further details about the group inverse, we refer to Campbell and Meyer [48, Chapters 7 and 8].

△

Lemma 1.15. *Assume that $\{X, \varphi\}$ is an MMBM with parameters $\boldsymbol{\mu}$, $\boldsymbol{\sigma}$, and Q. The expected value of $X(t)$ is*

$$\mathrm{E}[X(t)] = t\, \boldsymbol{\pi}^{\mathrm{T}} \boldsymbol{\mu} + \boldsymbol{\alpha}^{\mathrm{T}} (e^{Qt} - I) Q^\# \boldsymbol{\mu}, \qquad \text{for } t \ge 0, \tag{1.40}$$

where $\boldsymbol{\pi}$ is the stationary probability vector of $\{\varphi\}$ ($\boldsymbol{\pi}^{\mathrm{T}} Q = \mathbf{0}$, $\boldsymbol{\pi}^{\mathrm{T}} \mathbf{1} = 1$), $\boldsymbol{\alpha}$ is the distribution of $\varphi(0)$, and $Q^\#$ is the group inverse of Q.

For the stationary version of the process, $\boldsymbol{\alpha}$ is equal to $\boldsymbol{\pi}$ and

$$\mathrm{E}[X(t)] = t\, \boldsymbol{\pi}^{\mathrm{T}} \boldsymbol{\mu}. \tag{1.41}$$

Proof. Since μ_j is the rate of level increment at time s if $\varphi(s) = j$, and $(e^{Qs})_{ij}$ is the probability that $\varphi(s) = j$ given that $\varphi(0) = i$, we may write that

$$\mathrm{E}[X(t)] = \boldsymbol{\alpha}^{\mathrm{T}} \int_0^t e^{Qs} \boldsymbol{\mu} \, \mathrm{d}s.$$

Next,

$$\begin{aligned}\int_0^t e^{Qs} \, \mathrm{d}s &= \int_0^t (e^{Qs} - \mathbf{1} \cdot \boldsymbol{\pi}^{\mathrm{T}}) \mathrm{d}s + t \, \mathbf{1} \cdot \boldsymbol{\pi}^{\mathrm{T}} \\ &= (I - e^{Qt}) \int_0^\infty (e^{Qs} - \mathbf{1} \cdot \boldsymbol{\pi}^{\mathrm{T}}) \mathrm{d}s + t \, \mathbf{1} \cdot \boldsymbol{\pi}^{\mathrm{T}},\end{aligned}$$

and (1.40) follows from (1.36). Finally, (1.41) is a consequence of $\boldsymbol{\pi}^{\mathrm{T}} e^{Qt} = \boldsymbol{\pi}^{\mathrm{T}}$. □

The second term in (1.40) is bounded since e^{Qt} is a stochastic matrix. Thus, that equation tells us that $\mathrm{E}[X(t)]$ is asymptotically linear as $t \to \infty$, with slope equal to the asymptotic mean drift $\boldsymbol{\pi}^{\mathrm{T}} \boldsymbol{\mu}$. Furthermore,

$$\begin{aligned}\lim_{t \to \infty} (\mathrm{E}[X(t)] - t \, \boldsymbol{\pi}^{\mathrm{T}} \boldsymbol{\mu}) &= \lim_{t \to \infty} (\boldsymbol{\alpha}^{\mathrm{T}} (e^{Qt} - I) Q^{\#} \boldsymbol{\mu}) \\ &= \boldsymbol{\alpha}^{\mathrm{T}} (\mathbf{1} \cdot \boldsymbol{\pi}^{\mathrm{T}} - I) Q^{\#} \boldsymbol{\mu} \\ &= \boldsymbol{\alpha}^{\mathrm{T}} (-Q)^{\#} \boldsymbol{\mu} \end{aligned} \tag{1.42}$$

since $\boldsymbol{\pi}^{\mathrm{T}} Q^{\#} = \mathbf{0}$, so that the initial distribution $\boldsymbol{\alpha}$ plays a role in the asymptotics of $\{X\}$.

Remark 1.16. The expression in Asmussen [14, Chapter XI, Corollary 2.5] for the expected value of $X(t)$ is

$$\mathrm{E}_{0,i}[X(t)] = t\kappa^* + k_i^* - \boldsymbol{e}_i^{\mathrm{T}} e^{Qt} \boldsymbol{k}^*, \tag{1.43}$$

where $\boldsymbol{e}_i$ is the vector with all components equal to 0, except for the ith which is equal to 1, and κ^* and $\boldsymbol{k}^*$ are connected to the matrix polynomial $\mathcal{F}(s)$ of Lemma 1.13:

- $\kappa^* = \frac{\partial}{\partial s} \kappa(s)\big|_{s=0}$, where $\kappa(s)$ is the eigenvalue of $\mathcal{F}(s)$ of maximal real part, and
- $\boldsymbol{k}^* = \frac{\partial}{\partial s} \boldsymbol{k}(s)\big|_{s=0}$, where $\boldsymbol{k}(s)$ is the corresponding right eigenvector.

To show that the right-hand sides of (1.40) and (1.43) are identical, one determines κ^* and $\boldsymbol{k}^*$ by a standard procedure, such as the one used in Neuts [116, Page 17] for instance, and proceeds with simple algebraic manipulations.

For the second moment of $X(t)$, we restrict our attention to the stationary version of the process, to avoid dealing with uselessly messy expressions.

Lemma 1.17. *If $\varphi(0)$ has the distribution $\boldsymbol{\pi}$, then*

$$\mathrm{Var}(X(t)) = t\boldsymbol{\pi}^{\mathrm{T}} (\Delta_\sigma^2 - 2\Delta_\mu Q^{\#} \Delta_\mu) \mathbf{1} + 2\boldsymbol{\pi}^{\mathrm{T}} \Delta_\mu (e^{Qt} - I)(Q^{\#})^2 \Delta_\mu \mathbf{1}. \tag{1.44}$$

Proof. Rewrite (1.28) as

$$\mathrm{E}[e^{sX(t)}; \varphi(t)] = \sum_{\nu \geq 0} \frac{1}{\nu!} t^\nu \mathcal{F}(s)^\nu. \tag{1.45}$$

Take the second derivative of (1.45) with respect to s, evaluate it at $s = 0$, pre-multiply with $\boldsymbol{\pi}^\mathrm{T}$ and post-multiply with $\mathbf{1}$, and use $\boldsymbol{\pi}^\mathrm{T} Q^\# = \mathbf{0}$ and $Q^\# \mathbf{1} = \mathbf{0}$. After some tedious but simple manipulations, we find that

$$\mathrm{E}_{0,\pi}[X^2(t)] = t\boldsymbol{\pi}^\mathrm{T} \Delta_\sigma^2 \mathbf{1} + 2\boldsymbol{\pi}^\mathrm{T} \Delta_\mu \sum_{\nu \geq 2} \frac{1}{\nu!} t^\nu Q^{\nu-2} \boldsymbol{\mu}. \tag{1.46}$$

Now, $QQ^\# + \mathbf{1} \cdot \boldsymbol{\pi}^\mathrm{T} = I$ by (1.34) and $Q^n \mathbf{1} = \mathbf{0}$ for $n \geq 1$. Thus,

$$\begin{aligned}
\sum_{\nu \geq 2} \frac{1}{\nu!} t^\nu Q^{\nu-2} &= \sum_{\nu \geq 2} \frac{1}{\nu!} t^\nu Q^{\nu-1} Q^\# + \frac{1}{2} t^2 \mathbf{1} \cdot \boldsymbol{\pi}^\mathrm{T} \\
&= \sum_{\nu \geq 2} \frac{1}{\nu!} t^\nu Q^\nu (Q^\#)^2 + \frac{1}{2} t^2 \mathbf{1} \cdot \boldsymbol{\pi}^\mathrm{T} \\
&= (e^{Qt} - I)(Q^\#)^2 - tQ^\# + \frac{1}{2} t^2 \mathbf{1} \cdot \boldsymbol{\pi}^\mathrm{T};
\end{aligned} \tag{1.47}$$

to justify the second and third equalities, we use

$$Q^\# = Q^\# Q Q^\# = Q(Q^\#)^2$$

by (1.31).

The proof of the lemma follows from (1.41, 1.46, 1.47). □

The variance of $X(t)$ is asymptotically linear, and the coefficient of t has two different contributions: $\boldsymbol{\pi}^\mathrm{T} \Delta_\sigma^2 \mathbf{1}$ is merely a weighted average of the individual variances σ_i^2, $i \in \mathcal{E}$, while $\boldsymbol{\pi}^\mathrm{T} \Delta_\mu Q^\# \boldsymbol{\mu}$ depends on the dynamic behaviour of the phase process.

Example 1.18. The MMBM in this example has the parameters

$$Q = \begin{bmatrix} -10 & 5 & 5 \\ 10 & -10 & 0 \\ 10 & 0 & -10 \end{bmatrix}, \tag{1.48}$$

$\boldsymbol{\mu} = \begin{bmatrix} 3 & -3 & -15 \end{bmatrix}^\mathrm{T}$ and $\boldsymbol{\sigma} = \begin{bmatrix} 1 & 1 & 1 \end{bmatrix}^\mathrm{T}$. The process alternates between two subsets of phases, $\{1\}$ where the level increases, and $\{2, 3\}$ where it decreases. The expected sojourn time is 0.1 unit of time in each phase. In Phase 1, the process behaves like a BM(3,1) process. Upon leaving that phase, it starts behaving with equal probability like a BM(−3,1) or like a BM(−15,1) with steeper downward drift. The variance is 1 for all three phases.

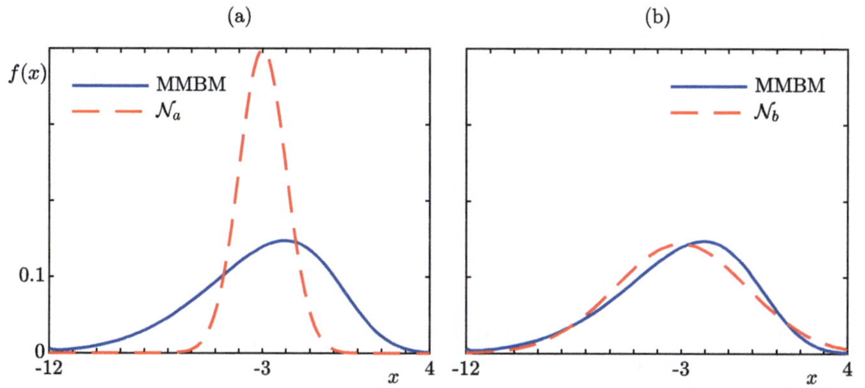

Fig. 1.4 The density function of $X(1)$ for Example 1.18 is given as the plain blue line. In (**a**) on the left, it is compared to the $\mathcal{N}(-3, 1)$ density (red dashed line). In (**b**) on the right, it is compared to the $\mathcal{N}(-3, 7.66)$ density (red dashed line)

The stationary probability vector of $\{\varphi\}$ is $\boldsymbol{\pi} = \begin{bmatrix}0.5 & 0.25 & 0.25\end{bmatrix}^{\mathrm{T}}$, so that $\mathrm{E}[X(1)] = \boldsymbol{\pi}^{\mathrm{T}}\boldsymbol{\mu} = -3$ if we choose $\boldsymbol{\alpha} = \boldsymbol{\pi}$. The density of $X(1)$ is shown as a continuous blue line in Fig. 1.4a on the left, and it is compared to the $\mathcal{N}(-3, 1)$ density shown as a dashed red line. The distribution of $X(1)$ is much more dispersed and, since $\sigma_i = 1$ for all i, we conclude that the greater dispersion of the probability mass is solely due to the random changes of drift.

Actually, $\mathrm{Var}(X(1)) = 7.66$, and the dominant term in the right-hand side of (1.44) is $-2\boldsymbol{\pi}^{\mathrm{T}}\Delta_\mu Q^{\#}\Delta_\mu \mathbf{1}$, equal to 7.20. In Fig. 1.4b on the right, the comparison is between $X(1)$ and $\mathcal{N}(-3, 7.66)$. The two density functions have roughly the same shape, confirming that the main effect is the increased variance.

Example 1.19. This is an MMBM with two phases. The drift is positive in Phase 1 and negative in Phase 2, and the variance is 1 in both: $\boldsymbol{\mu} = \begin{bmatrix}3 & -4\end{bmatrix}$ and $\boldsymbol{\sigma} = \begin{bmatrix}1 & 1\end{bmatrix}$. The Markov chain generator is

$$Q = \begin{bmatrix}-1 & 1 \\ 0.5 & -0.5\end{bmatrix},$$

and $\boldsymbol{\pi} = \begin{bmatrix}1/3 & 2/3\end{bmatrix}^{\mathrm{T}}$, so that the stationary drift is $\boldsymbol{\pi}^{\mathrm{T}}\boldsymbol{\mu} = -5/3$ and $\mathrm{Var}(X(1)) = 8.00$ if $\varphi(0)$ has the distribution $\boldsymbol{\pi}$.

The distribution of $X(1)$ is significantly different from the Gaussian, as we see clearly in Fig. 1.5: We display on the left the cumulative distribution function of $X(1)$ as a continuous blue line and of the $\mathcal{N}(-3/5, 8.00)$ random variable as a dashed red line, and we show their density in Fig. 1.5b on the right.

Actually, the environment in this example does not change very fast and it is likely that there is no change of phase before time $t = 1$. To be precise, the probability that there is no change before time 1 is $\mathrm{P}[\varphi(s) = \varphi(0) : 0 \le s \le 1] = 0.527$. This explains the accumulation of probability mass around the values $X(1) = -4$ and

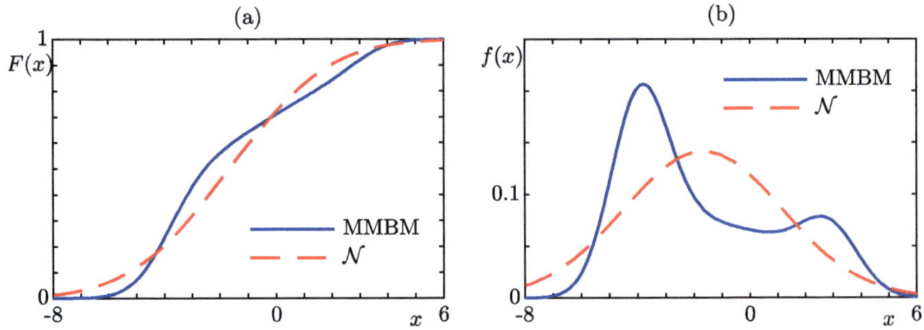

Fig. 1.5 Distribution of $X(1)$ for the MMBM of Example 1.19 (plain blue line), compared to the $\mathcal{N}(-5/3, 8.00)$ (dashed red line). The distribution function is shown in (**a**) on the left and the density function in (**b**) on the right

$X(1) = 3$. For Example 1.18 on the contrary, the probability that there is no change of phase before $t = 1$ is $4.54\,10^{-5}$, there are 10 changes on average before time 1, more mixing of intervals with positive and with negative drift, and this explains why the density in Fig. 1.4 has a single mode.

To conclude this section, we briefly examine the dependence of level increments over disjoint intervals.

Lemma 1.20. *Assume that* $X(0) = 0$ *and that* $\varphi(0)$ *has the distribution* $\boldsymbol{\pi}$. *The covariance* $C(t, h)$ *between* $X(t)$ *and* $X(t+h) - X(t)$ *is given by*

$$C(t, h) = C(h, t) = \boldsymbol{\pi}^{\mathrm{T}} \Delta_\mu Q^{\#}(e^{Qt} - I)(e^{Qh} - I)Q^{\#}\Delta_\mu \mathbf{1}. \tag{1.49}$$

Proof. We start from

$$C(t, h) = \mathrm{E}_{0,\pi}[X(t)(X(t+h) - X(t))] - \mathrm{E}_{0,\pi}[X(t)]\mathrm{E}_{0,\pi}[X(t+h) - X(t)]. \tag{1.50}$$

The distribution of $\varphi(t)$ is $\boldsymbol{\pi}^{\mathrm{T}} e^{Qt} = \boldsymbol{\pi}^{\mathrm{T}}$ for any given t, and so

$$\mathrm{E}_{0,\pi}[X(t)]\mathrm{E}_{0,\pi}[X(t+h) - X(t)] = \mathrm{E}_\pi[X(t)]\mathrm{E}_\pi[X(h)] \tag{1.51}$$

by the Markov property. We condition on the phase at time t and the first term in the right-hand side of (1.50) becomes

$$\begin{aligned} &\mathrm{E}_{0,\pi}[X(t)(X(t+h) - X(t))] \\ &\quad = \sum_{k \in \mathcal{E}} \mathrm{E}_{0,\pi}[X(t)\mathbb{1}[\varphi(t) = k]]\mathrm{E}[X(h)|X(0) = 0, \varphi(0) = k] \\ &\quad = \boldsymbol{\pi}^{\mathrm{T}}\mathrm{E}[X(t)\,;\,\varphi(t)]\mathrm{E}[X(h)\,;\,\varphi(h)]\mathbf{1} \end{aligned} \tag{1.52}$$

using the notation (1.26). Now,

$$\begin{aligned}
\mathrm{E}[X(t)\,;\,\varphi(t)] &= \frac{\partial}{\partial s}\mathrm{E}[e^{sX(t)}\,;\,\varphi(t)]\bigg|_{s=0} \\
&= \frac{\partial}{\partial s}\sum_{\nu\ge 0}\frac{1}{\nu!}t^{\nu}\mathcal{F}(s)^{\nu}\bigg|_{s=0} && \text{by (1.45)} \\
&= \sum_{\nu\ge 1}\frac{1}{\nu!}t^{\nu}\sum_{0\le k\le \nu-1} Q^{k}\Delta_{\mu}Q^{\nu-1-k} && \text{by (1.29),}
\end{aligned}$$

from which we find that

$$\boldsymbol{\pi}^{\mathrm{T}}\mathrm{E}[X(t)\,;\,\varphi(t)] = \boldsymbol{\pi}^{\mathrm{T}}\Delta_{\mu}\sum_{\nu\ge 1}\frac{1}{\nu!}t^{\nu}Q^{\nu-1}, \tag{1.53}$$

and

$$\mathrm{E}[X(h)\,;\,\varphi(h)]\mathbf{1} = \sum_{\nu\ge 1}\frac{1}{\nu!}h^{\nu}Q^{\nu-1}\Delta_{\mu}\mathbf{1}. \tag{1.54}$$

We bring together (1.51–1.54) and obtain

$$\begin{aligned}
C(t,h) &= \boldsymbol{\pi}^{\mathrm{T}}\Delta_{\mu}\sum_{n\ge 1}\frac{1}{n!}t^{n}Q^{n-1}\sum_{n'\ge 1}\frac{1}{n'!}h^{n'}Q^{n'-1}\Delta_{\mu}\mathbf{1} - th(\boldsymbol{\pi}^{\mathrm{T}}\boldsymbol{\mu})^{2} \\
&= \boldsymbol{\pi}^{\mathrm{T}}\Delta_{\mu}\sum_{\nu\ge 2}\frac{1}{\nu!}((t+h)^{\nu} - t^{\nu} - h^{\nu})Q^{\nu-2}\Delta_{\mu}\mathbf{1} - th(\boldsymbol{\pi}^{\mathrm{T}}\boldsymbol{\mu})^{2}.
\end{aligned}$$

Using (1.47), this shows that

$$\begin{aligned}
C(t,h) &= \boldsymbol{\pi}^{\mathrm{T}}\Delta_{\mu}(e^{Q(t+h)} - e^{Qt} - e^{Qh} + I)(Q^{\#})^{2}\Delta_{\mu}\mathbf{1} \\
&= \boldsymbol{\pi}^{\mathrm{T}}\Delta_{\mu}(e^{Q(t)} - I)(e^{Qh} - I)(Q^{\#})^{2}\Delta_{\mu}\mathbf{1}.
\end{aligned} \tag{1.55}$$

As Q and $Q^{\#}$ commute by (1.31), that last equation is equivalent to (1.49) and the lemma is proved. □

Remark 1.21. Notice that the covariance of $X(t)$ and $X(t+h) - X(t)$ is independent of the volatility parameter $\boldsymbol{\sigma}$. Also, if the drift parameters μ_i are all equal, then the increments $X(t)$ and $X(t+h) - X(t)$ are uncorrelated since $\Delta_{\mu}\mathbf{1}$ is proportional to $\mathbf{1}$ in that case, and $Q^{\#}\mathbf{1} = \mathbf{0}$.

Example 1.22. This example is based on a singularly perturbed system, analysed in Dendievel et al. [61, Section 8]. There are two sets of two phases each and $\{\varphi\}$ moves slowly from one set to the other and back. The parameters are

$$Q = \begin{bmatrix} -a_1 & a_1 & 0 & 0 \\ a_2 & -a_2 & 0 & 0 \\ 0 & 0 & -a_3 & a_3 \\ 0 & 0 & a_4 & -a_4 \end{bmatrix} + \varepsilon\begin{bmatrix} -1 & 0 & 1 & 0 \\ 0 & -1 & 0 & 1 \\ 1 & 0 & -1 & 0 \\ 0 & 1 & 0 & -1 \end{bmatrix}$$

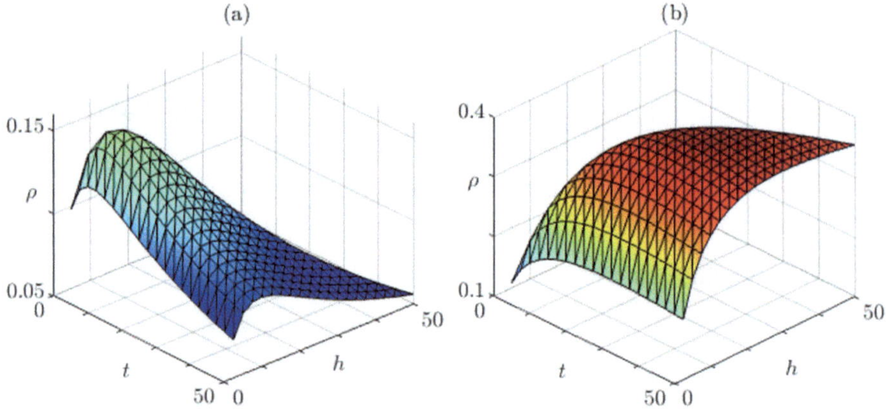

Fig. 1.6 For the MMBM of Example 1.22, correlation coefficient of $X(t)$ and $X(t+h) - X(t)$ in the intervals $0 \le h \le t \le 50$. In (**a**) on the left, $\varepsilon = 0.05$, in (**b**) on the right, $\varepsilon = 0.01$

with

$$\boldsymbol{\mu} = \begin{bmatrix} 1 & -1 & 1 & -1 \end{bmatrix}^{\mathrm{T}}, \qquad \boldsymbol{\sigma} = \begin{bmatrix} 1 & 1 & 1 & 1 \end{bmatrix}^{\mathrm{T}},$$

and $a_1 = 2$, $a_2 = 3$, $a_3 = 3$, $a_4 = 1.5$. In the first set of phases, the process spends more time in Phase 1, and the tendency is for the level to increase; in the second set, the process spends more time in Phase 4, and the tendency is to move down.

We give in Fig. 1.6 the correlation coefficients

$$\rho(t,h) = C(t,h)/\sqrt{\mathrm{Var}(X(t))\mathrm{Var}(X(h))}, \qquad 0 \le h \le t \le 50$$

for two values of ε. In Fig. 1.6a on the left, $\varepsilon = 0.05$, the process spends on average $1/\varepsilon = 20$ units of time in either of the two sets $\{1,2\}$ and $\{3,4\}$ before switching to the other, and the stationary drift is -0.0664. In Fig. 1.6b on the right, $\varepsilon = 0.01$, the expected time spent in a set is 100 units of time, and the stationary drift is -0.0666, nearly the same as with $\varepsilon = 0.05$. The shapes of the two surfaces are very different, in all likelihood because on the right, with ε being so small, it is more likely that the system parameters have not changed before time t and are the same in the two intervals $(0,t)$ and $(t, t+h)$. Note that the range on the vertical axis is 0.1–0.4 on the right, while it is 0.05–0.15 on the left, so that ρ is significantly closer to 0 for $\varepsilon = 0.05$.

1.10 Markov-Modulated Lévy Processes

A next step would be to combine the Markov modulation from Sect. 1.7 and the sample paths with jumps from Sect. 1.6: $\{X\}$ would behave like a Lévy process with parameters a_i, σ_i and jump intensity $\nu_i(\cdot)$ during intervals of time when $\varphi(t)$ remains equal to i and, in addition, there might be jumps when $\varphi(t)$ switches from one state to another. Markov-modulated Lévy processes are also known as Markov additive processes (MAPs), see Asmussen [14, Chapter 11] for an introductory presentation.

MAPs are analysed in Asmussen and Kella [16], Breuer [44], D'Auria et al. [55], and Ivanovs [89], to cite but a few authors. Ivanovs et al. [91] illustrate some differences between the analysis when $\{X\}$ takes real and when it takes integer values.

1.11 Fast-Switching Fluid Flows

Linear fluid flows, or Fluid flows for short, form a family of MMBMs characterised by the fact that $\sigma_i = 0$ for all i in $\mathcal{E}$. Thus, the level is a stepwise linear function of time, with slope μ_i during every interval of sojourn in phase i, i in $\mathcal{E}$. Fluid flows have been used as models for dams and reservoirs (Loynes [110]) and for buffers in telecommunication systems (Anick et al. [12]); we shall use the popular name *Fluid queues* for their regulated version, when the level may not become negative.

Fast-switching Fluid flows have been suggested in Ramaswami [131] as approximations to Brownian motion and may be used as a bridge between the results obtained by Matrix-analytic methods for Fluid flows and properties of MMBMs. We start with a two-state Fluid flow indexed by the parameter ε: $\{X_\varepsilon, \beta_\varepsilon\} = \{X_\varepsilon(t), \beta_\varepsilon(t) : t \in \mathbb{R}^+\}$, β_ε in $\{1, 2\}$, with fluid rate vector $\boldsymbol{\mu}_\varepsilon = \begin{bmatrix}1/\varepsilon & -1/\varepsilon\end{bmatrix}^{\mathrm{T}}$ and transition matrix

$$T_\varepsilon = \begin{bmatrix} -1/\varepsilon^2 & 1/\varepsilon^2 \\ 1/\varepsilon^2 & -1/\varepsilon^2 \end{bmatrix}. \tag{1.56}$$

At time 0, $X_\varepsilon(0) = 0$ and $\beta_\varepsilon(0)$ has the uniform distribution $\begin{bmatrix}0.5 & 0.5\end{bmatrix}$.

In this model, the fluid level changes direction faster at the rate $1/\varepsilon^2$ as ε takes steadily smaller values, and the slope keeps increasing as well, at the rate $1/\varepsilon$. In the limit as $\varepsilon \to 0$, the marginal process $\{X_\varepsilon\}$ converges to the standard Brownian motion $\mathcal{B}$. The process $\{X_\varepsilon, \beta_\varepsilon\}$ is named the *flip-flop* in Latouche and Nguyen [103], we use it in the proof of Theorem 5.2.

Starting from an MMBM with parameters $\boldsymbol{\mu}$, $\boldsymbol{\sigma}$, and Q, we construct the fast-switching Fluid process $\{X_\varepsilon(t), \beta_\varepsilon(t), \varphi_\varepsilon(t) : t \geq 0\}$ as follows: Here, the phase process is a two-dimensional Markov chain $\{\beta_\varepsilon, \varphi_\varepsilon\}$ with state space $\mathcal{S} = \{(k, i) : k \in \{1, 2\} \text{ and } i \in \mathcal{E}\}$ and generator

$$T_\varepsilon = \begin{bmatrix} Q - (1/\varepsilon^2)I & (1/\varepsilon^2)I \\ (1/\varepsilon^2)I & Q - (1/\varepsilon^2)I \end{bmatrix}.$$

The matrix T_ε is also written as

$$T_\varepsilon = I_2 \otimes Q + (1/\varepsilon^2)G \otimes I_m, \tag{1.57}$$

where I_n denotes the $n \times n$ identity matrix, $\otimes$ is the Kronecker product operator[1] and

$$G = \begin{bmatrix} -1 & 1 \\ 1 & -1 \end{bmatrix}. \tag{1.58}$$

The fluid rate vector is

$$\begin{aligned} \boldsymbol{\mu}_\varepsilon &= \begin{bmatrix} \boldsymbol{\mu} + (1/\varepsilon)\boldsymbol{\sigma} \\ \boldsymbol{\mu} - (1/\varepsilon)\boldsymbol{\sigma} \end{bmatrix} \\ &= \begin{bmatrix} 1 \\ 1 \end{bmatrix} \otimes \boldsymbol{\mu} + (1/\varepsilon) \begin{bmatrix} 1 \\ -1 \end{bmatrix} \otimes \boldsymbol{\sigma}. \end{aligned} \tag{1.59}$$

Simply stated, we duplicate the state space $\mathcal{E}$ in the Markov-modulated Brownian motion $\{X, \varphi\}$, and we introduce the auxiliary process $\{\beta_\varepsilon\}$ to keep track of which copy is in use at any given time. For ε sufficiently small, the components $(\mu_\varepsilon)_{1i}$ in the first copy are positive, while the components $(\mu_\varepsilon)_{2i}$ in the other copy are negative, for all i in $\mathcal{E}$.

The next theorem is proved in Latouche and Nguyen [101].

Theorem 1.23. *Assume that $X_\varepsilon(0) = 0$ and that the initial phases $\beta_\varepsilon(0)$ and $\varphi_\varepsilon(0)$ have their respective stationary distributions $\begin{bmatrix} 0.5 & 0.5 \end{bmatrix}$ and $\boldsymbol{\pi}$.*

The processes $\{X_\varepsilon(t), \varphi_\varepsilon(t) : t \geq 0\}$ converge weakly to the Markov-modulated Brownian motion $\{X(t), \varphi(t) : t \geq 0\}$ as $\varepsilon \to 0$. □

Example 1.24. As an illustration, we compare $X(1)$ and $X_\varepsilon(1)$, for a few values of ε, for the MMBMs in Examples 1.18 and 1.19. We do not plot the density functions themselves, or the cumulative distribution functions, as they nearly coincide, even with the larger value that we use for ε. Instead, we estimate the maximum difference

$$\delta_\varepsilon = \max_{a \leq x \leq b} |f(x) - f_\varepsilon(x)| \tag{1.60}$$

between the density functions of $X(1)$ and $X_\varepsilon(1)$, taken over an interval (a, b) which covers most of the probability mass of $X(1)$; for Example 1.18, the interval is $(-12, 4)$, and it is $(-8, 6)$ for Example 1.19—these are the same intervals as in Figs. 1.4 and 1.5.

We give $\log_{10} \delta_\varepsilon$ in Table 1.1 for $\varepsilon = 10^{-1}$, 10^{-2}, 10^{-3}, and 10^{-4}. As we can see, δ_ε is very nearly a linear function of ε for both examples, and it is quite small even for $\varepsilon = 0.1$. The difference between the cumulative distribution functions is of similar order of magnitude and is not reported. The illustrations in this book are obtained from simulations of fast-switching Fluid flows with $\varepsilon = 10^{-3}$.

---

[1] If A is an $n \times m$ matrix and B is an $n' \times m'$ matrix, the Kronecker product $A \otimes B$ is an $nn' \times mm'$ matrix defined by

$$A \otimes B = \begin{bmatrix} a_{11}B & \cdots & a_{1m}B \\ \vdots & & \vdots \\ a_{n1}B & \cdots & a_{nm}B \end{bmatrix}.$$

See Graham [71] for further details.

$\log_{10} \varepsilon$		-1	-2	-3	-4
$\log_{10} \delta_\varepsilon$	Ex. 1.18	-3.3238	-4.3378	-5.3391	-6.3392
	Ex. 1.19	-2.5234	-3.5526	-4.5554	-5.5557

Table 1.1 Comparison of the density functions of $X(1)$ and $X_\varepsilon(1)$ for the MMBMs in Examples 1.18 and 1.19, and for $\varepsilon = 10^{-i}$, $i = 1, \ldots, 4$; δ_ε measures the difference and is defined in (1.60)

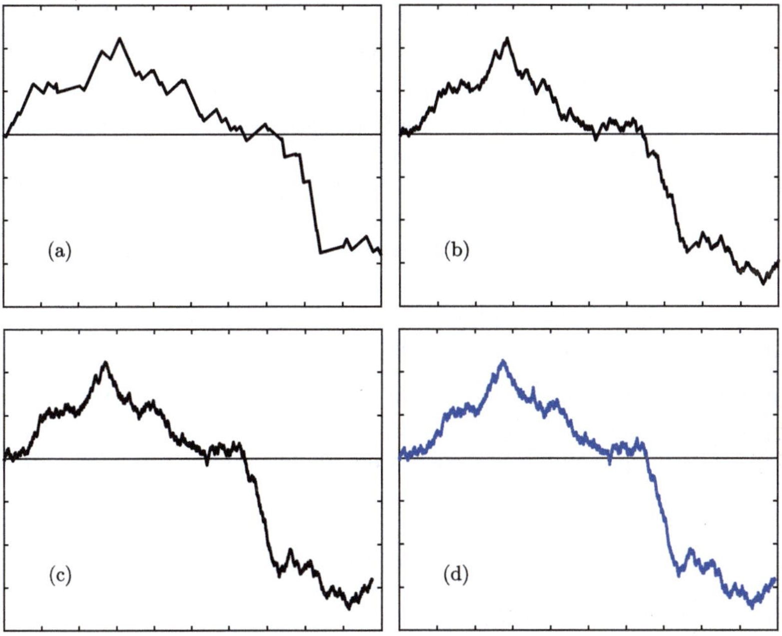

Fig. 1.7 Sample path for $0 \le t \le 1$ of a fast-switching Fluid flow with decreasing values of ε. The MMBM is the same as in Example 1.18. From approximation **(a)** to approximation **(d)**, ε takes successively the values $1/(2^2 10)$, $1/(2^4 10)$, $1/(2^6 10)$, and $1/(2^8 10)$

Nguyen and Peralta [119] construct a probability space and a sequence of fast-switching Fluid flows $\{X_{\varepsilon(n)}, \varphi_{\varepsilon(n)}\}$, $n \ge 1$, with $\varepsilon(n) < \varepsilon(n-1)$, $\lim_{n\to\infty} \varepsilon(n) = 0$, such that the sequence converges strongly to $\{X, \varphi\}$. The authors' construction is based on a set of embedded Poisson processes with increasing rates $\{\lambda_n\}$, $n \ge 1$, with $\lim_{n\to\infty} \lambda_n = \infty$, and it makes use of Lemma 2.6. That property, known as the *Wiener-Hopf factorisation* of Brownian motion, makes it possible to track both the value of $\{X\}$ at epochs determined by the Poisson processes and the minimum over intervals between successive epochs. Lemma 2.6 is stated in the next chapter when we first make use of it. We shall repeatedly refer to it in the book,

Example 1.25. To illustrate Theorem 1.23, we take the MMBM of Example 1.18, and we show in Fig. 1.7 four trajectories of $\{X_\varepsilon\}$ for decreasing values of ε. The four simulations have been implemented with the construction in [119] so that the strong convergence is made visible. The parameters μ_ε and T_ε are defined in (1.59) and (1.57), and we give to the parameter ε the four values $1/(2^i 10)$, $i = 2, 4, 6$, and 8. With $\varepsilon = 1/40$, there is an average of four changes of β during a sojourn in any of the phases, while there are more than 250 such changes for the smallest value.

The nth approximated trajectory is obtained by linear interpolation of a sequence of points $\{X_\varepsilon(t_i^{(n)}) = x_i^{(n)} : i \in \mathbb{N}\}$ and we draw the attention on three features. Firstly, $x_i^{(n)} > x_{i+1}^{(n)} < x_{i+2}^{(n)}$ for even values of i: The trajectory alternates between intervals of linear growth and intervals of linear decrease, as we clearly see in Fig. 1.7a.

Next, the set $\{x_i^{(n)} : i \in \mathbb{N}\}$ of levels at step n is a subset of the set $\{x_i^{(n+1)} : i \in \mathbb{N}\}$ of levels at the next iteration: Once a value of $X(\cdot)$ has been determined, it is kept for all subsequent approximations. This is why the general shape of the trajectory is preserved from one simulation to the next, even though the relative smoothness of the trajectory in Fig. 1.7a gives way to steadily faster oscillations as ε decreases.

Finally, the sequence of epochs $\{t_i^{(n)}) : i \in \mathbb{N}\}$ is not preserved. For instance, the sharp drop of level in the last part of the graphs does not occur at the same time over the four figures, but the difference in timing becomes smaller with ε.

Remark 1.26. A classic construction of the standard Brownian motion is based on Donsker's theorem ([33, Theorem 8.2]) which goes as follows: Take a sequence $\{\xi_1, \xi_2, \ldots\}$ of independent and identically distributed random variables with mean 0 and variance 1 and define

$$Z_n(t) = \frac{1}{\sqrt{n}}\{ \sum_{1 \le i \le \lfloor nt \rfloor} \xi_i + (nt - \lfloor nt \rfloor)\xi_{\lfloor nt \rfloor + 1} \},$$

where $\lfloor \cdot \rfloor$ is the floor function. The sequence $\{Z_n\}$ of processes converges weakly to $\mathcal{B}$ as $n \to \infty$. This means that we may approximate a trajectory of the standard Brownian motion with linear interpolation between the points of a random walk $\{S_{n\tau} : n \ge 0\}$, with $S_0 = 0$ and $S_{n\tau} = S_{(n-1)\tau} + \xi_n$, if we choose τ sufficiently small.

Another construction is based on the Haar wavelet,[2] that is, the system of functions $\{H_n(\cdot) : n \ge 0\}$ with

$$H_{2^j+k}(t) = \begin{cases} 2^{j/2} & \text{for } t \in [\frac{k}{2^j}, \frac{k+1/2}{2^j}) \\ -2^{j/2} & \text{for } t \in [\frac{k+1/2}{2^j}, \frac{k}{2^j}) \\ 0 & \text{elsewhere,} \end{cases}$$

[2] Wavelets are used in data analysis. Each family forms a complete orthonormal system in the Hilbert space $L^2(\mathbb{R})$. Two references are Mallat [111] and Meyer [114].

for $j \geq 0, 0 \leq k \leq 2^j - 1$. Define

$$W_n(t) = \sum_{0 \leq \nu \leq n} \xi_\nu \int_0^t H_\nu(s)\,\mathrm{d}s,$$

for $n \geq 0$. The sequence $\{W_n\}$ converges weakly to $\mathcal{B}$ (Schilling and Pastzsch [136, Chapter 3]).

Donsker's theorem and most of the classic constructions are rooted in the central limit theorem and are inspired by Definition 1, which requires that increments over an interval are normally distributed. Ramaswami's approximation described in this section is more attuned to Theorem 1.2 and its qualitative characterisation of Brownian motion.

Parenthesis 1.27. About Simulation

One is tempted, when exact calculations are not feasible, to use simulation as an alternative computational approach. For instance, Mathews [112] investigates the use of fast-switching linear fluid to evaluate stochastic integrals and analyses convergence rates as $\varepsilon \to 0$. The author determines that it is better to use a four-state Fluid flow process with parameters $\boldsymbol{\mu}'_\varepsilon = (1/\varepsilon)\begin{bmatrix}1 & 1 & -1 & -1\end{bmatrix}$ and

$$T'_\varepsilon = \begin{bmatrix} -2/\varepsilon^2 & 1/\varepsilon^2 & 1/\varepsilon^2 & 0 \\ 1/\varepsilon^2 & -2/\varepsilon^2 & 0 & 1/\varepsilon^2 \\ 1/\varepsilon^2 & 0 & -2/\varepsilon^2 & 1/\varepsilon^2 \\ 0 & 1/\varepsilon^2 & 1/\varepsilon^2 & -2/\varepsilon^2 \end{bmatrix},$$

instead of the two-phase process defined at the beginning of Sect. 1.11. With this four-phase Fluid flow, an interval of constant direction, A say, is the sum $A = \sum_{1 \leq i \leq N} \xi_i$ where N is geometric with parameter $1/2$, and the ξ_is are iid random variables exponentially distributed with parameter $2/\varepsilon^2$, so that A is exponential with parameter $1/\varepsilon^2$, exactly like the two-phase process.

The reason why the four-state process is more efficient is that the two-phase process suffers from a lack of independence in time, as an interval with positive fluid rate is automatically followed with one of negative fluid rate and vice versa. With the four-state process, any interval may be followed with equal probability with a positive or a negative rate, and successive intervals are thoroughly independent.

△

1.12 Notes

I should have made, somewhere, the explicit assumption that

- there should be at least one i for which σ_i is strictly positive,
- or, if all σ_is are equal to 0, there should be at least one pair of phases i and j such that $\mu_i < 0 < \mu_j$,

thereby making sure that trajectories are not monotone and that we avoid Markov-modulated continuous subordinators. That I did not find a suitable place to write down this assumption earlier in the chapter is entirely due to my lack of imagination.

For my presentation of Brownian motion, I generally took my inspiration in Harrison [80, Chapter 1]. Other good places for the reader to explore are Breiman [43, Chapter 12], Freedman [66, Chapter 1.3] where I have found much about sample path properties, and Resnick [132, Chapter 6.6] which has some very nice arguments and proofs.

There are several path properties of Brownian motion that I have not discussed, such as the trajectories being monotone in no interval, *or* the fact that Brownian motion almost surely has infinite variation over $(0, t)$ for any positive t, *or* ... One weak excuse is that I do not use such properties later in the book. The real reason, probably, being that the present piece of work is about the Matrix-analytic approach to the study of Markov-modulated Brownian motion, and that I have not (not yet ?) found how these fundamental properties find a place there.

For Lévy processes, I like to mention Bertoin [32], Kyprianou [98], and Papa-pantoleon [125]. The physical description given at the end of Sect. 1.6 is based on an argument I have seen on the Web. It may be deduced from the more common (but somewhat less transparent) description given, for instance, in [32, Page 13].

Nguyen and Peralta [119] mention that the fast-switching linear fluid is also known as the telegraph process and is analysed in Goldstein [68], Kac [94], Pinsky [126], and Watanabe [143]. Its definition is given there as

$$X_\lambda(t) = \sqrt{\lambda} \int_0^t (-1)^{N_\lambda(s)} \, \mathrm{d}s,$$

where $\{N_\lambda\}$ is a Poisson process with rate λ. It is clearly equivalent to our definition if $\lambda = 1/\varepsilon^2$.

Chapter 2
First Passage Across a Level

⋄ Ladder time ⋄ Ladder process ⋄ Stationary drift ⋄ Purely Brownian MMBM ⋄ Linear fluid ⋄ General MMBM ⋄ Distribution of $\varphi(\tau_x)$ ⋄ Distribution of τ_x ⋄

We deal in this chapter with the random times τ_x, $x \geq 0$, defined as

$$\begin{aligned} \tau_x &= \inf\{t \geq 0 : X(t) \geq x\}, \\ &= \infty \qquad\qquad \text{if } X(t) < x \text{ for all } t. \end{aligned}$$

If $X(0) = 0$, as the trajectories of the MMBM process are continuous, τ_x is the first epoch t where the level of the MMBM is *equal* to x, if there is such an epoch; in case the process $\{X\}$ remains below level x during its whole history, then τ_x is arbitrarily given the value ∞. The process $\{\tau_x : x \geq 0\}$ is known as the *ascending ladder time process*. Its twin $\{\tau_{-x} : x \geq 0\}$ is the *descending* ladder time process, with

$$\begin{aligned} \tau_{-x} &= \inf\{t \geq 0 : X(t) \leq -x\}, \\ &= \infty \qquad\qquad \text{if } X(t) > -x \text{ for all } t. \end{aligned}$$

We are interested in both the first passage times to higher levels and those to lower levels; their analysis is similar, but the results are slightly different and it is worth our while to give the two separately.

In analogy with Bertoin [32, Section VI.1], we call $\{\varphi(\tau_x) : x \geq 0, \tau_x < \infty\}$ and $\{\varphi(\tau_{-x}) : x \geq 0, \tau_{-x} < \infty\}$ the ascending and descending *ladder processes*, respectively. Remember that "level 0" is a purely conventional mark, and that the evolution of $X(t)$ is defined in terms of increments $X(t + h) - X(t)$ which are independent of $X(t)$ and only depend on h, conditionally given $\varphi(t)$. Thus,

$$\mathrm{P}[\tau_x \leq t | X(0) = 0, \varphi(0) = i] = \mathrm{P}[\tau_{a+x} \leq t | X(0) = a, \varphi(0) = i]$$

for any starting level a in $\mathbb{R}$. We assume most often that $X(0) = 0$.

© The Author(s), under exclusive license to Springer Nature Switzerland AG 2026

G. Latouche, *Markov-Modulated Brownian Motion*, Springer Series in Operations Research and Financial Engineering, https://doi.org/10.1007/978-3-031-98576-8_2

2.1 Stationary Drift

We begin with a representation of $X(t)$ slightly different from (1.23). Define $\{T_n : n \geq 0\}$ as the successive jump epochs of the phase process $\{\varphi\}$, with $T_0 = 0$, and define $K = \sup\{n : T_n < t\}$. Time T_K is the last jump epoch before t and we have

$$\begin{aligned} X(t) &= X(T_K) + \sum_{i\in\mathcal{E}} \mathbb{1}\{\varphi(T_K) = i\}(X(t) - X(T_K)) \\ &= \sum_{n=0}^{K-1} \sum_{i\in\mathcal{E}} \mathbb{1}\{\varphi(T_n) = i\}(X(T_{n+1}) - X(T_n)) \\ &\quad + \sum_{i\in\mathcal{E}} \mathbb{1}\{\varphi(T_K) = i\}(X(t) - X(T_K)) \\ &= \sum_{i\in\mathcal{E}} X_i^*(t), \end{aligned} \tag{2.1}$$

where

$$\begin{aligned} X_i^*(t) = &\sum_{n=0}^{K-1} \mathbb{1}\{\varphi(T_n) = i\}(X_i(T_{n+1}) - X_i(T_n)) \\ &+ \mathbb{1}\{\varphi(T_K) = i\}(X_i(t) - X_i(T_K)), \qquad i \in \mathcal{E} \end{aligned}$$

is the contribution during the interval of time $(0, t)$ from the process $\{X_i\}$ that was part of Definition 1.11. There is some similarity between (1.23) and (2.1) in that the right-hand side is based in both cases on the superposition of m continuous processes; the difference is that in (1.23) each original Brownian motion $\{X_i\}$, $1 \leq i \leq n$, is allowed to evolve independently of everything else during intervals where $\varphi(t) \neq i$, while in (2.1), the artificial processes $\{X_i^*\}$ remain constant in those intervals.

The time spent in phase i during the interval $(0, t)$ is

$$\begin{aligned} \theta_i(t) &= \sum_{n=0}^{K-1} \mathbb{1}\{\varphi(T_n) = i\}(T_{n+1} - T_n) + \mathbb{1}\{\varphi(T_K) = i\}(t - T_K) \\ &= \int_0^t \mathbb{1}\{\varphi(u) = i\}\,\mathrm{d}u. \end{aligned} \tag{2.2}$$

We give in Fig. 2.1 an illustration of the pair of functions $X_i^*(\cdot)$ and $\theta_i(\cdot)$ for a given phase i, the dashed segments correspond to intervals when the process $\{\varphi\}$ visits states different from i. The piecewise linear function in red is the trajectory of θ_i, it is continuous and is like a stopwatch which is stopped when the phase is different from i. The blue line is the trajectory of X_i^*, it behaves like the Brownian motion over the intervals of time when $\varphi(t) = i$ and it is constant across intervals when $\varphi(t) \neq i$.

Next, we define for each phase i the inverse function ρ_i of θ_i:

$$\rho_i(t) = \max\{s : \theta_i(s) = t\},$$

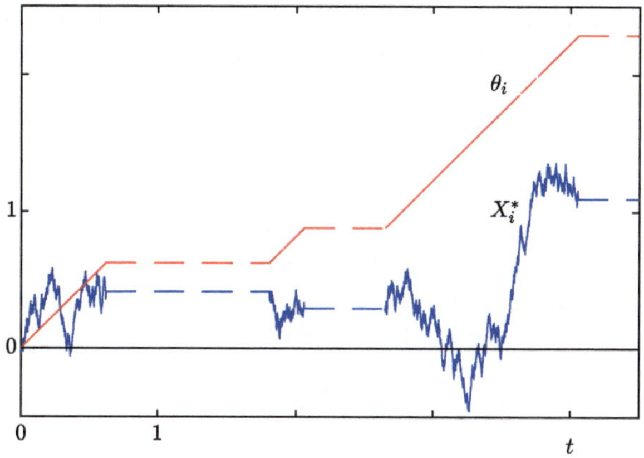

Fig. 2.1 The piecewise linear red line is thesample trajectory of the time spent in phase i, assuming that $\varphi(0) = i$, and the blue line shows the increments of the MMBM during the sojourn intervals in phase i

and we call $\rho_i(\cdot)$ the local clock associated with i. It is strictly increasing and continuous from the right and its jumps are equal to the intervals of time spent away from phase i.

Finally, we define for each i the restricted process $\{V_i(t) : t \geq 0\}$ as

$$V_i(t) = X_i^*(\rho_i(t)).$$

The effect is to remove the intervals of time when $\{X_i^*\}$ does not evolve, and to create thereby a new Brownian motion with parameters (μ_i, σ_i), $i \in \mathcal{E}$. The strong Markov property comes into play here, to justify that $\{V_i\}$ is a Brownian motion indeed, and that $V_i(t) \overset{\mathrm{d}}{=} X_i(t)$ for all t.

With this, (2.1) may be written as

$$X(t) = \sum_{i \in \mathcal{E}} V_i(\theta_i(t)) \overset{\mathrm{d}}{=} \sum_{i \in \mathcal{E}} \theta_i(t) \frac{X_i(\theta_i(t))}{\theta_i(t)}.$$

In the limit as $t \to \infty$,

$$\lim_{t \to \infty} \theta_i(t) = \infty, \quad \lim_{t \to \infty} \frac{\theta_i(t)}{t} = \pi_i, \quad \text{and} \quad \lim_{t \to \infty} \frac{X_i(t)}{t} = \mu_i$$

with probability one, so that the stationary expected drift is

$$\lim_{t \to \infty} \frac{X(t)}{t} = \boldsymbol{\pi}^{\mathrm{T}} \boldsymbol{\mu} \qquad \text{a.s.,} \tag{2.3}$$

and the process $\{X\}$ drifts to $+\infty$ if $\boldsymbol{\pi}^{\mathrm{T}} \boldsymbol{\mu} > 0$, to $-\infty$ if $\boldsymbol{\pi}^{\mathrm{T}} \boldsymbol{\mu} < 0$.

To determine the limiting behaviour of $X(t)$ when $\boldsymbol{\pi}^{\mathrm{T}}\boldsymbol{\mu} = 0$, we turn to the renewal process $\{t_n : n \geq 0\}$ of epochs at which $\{\varphi\}$ returns to its initial value $\varphi(0)$:

$$\begin{aligned} t_0 &= 0 \\ t_n &= \inf\{T_k > t_{n-1} : \varphi_k = \varphi_0\}, \end{aligned}$$

where $\{\varphi_k : k \geq 0\}$ is the jump chain and $\{T_k : k \geq 0\}$ is the sequence of jump epochs introduced at the beginning of this section.

Define $S_n = X(t_n)$, $n \geq 0$. Since $\varphi(t_n) = \varphi(0)$ for all n, the differences $S_n - S_{n-1}$ are independent by the strong Markov properties, and

$$S_n - S_{n-1} \overset{\mathrm{d}}{=} \sum_{j \in \mathcal{E}} X_j(\theta_j(t_1)) \overset{\mathrm{d}}{=} S_1$$

so that $\{S_n : n \geq 0\}$ is a standard random walk on $\mathbb{R}$, with well-known properties.

For any initial phase i, we have

$$\begin{aligned} \mathrm{E}_{0,i}[S_n - S_{n-1}] &= \sum_{j \in \mathcal{E}} \mathrm{E}_{0,i}[X_j(\theta_j(t_1))] \\ &= \sum_{j \in \mathcal{E}} \mu_j \mathrm{E}_i[\theta_j(t_1)]. \end{aligned}$$

The expected sojourn time in phase j between two successive visits to phase i is $c\pi_j$ for some c independent of i (see Çinlar [49, Eq. 8.5.10]). Therefore,

$$\mathrm{E}_{0,i}[S_n - S_{n-1}] = c \sum_{j \in \mathcal{E}} \mu_j \pi_j = 0$$

in this case, and

$$\limsup_n S_n = +\infty, \qquad \liminf_n S_n = -\infty$$

(see Chung [51, Theorems 8.3.4, 8.4.4]).

Altogether, our analysis is summarised as follows.

Lemma 2.1. *Assuming as we do that $\{X\}$ is not constant, we recognise three cases:*

(a) If $\boldsymbol{\pi}^{\mathrm{T}}\boldsymbol{\mu} > 0$, then $\lim_{t\to\infty} X(t) = +\infty$ and $\liminf_t X(t) > -\infty$.
(b) If $\boldsymbol{\pi}^{\mathrm{T}}\boldsymbol{\mu} < 0$, then $\limsup_t X(t) < +\infty$ and $\lim_{t\to\infty} X(t) = -\infty$.
(c) If $\boldsymbol{\pi}^{\mathrm{T}}\boldsymbol{\mu} = 0$, then $\limsup_t X(t) = +\infty$ and $\liminf_t X(t) = -\infty$. □

Lemma 2.1 has a number of immediate consequences about the first passage time across levels.

Corollary 2.2. *Assume that $X(0) = 0$, $\varphi(0) = i$ in $\mathcal{E}$. Take $x > 0$ arbitrary:*

(a) If $\boldsymbol{\pi}^{\mathrm{T}}\boldsymbol{\mu} > 0$, then $\tau_x < \infty$ a.s. and $\mathrm{E}[\tau_x] < \infty$, while $\mathrm{P}[\tau_{-x} < \infty] < 1$.
(b) If $\boldsymbol{\pi}^{\mathrm{T}}\boldsymbol{\mu} < 0$, then $\mathrm{P}[\tau_x < \infty] < 1$, while $\tau_{-x} < \infty$ a.s. and $\mathrm{E}[\tau_{-x}] < \infty$.

(c) If $\boldsymbol{\pi}^{\mathrm{T}}\boldsymbol{\mu} = 0$, *then both* τ_x *and* τ_{-x} *are finite a.s., and both* $\mathrm{E}[\tau_x]$ *and* $\mathrm{E}[\tau_{-x}]$ *are infinite.*

We write that $\{X\}$ *is transient when* $\boldsymbol{\pi}^{\mathrm{T}}\boldsymbol{\mu} \neq 0$ *and null recurrent when* $\boldsymbol{\pi}^{\mathrm{T}}\boldsymbol{\mu} = 0$.

Proof. In Case (a), $\tau_x < \infty$ a.s. and $\mathrm{P}[\tau_{-x} < \infty] < 1$, in direct consequence of Lemma 2.1, and we only need to focus on the expectation of τ_x.

In proving Lemma 2.1, we have defined the epochs $\{t_n : n \geq 0\}$ when $\{X, \varphi\}$ returns to the initial phase, and the sequence $\{S_n : n \geq 0\}$ of levels at these times. We now extract two subprocesses characterised by increasing values of the level:

$$t_0^* = 0, \qquad S_0^* = 0,$$
$$t_k^* = \inf\{t_\nu > t_{k-1}^* : S_\nu > S_{k-1}^*\}, \qquad S_k^* = X(t_k^*)$$

for $k \geq 1$. The differences $t_k^* - t_{k-1}^*$ and $S_k^* - S_{k-1}^*$ are i.i.d. strictly positive random variables, and both $\{t^*\}$ and $\{S^*\}$ are strictly increasing renewal processes.

The index τ^* when $\{S^*\}$ gets over x, defined with $S_{\tau^*-1}^* < x \leq S_{\tau^*}^*$, is finite and $\mathrm{E}[\tau^*] < \infty$ by [49, Proposition 9.1.14]. Clearly, $\tau_x \leq t_{\tau^*}^*$, and it suffices to show that $\mathrm{E}[t_{\tau^*}^*] < \infty$. We use Wald's identity [14, Page 414] and write that

$$\mathrm{E}[t_{\tau^*}^*] = \mathrm{E}[\sum_{1 \leq \nu \leq \tau^*} (t_\nu^* - t_{\nu-1}^*)] = \mathrm{E}[\tau^*]\mathrm{E}[t_1^*] < \infty$$

because $\mathrm{E}[t_1^*]$ is finite by [51, Theorem 8.4.4].

The argument for Case (b) is identical. For Case (c), τ_x and τ_{-x} are finite by direct consequence of Lemma 2.1, and we prove by contradiction that their expectations are infinite. It results from (1.41) that

$$\mathrm{E}_{0,\pi}[X(T)] = \boldsymbol{\pi}^{\mathrm{T}}\boldsymbol{\mu}\, \mathrm{E}_{0,\pi}[T] = 0 \tag{2.4}$$

if T is a nonnegative random variable with finite mean, and so $\mathrm{E}_{0,\pi}[X(\tau_x)]$ would be equal to 0 if $\mathrm{E}[\tau_x]$ were finite. However, this would be contradictory with $X(\tau_x) = x$ by definition of τ_x. Thus, we conclude that $\mathrm{E}_{0,\pi}[\tau_y]$ is infinite for all y. This implies that for all y, there exists j such that $\mathrm{E}_{0,j}[\tau(y)] = \infty$.

Now, take any y with $0 < y < x$, and any initial phase i. We have

$$\mathrm{E}_{0,i}[\tau_x] \geq \mathrm{P}_{0,i}[A]\, \mathrm{E}_{0,j}[\tau_y],$$

where A is the event that the process $\{X(t), \varphi(t)\}$ reaches $(x - y, j)$ before reaching level x. By our irreducibility assumption, $\mathrm{P}_{0,i}[A] > 0$ and so $\mathrm{E}_{0,i}[\tau_x] = \infty$ for all x and all i. □

Even though $X(t)$ oscillates forever without bounds when $\boldsymbol{\pi}^{\mathrm{T}}\boldsymbol{\mu} = 0$, still its expected value does converge to a finite value as $t \to \infty$. Equation (2.5) follows as a direct consequence of (1.40) with $\boldsymbol{\alpha} = \boldsymbol{e}_i$. We shall use Proposition 2.3 in the proof of Lemma 3.7.

Proposition 2.3. *If* $\pi^{\mathrm{T}}\mu = 0$, *then*

$$\lim_{t\to\infty} \mathrm{E}_{0,i}[X(t)] = (-Q^{\#}\mu)_i \tag{2.5}$$

for $i \in \mathcal{E}$. □

2.2 Ladder Processes

It is useful to organise the phases into different sets according to the dynamics of the evolution of the corresponding Brownian motions:

- $\mathcal{E}_\sigma = \{i : \sigma_i > 0\}$, this is the set of phases for which $\{X_i\}$ exhibits a truly Brownian behaviour.
- $\mathcal{E}_+ = \{i : \sigma_i = 0, \mu_i > 0\}$ and
  $\mathcal{E}_- = \{i : \sigma_i = 0, \mu_i < 0\}$, these are the phases for which X_i is a linear flow, with positive slopes for phases in $\mathcal{E}_+$ and negative slopes in $\mathcal{E}_-$.
- $\mathcal{E}_0 = \{i : \sigma_i = 0, \mu_i = 0\}$, the level remaining constant as long as φ remains in $\mathcal{E}_0$.

These subsets determine a partition of $\mathcal{E}$. In addition, we define

$$\mathcal{E}_{\mathrm{u}} = \mathcal{E}_\sigma \cup \mathcal{E}_+ \qquad \text{and} \qquad \mathcal{E}_{\mathrm{d}} = \mathcal{E}_\sigma \cup \mathcal{E}_- :$$

When φ is in $\mathcal{E}_{\mathrm{u}}$, the level may go up, either linearly or in Brownian fashion, and in $\mathcal{E}_{\mathrm{d}}$, the level may go down.

The generator Q and the vector μ are organised in a manner conformant with the partition of $\mathcal{E}$, and we write, possibly after a permutation of rows and columns, that

$$Q = \begin{bmatrix} Q_{\sigma\sigma} & Q_{\sigma+} & Q_{\sigma-} & Q_{\sigma 0} \\ Q_{+\sigma} & Q_{++} & Q_{+-} & Q_{+0} \\ Q_{-\sigma} & Q_{-+} & Q_{--} & Q_{-0} \\ Q_{0\sigma} & Q_{0+} & Q_{0-} & Q_{00} \end{bmatrix} \qquad \text{and} \qquad \mu = \begin{bmatrix} \mu_\sigma \\ \mu_+ \\ \mu_- \\ \mu_0 \end{bmatrix}.$$

More generally, we use the notation M_{ab} for a matrix, or a block in a matrix, with rows and columns indexed by states in $\mathcal{E}_a$ and $\mathcal{E}_b$, respectively, and we denote by m_a the number of states in $\mathcal{E}_a$.

We start our analysis with the descending ladder process $\{\varphi(\tau_{-x}) : x \geq 0\}$ of phases visited by the MMBM as it reaches levels further and further down, and we define the transition probability

$$G_{ij}(x) = \mathrm{P}_{0,i}[\tau_{-x} < \infty, \varphi(\tau_{-x}) = j], \qquad \text{for } x \geq 0, i, j \in \mathcal{E}. \tag{2.6}$$

It is clear that upon a *first* passage to a lower level, the phase has to belong to $\mathcal{E}_{\mathrm{d}}$, and so we may write by the strong Markov property that

$$G_{\mathrm{dd}}(x + y) = G_{\mathrm{dd}}(x) G_{\mathrm{dd}}(y).$$

Since $G_{dd}(0) = I$, we conclude that there exists a matrix U indexed by $\mathcal{E}_d$ such that

$$G_{dd}(x) = e^{Ux}. \tag{2.7}$$

In other words, if, in addition to $X(0) = 0$, we assume that $\varphi(0)$ is in $\mathcal{E}_d$, then $\{\varphi(\tau_{-x} : x \geq 0)\}$ is a continuous-parameter Markov chain on $\mathcal{E}_d$ with infinitesimal generator U.

If $\varphi(0)$ is in $\mathcal{E}_+$, the level initially increases during a strictly positive interval before it gets a chance to decrease, and the level will have to return to 0, in a phase of $\mathcal{E}_d$, before being able to reach level $-x$. If $\varphi(0)$ is in $\mathcal{E}_0$, the level remains equal to 0 until the phase jumps to a value in $\mathcal{E}_d$ or to a phase in $\mathcal{E}_+$; in the second case, the process will have to return to 0 as we have just discussed. Therefore,

$$\begin{bmatrix} G_{+d}(x) \\ G_{0d}(x) \end{bmatrix} = \begin{bmatrix} \Psi_{+d} \\ \Psi_{0d} \end{bmatrix} e^{Ux},$$

with

$$\Psi_{ij} = \mathrm{P}_{0,i}[\tau_0^* < \infty, \varphi(\tau_0^*) = j], \qquad i \in \mathcal{E}_+ \cup \mathcal{E}_0,\ j \in \mathcal{E}_d, \tag{2.8}$$

where $\tau_0^* = \inf\{t > 0 : X(t) = 0, \varphi(t) \in \mathcal{E}_d\}$ is the first passage time to level 0 in a phase of $\mathcal{E}_d$.

For similar reasons, the process $\{\varphi(\tau_x) : x \geq 0\}$ of first passages to higher levels is a Markov chain on $\mathcal{E}_u$. We denote by $\widehat{U}$ its generator, by $\widehat{G}(x)$ the transition probability matrix

$$\widehat{G}_{ij}(x) = \mathrm{P}_{0,i}[\tau_x < \infty, \varphi(\tau_x) = j], \qquad \text{for } x \geq 0, \tag{2.9}$$

and we have

$$\widehat{G}_{uu}(x) = e^{\widehat{U}x}, \qquad \begin{bmatrix} \widehat{G}_{-u}(x) \\ \widehat{G}_{0u}(x) \end{bmatrix} = \begin{bmatrix} \widehat{\Psi}_{-u} \\ \widehat{\Psi}_{0u} \end{bmatrix} e^{\widehat{U}x}, \tag{2.10}$$

where

$$\widehat{\Psi}_{ij} = \mathrm{P}_{0,i}[\widehat{\tau}_0^* < \infty, \varphi(\widehat{\tau}_0^*) = j], \qquad i \in \mathcal{E}_- \cup \mathcal{E}_0,\ j \in \mathcal{E}_u,$$

and $\widehat{\tau}_0^* = \inf\{t > 0 : X(t) = 0, \varphi(t) \in \mathcal{E}_u\}$ is the first passage time to level 0 in a phase of $\mathcal{E}_u$.

The four matrices so defined, U and Ψ on the one hand, $\widehat{U}$ and $\widehat{\Psi}$ on the other hand, are keys to the analysis of Markov-modulated Brownian motion. Their most salient properties are given now.

Lemma 2.4. *The matrices U and $\widehat{U}$ are irreducible, all entries of Ψ and $\widehat{\Psi}$ are strictly positive. Furthermore:*

(a) If $\boldsymbol{\pi}^{\mathrm{T}}\boldsymbol{\mu} > 0$, then $\widehat{U}\mathbf{1} = \mathbf{0}$ and $e^{\widehat{U}x}$ is stochastic, $U\mathbf{1} < \mathbf{0}$ and e^{Ux} is substochastic, $\widehat{\Psi}\mathbf{1} = \mathbf{1}$, and $\Psi\mathbf{1} < \mathbf{1}$.
(b) If $\boldsymbol{\pi}^{\mathrm{T}}\boldsymbol{\mu} < 0$, then $U\mathbf{1} = \mathbf{0}$ and e^{Ux} is stochastic, $\widehat{U}\mathbf{1} < \mathbf{0}$ and $e^{\widehat{U}x}$ is substochastic, $\Psi\mathbf{1} = \mathbf{1}$, and $\widehat{\Psi}\mathbf{1} < \mathbf{1}$.

(c) If $\boldsymbol{\pi}^{\mathrm{T}}\boldsymbol{\mu} = 0$*, then* $U\mathbf{1} = \mathbf{0}$ *and* e^{Ux} *is stochastic,* $\widehat{U}\mathbf{1} = \mathbf{0}$ *and* $e^{\widehat{U}x}$ *is stochastic,* $\Psi\mathbf{1} = \mathbf{1}$*, and* $\widehat{\Psi}\mathbf{1} = \mathbf{1}$.

Proof. The first claim results from Definition 1.11 where we made the assumption that $\{\varphi\}$ is irreducible.

The other statements are direct consequences of Corollary 2.2. In Case (a) for instance, $\tau_x < \infty$ with probability 1 for positive x, which implies that $e^{\widehat{U}x}$ is stochastic and that $\widehat{\Psi}\mathbf{1} = \mathbf{1}$. On the contrary, $\tau_{-x} < \infty$ with probability strictly less than one, and so e^{Ux} is strictly substochastic, etc. □

We show below that

$$\Psi_{0\mathrm{d}} = (-Q_{00})^{-1}(Q_{0\mathrm{d}} + Q_{0+}\Psi_{+\mathrm{d}}). \tag{2.11}$$

We condition on the time u when the process leaves $\mathcal{E}_0$, and on the first state visited at time u, and obtain

$$\Psi_{0\mathrm{d}} = \int_0^\infty e^{Q_{00}u}(Q_{0\mathrm{d}} + Q_{0+}\Psi_{+\mathrm{d}})\,\mathrm{d}u.$$

As $\{\varphi\}$ is irreducible, the expected uninterrupted sojourn time in the states of $\mathcal{E}_0$ is finite[1] and given by

$$\int_0^\infty e^{Q_{00}u}\,\mathrm{d}u = (-Q_{00})^{-1}$$

(see Sericola [137, Sections 1.11 and 2.13]) from which (2.11) follows. By the same argument,

$$\widehat{\Psi}_{0\mathrm{u}} = (-Q_{00})^{-1}(Q_{0\mathrm{u}} + Q_{0-}\widehat{\Psi}_{-\mathrm{u}}). \tag{2.12}$$

Remark 2.5. We denote by $\rho(M)$ the spectrum of a matrix M, that is, the set of all of its eigenvalues, counting their algebraic multiplicity. We also denote as $\mathbb{C}_{\leq}$ and $\mathbb{C}_{<}$ the sets of complex numbers with negative real parts, and with strictly negative real parts, respectively, and $\mathbb{C}_{\geq}$ and $\mathbb{C}_{>}$ are similarly defined. The real part of the complex number z is written as $\mathrm{Re}(z)$.

If M is the generator of a continuous-parameter Markov process on a finite state space, then $\rho(M) \subset \mathbb{C}_{\leq}$. If M is an irreducible generator, then there exists an eigenvalue λ which is real, has multiplicity 1, and strictly dominates all others: $\mathrm{Re}(\lambda') < \lambda$ for any eigenvalue λ' of M different from λ. Moreover,

- if $M\mathbf{1} = \mathbf{0}$, then $\lambda = 0$ and all other eigenvalues are in $\mathbb{C}_{<}$,
- if $M\mathbf{1} \leq \mathbf{0}$ with at least one strict inequality, then $\lambda < 0$ and all eigenvalues are in $\mathbb{C}_{<}$.

This applies to U and $\widehat{U}$, and Lemma 2.4 gives us some basic properties of their eigenvalues, information that we shall use in the proof of Theorem 2.11 and later.

We develop in three steps our analysis of the ladder processes $\{\varphi(\tau_x) : x \geq 0\}$ and $\{\varphi(\tau_{-x}) : x \geq 0)\}$. In Sect. 2.3, we assume that $\sigma_i > 0$ for all i, so that $\mathcal{E}_\sigma = \mathcal{E}$ and

[1] Assuming as mentioned at the beginning of Sect. 1.12 that there are states outside of $\mathcal{E}_0$.

$\{X_i\}$ is a Brownian motion for each phase i. In Sect. 2.4, $\sigma_i = 0$ for all i, meaning that $\{X_i\}$ is a Linear fluid for all i. The general case is analysed in Sect. 2.5.

In so doing, we pursue two objectives. On the one hand, the key equations for the two special cases have different structures, identified in Theorems 2.8 and 2.19, respectively, well adapted to the physical behaviour of their corresponding processes, and easier to read than the structure in Theorem 2.22 for the general case.

In addition, the Matrix-analytic approach to Fluid flows is well documented (see in particular Latouche and Nguyen [104] and the references therein), and arguments in the general case are easy to follow once one has dealt separately with the specific aspects of the purely Brownian and the Fluid flow processes. We focus in this book on the analysis of Brownian processes, and this is why we feel justified in assuming in Chaps. 3–6 that $\sigma_i > 0$ for all i.

2.3 Purely Brownian MMBM

Assume that $\mathcal{E} = \mathcal{E}_\sigma$. In that case, there is no matrix Ψ or $\widehat{\Psi}$, and we only need to determine the matrices U and $\widehat{U}$. We make in Theorem 2.8 our first use of the following lemma, called the *Wiener-Hopf factorisation* of the Brownian motion in Bertoin [32, Section 6.2]. Details are also to be found in Sato [135, Section 45], in particular with Theorem 45.8 and with Corollary 45.9.

Lemma 2.6. *Assume that $\{X\}$ is a Brownian motion with parameters μ and σ and that $X(0) = 0$. Assume that T is an exponential random variable with parameter λ, independent of $\{X\}$. Define*

$$M(T) = \max\{X(s) : 0 \le s \le T\} \qquad \text{and} \qquad m(T) = \min\{X(s) : 0 \le s \le T\}.$$

Define

$$\alpha_-(\mu, \sigma, \lambda) = \frac{\sqrt{\mu^2 + 2\lambda\sigma^2} - \mu}{\sigma^2} \tag{2.13}$$

$$\alpha_+(\mu, \sigma, \lambda) = \frac{\sqrt{\mu^2 + 2\lambda\sigma^2} + \mu}{\sigma^2}. \tag{2.14}$$

The random variables $M(T)$ and $M(T) - X(T)$ are independent and exponentially distributed, respectively with parameters $\alpha_-(\mu, \sigma, \lambda)$ and $\alpha_+(\mu, \sigma, \lambda)$.

The random variables $|m(T)|$ and $X(T) - m(T)$ are independent and exponentially distributed, respectively with parameters $\alpha_+(\mu, \sigma, \lambda)$ and $\alpha_-(\mu, \sigma, \lambda)$.

We note for later reference that

$$\alpha_+(\mu, \sigma, \lambda) - \alpha_-(\mu, \sigma, \lambda) = 2\mu/\sigma^2 \tag{2.15}$$

$$\alpha_-(\mu, \sigma, \lambda)\,\alpha_+(\mu, \sigma, \lambda) = 2\lambda/\sigma^2. \tag{2.16}$$

□

We shall often apply this lemma to intervals $[0, \tau^*]$ where τ^* is the first jump time of the Markov process $\{\varphi\}$, it is exponentially distributed with parameter $\lambda_i = |Q_{ii}|$ if $\varphi(0) = i$. We use the density

$$V_{ab}(x, y) = \Delta_a e^{-\Delta_a x} \Delta_b e^{-\Delta_b y}, \qquad x, y \geq 0, \tag{2.17}$$

where $\Delta_x = \mathrm{diag}(\boldsymbol{x})$ for any vector $\boldsymbol{x}$,

$$a_i = \alpha_-(\mu_i, \sigma_i, \lambda_i), \qquad b_i = \alpha_+(\mu_i, \sigma_i, \lambda_i), \tag{2.18}$$

for i in $\mathcal{E}$. Both Δ_a and Δ_b are diagonal matrices, $V_{ab}(x, y)$ is diagonal as well, and the products in (2.17) commute.

Remark 2.7. The matrix function V_{ab} will be used in two different contexts. On the one hand, the entry $(V_{ab}(x, y))_{ii}$ is the joint density of $(M(\tau^*), M(\tau^*) - X(\tau^*))$ conditionally given $\varphi(0) = i$, it is also the joint density of $(X(\tau^*) - m(\tau^*), |m(\tau^*)|)$. In both interpretations, $X(\tau^*) = X(0) + x - y$.

Theorem 2.8. *If $\sigma_i > 0$ for all i, then the matrix U is a solution of the fixed point equation*

$$U = -\Delta_b + \Delta_b \int_0^\infty \Delta_a e^{-\Delta_a y} \widetilde{P} e^{Uy}\, \mathrm{d}y, \tag{2.19}$$

and the matrix $\widehat{U}$ is a solution of the fixed point equation

$$\widehat{U} = -\Delta_a + \Delta_a \int_0^\infty \Delta_b e^{-\Delta_b y} \widetilde{P} e^{\widehat{U}y}\, \mathrm{d}y. \tag{2.20}$$

Proof. For i and j fixed, $(e^{Ux})_{ij}$ is the probability that, starting from level 0 in phase i, the process crosses level $-x$ in finite time and does so for the first time in phase j. Like in the proof of Lemma 1.13, we condition on the epoch T when the phase process leaves i for the first time. The random variable T has an exponential distribution with parameter $\lambda_i = |Q_{ii}|$ and, by Lemma 2.6, $|m_X(T)|$ and $X(T) - m_X(T)$ are exponentially distributed with parameters b_i and a_i given by (2.18).

We may write

$$(e^{Ux})_{ij} = e^{-b_i x}\delta_{ij} + \int_0^x e^{-b_i v} b_i \int_0^\infty e^{-a_i y} a_i \sum_{k \in \mathcal{E}} \widetilde{P}_{ik} (e^{U(x+y-v)})_{kj}\, \mathrm{d}v\, \mathrm{d}y, \tag{2.21}$$

where δ_{ij} is the Kronecker delta function, equal to 1 if $i = j$, equal to 0 otherwise. Recall that $\widetilde{P} = I + \Delta_\lambda^{-1} Q$ defined in (1.30) is the transition matrix of the jump chain for the phase process. The justification of (2.21) goes as follows. The first term in the right-hand side is the probability that $|m_X(T)| > x$, in which case $\tau_x < T$ and $\varphi(\tau_x)$ is necessarily equal to i. The second term (see an illustration in Fig. 2.2) is the probability that

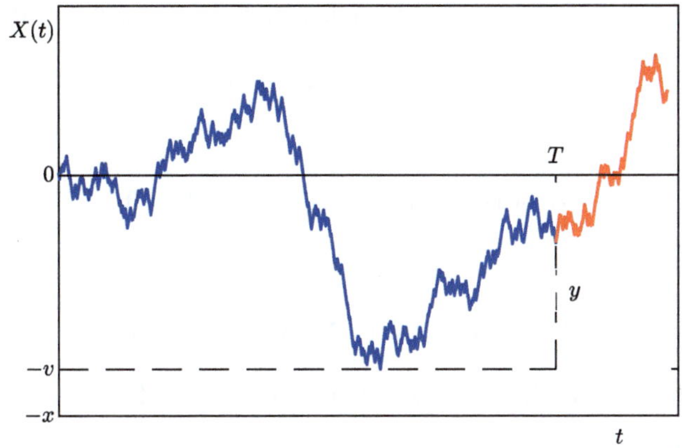

Fig. 2.2 Illustration or the conditioning argument for Eq. (2.21). At time 0, the process is in state $(0, i)$, and at time T, it is in state $(y - v, k)$

- the process goes down to level $-v$ during $(0, T)$, for some $v < x$, but not lower,
- at time T, the process is y units above its minimum, so that $X(T) = y - v$,
- the phase then jumps to the new value k,
- and the process is eventually in phase j after it will have gone down $y - v + x$ units below its value at time T.

In matrix notation,

$$e^{Ux} = e^{-\Delta_b x} + \int_0^x \int_0^\infty V_{ab}(y, v)\widetilde{P}\, e^{U(x+y-v)}\, dv\, dy. \tag{2.22}$$

We take the derivative with respect to x on both sides of (2.22), evaluate it at $x = 0$, and obtain (2.19).

The proof that $\widehat{U}$ is a solution of (2.20) follows through the same steps, using $M_X(T)$ and $M_X(T) - X(T)$ instead of $|m_X(T)|$ and $X(T) - m_X(T)$. □

Corollary 2.9. *If $\sigma_i > 0$ for all i, then the matrix U is a solution of the matrix-quadratic equation*

$$\frac{1}{2}\Delta_\sigma^2 X^2 + \Delta_\mu X + Q = 0, \tag{2.23}$$

and the matrix $\widehat{U}$ is a solution of

$$\frac{1}{2}\Delta_\sigma^2 X^2 - \Delta_\mu X + Q = 0. \tag{2.24}$$

Proof. We integrate by part the integral in (2.19) and obtain that

$$U = -\Delta_b + \Delta_b[-e^{-\Delta_a y}\widetilde{P}e^{Uy}]_0^\infty + \Delta_b \int_0^\infty e^{-\Delta_a y}\widetilde{P}e^{Uy}\, U\, dy.$$

The matrix e^{Ux}, being stochastic, is bounded, and $\lim_{y\to\infty} e^{-\Delta_a y} = 0$ since $a_i > 0$ for all i. Therefore,

$$U = -\Delta_b + \Delta_b \widetilde{P} + \Delta_a^{-1}(U + \Delta_b)U,$$

which we rewrite as

$$U^2 + (\Delta_b - \Delta_a)U + \Delta_a \Delta_b(\widetilde{P} - I) = 0$$

or, using (2.15, 2.16),

$$U^2 + 2\Delta_\mu \Delta_\sigma^{-2} U + 2\Delta_\sigma^{-2}\Delta_\lambda(\widetilde{P} - I) = 0.$$

This proves that U is a solution of (2.23), the proof of (2.24) is similar. □

Remark 2.10. If we bring (1.29) and (2.23) together, it looks like we might write that U is a solution of the matrix equation $\mathcal{F}(X) = 0$, where $\mathcal{F}(\cdot)$ is the Laplace exponent of the MMBM. However, we are dealing with matrices, and their products usually do not commute. Therefore, we should specify that X is here a factor on the right of the linear and the quadratic terms of $\mathcal{F}(\cdot)$. This is all the more necessary since we shall have in (2.37) an equation in which the unknown matrix is a factor on the left. To avoid ambiguities, we shall write that U and $-\widehat{U}$ are *right solutions* of $\mathcal{F}(\cdot) = 0$.

It follows from the fixed point equation (2.19) that

$$U = -\Delta_b + \Delta_b \Phi, \tag{2.25}$$

where

$$\Phi = \int_0^\infty \Delta_a e^{-\Delta_a y} \widetilde{P} e^{Uy}\, \mathrm{d}y. \tag{2.26}$$

The matrix Φ is (sub)stochastic;[2] to see this, assume that $X(0) = 0$ and $\varphi(0) = i$, and let us temporarily forget about the detailed oscillations of the Brownian motion. Assume that the minimal value $m(\tau^*)$ until the first change of phase is equal to some value $-v$. The entry Φ_{ij} is the probability that the level returns to $-v$ after the change of phase at time τ^*, and that the phase is j upon return to $-v$. Indeed,

- the expression $a_i e^{-a_i y}\, \mathrm{d}y$ in (2.26) is the probability that $X(\tau^*)$ takes a value in $(y, y + \mathrm{d}y)$ units above $-v$,
- the factor $\widetilde{P}_{ik}$ is the probability that the phase jumps from i to k at time τ^*,
- and the factor $(e^{Uy})_{kj}$ is the probability that the phase is j when the level returns to the level $-v$ for the first time thereafter.

With this, the right-hand side of (2.25) is easy to interpret: If $\varphi(t) = i$ at some time t, then the running minimum ceases at rate b_i to get smaller, and if it resumes its descent at a later time, it will do so in phase j with probability Φ_{ij}.

---

[2] A nonnegative matrix M is (sub)stochastic if $M\mathbf{1} \le \mathbf{1}$, it is substochastic if there is at least one strict inequality, it is stochastic if $M\mathbf{1} = \mathbf{1}$.

We integrate by part the right-hand side of (2.26) and find after a few simple manipulations that Φ is a solution of the Riccati equation

$$\Delta_a \widetilde{P} - \Delta_a \Phi - \Phi \Delta_b + \Phi \Delta_b \Phi = 0, \tag{2.27}$$

a quadratic equation of a type extensively studied in Bini et al. [40].

Wiener-Hopf Factorisation We embark now on a side trip into linear algebra. The reason is that Eq. (2.23) may have several solutions and we want to identify precisely which one is the matrix U and which solution of (2.24) is $\widehat{U}$. As shown in Theorem 2.13, this is done through an analysis of the eigenvalues of the two matrices. We only give a high-level presentation of the argument, and for detailed justifications, we refer to [40, Chapter 2].

As $\sigma_i > 0$ for all i, Δ_σ is nonsingular, and we rewrite (2.23) as

$$X^2 + 2\Delta_\sigma^{-2}\Delta_\mu X + 2\Delta_\sigma^{-2} Q = 0. \tag{2.28}$$

We show by direct verification that X is a solution of (2.28), and therefore a solution of (2.23), if and only if

$$D \begin{bmatrix} I & 0 \\ X & I \end{bmatrix} = \begin{bmatrix} I & 0 \\ X & I \end{bmatrix} \begin{bmatrix} X & I \\ 0 & -X - 2\Delta_\sigma^{-2}\Delta_\mu \end{bmatrix}, \tag{2.29}$$

with

$$D = \begin{bmatrix} 0 & I \\ -2\Delta_\sigma^{-2} Q & -2\Delta_\sigma^{-2}\Delta_\mu \end{bmatrix}. \tag{2.30}$$

The matrix D is a version of the companion matrix associated with the matrix polynomial

$$A(z) = 2\Delta_\sigma^{-2}\mathcal{F}(z) = z^2 I + 2z\Delta_\sigma^{-2}\Delta_\mu + 2\Delta_\sigma^{-2} Q, \tag{2.31}$$

so that the eigenvalues of D coincide with the roots of the polynomial

$$a(z) = \det(A(z)), \tag{2.32}$$

meaning that λ is a root of $a(z)$ if and only if it is an eigenvalue of D, with the same multiplicity. In addition, D and $A(z)$ have the same invariant subspaces.

The next theorem, therefore, is a statement about the eigenvalues of D as much as it is a statement about the roots of $a(z)$.

Theorem 2.11. *The roots $\lambda_1, \lambda_2, \ldots, \lambda_{2m}$ of $a(z)$, indexed in increasing order of their real parts, are the eigenvalues of D. They satisfy the inequalities*

$$\mathrm{Re}(\lambda_1) \le \cdots \le \mathrm{Re}(\lambda_{m-1}) < \lambda_m \le \lambda_{m+1} < \mathrm{Re}(\lambda_{m+2}) \le \cdots \le \mathrm{Re}(\lambda_{2m}). \tag{2.33}$$

The eigenvalues of U are $\{\lambda_1, \ldots, \lambda_m\}$ and those of $\widehat{U}$ are $\{-\lambda_{2m}, \ldots, -\lambda_{m+1}\}$.

Furthermore:

(a) If $\boldsymbol{\pi}^{\mathrm{T}}\boldsymbol{\mu} > 0$, then $\lambda_m < 0 = \lambda_{m+1}$.
(b) If $\boldsymbol{\pi}^{\mathrm{T}}\boldsymbol{\mu} < 0$, then $\lambda_m = 0 < \lambda_{m+1}$.

(c) If $\boldsymbol{\pi}^{\mathrm{T}}\boldsymbol{\mu} = 0$*, then* $\lambda_m = 0 = \lambda_{m+1}$.

Proof. We rewrite (2.29) as

$$\begin{bmatrix} I & 0 \\ X & I \end{bmatrix}^{-1} D \begin{bmatrix} I & 0 \\ X & I \end{bmatrix} = \begin{bmatrix} X & I \\ 0 & -X - 2\Delta_\sigma^{-2}\Delta_\mu \end{bmatrix}. \tag{2.34}$$

Thus, D is similar to the matrix in the right-hand side of (2.34), and we conclude that $\rho(X) \subset \rho(D)$ for any solution X of (2.23). In particular, both $\rho(U)$ and $\rho(-\widehat{U})$ are subsets of $\rho(D)$. As U and $\widehat{U}$ are generators, their spectra are in $\mathbb{C}_{\leq}$, and we see that $\rho(D)$ is about evenly split between $\mathbb{C}_{\leq}$ and $\mathbb{C}_{\geq}$, with $\rho(U) = \{\lambda_1, \ldots, \lambda_m\}$ and $\rho(\widehat{U}) = \{-\lambda_{2m}, \ldots, -\lambda_{m+1}\}$.

Case (a). Refer to Lemma 2.4. If $\boldsymbol{\pi}^{\mathrm{T}}\boldsymbol{\mu} > 0$, then e^{Ux} is substochastic and the m eigenvalues of U are all in $\mathbb{C}_<$. In addition, its maximal eigenvalue λ_m is real and is isolated from the others. At the same time, the matrix $e^{\widehat{U}x}$ is stochastic, meaning that $\widehat{U}$ has the eigenvalue 0 with multiplicity one, while its $m-1$ other eigenvalues are in $\mathbb{C}_<$, so that $-\widehat{U}$ has one eigenvalue equal to 0 and the others are in $\mathbb{C}_>$. With this argument, we have proved that

$$\mathrm{Re}(\lambda_{m-1}) < \lambda_m < 0 = \lambda_{m+1} < \mathrm{Re}(\lambda_{m+2}).$$

Case (b). If $\boldsymbol{\pi}^{\mathrm{T}}\boldsymbol{\mu} < 0$, then e^{Ux} is stochastic, $e^{\widehat{U}x}$ is substochastic, and the same argument leads to the conclusion that

$$\mathrm{Re}(\lambda_{m-1}) < \lambda_m = 0 < \lambda_{m+1} < \mathrm{Re}(\lambda_{m+2}).$$

Case (c). If $\boldsymbol{\pi}^{\mathrm{T}}\boldsymbol{\mu} = 0$, both U and $\widehat{U}$ have an eigenvalue equal to 0, the others being in $\mathbb{C}_<$. Thus, we know that $a(z)$ has $m-1$ roots in $\mathbb{C}_<$ and $m-1$ roots in $\mathbb{C}_>$ but to determine the location of the two remaining roots require a bit more work. It is conceivable, after all, that U and $-\widehat{U}$ share the single root 0 and that some additional root would exist, that does not belong to either U or $-\widehat{U}$.[3]

Actually, D has two eigenvalues equal to 0 with only one eigenvector, and this will complete the proof that

$$\mathrm{Re}(\lambda_{m-1}) < \lambda_m = 0 = \lambda_{m+1} < \mathrm{Re}(\lambda_{m+2}).$$

Take $\boldsymbol{v}_1^{\mathrm{T}} = \begin{bmatrix} \mathbf{1}^{\mathrm{T}} & \mathbf{0}^{\mathrm{T}} \end{bmatrix}$. We easily verify that $D\boldsymbol{v}_1 = \mathbf{0}$, which accounts for one eigenvalue 0. If we find a vector $\boldsymbol{v}_2$ such that $D\boldsymbol{v}_2 = \boldsymbol{v}_1$, then $(\boldsymbol{v}_1, \boldsymbol{v}_2)$ will form a Jordan chain for the eigenvalue 0 (Gantmacher [67, Chapter 5]), with $\boldsymbol{v}_1$ as the only proper eigenvector, and the proof will be complete.

The system $D\boldsymbol{x} = \boldsymbol{v}_1$ is written in expanded form as

$$\begin{bmatrix} 0 & I \\ -2\Delta_\sigma^{-2}Q & -2\Delta_\sigma^{-2}\Delta_\mu \end{bmatrix} \begin{bmatrix} \boldsymbol{x}_1 \\ \boldsymbol{x}_2 \end{bmatrix} = \begin{bmatrix} \mathbf{1} \\ \mathbf{0} \end{bmatrix}.$$

---

[3] As a matter of fact, this is what happens, to some extent: There are two roots equal to 0, so that U and $\widehat{U}$ each have its own copy, but they share the single associated eigenvector $\mathbf{1}$.

The first block of rows gives us $\boldsymbol{x}_2 = \mathbf{1}$, and the second block of rows leaves us with the system

$$-2\Delta_\sigma^{-2} Q \boldsymbol{x}_1 - 2\Delta_\sigma^{-2}\Delta_\mu \mathbf{1} = \mathbf{0},$$

that is, $Q\boldsymbol{x}_1 = -\boldsymbol{\mu}$. The matrix Q being a generator, this is a singular system. As we assume that $\boldsymbol{\pi}^{\mathrm{T}}\boldsymbol{\mu} = 0$, the system has the solutions $\boldsymbol{x} = -Q^{\#}\boldsymbol{\mu} + c\mathbf{1}$, where c is an arbitrary scalar, see (1.32). In summary, the vectors

$$\boldsymbol{v}_1 = \begin{bmatrix} \mathbf{1} \\ \mathbf{0} \end{bmatrix} \quad \text{and} \quad \boldsymbol{v}_2 = \begin{bmatrix} -Q^{\#}\boldsymbol{\mu} \\ \mathbf{1} \end{bmatrix} \tag{2.35}$$

form a Jordan chain for the eigenvalue 0 of D and the proof is complete. □

We state for later reference the corollary below, its proof is embedded in the proof of Theorem 2.11.

Corollary 2.12. *The matrix D defined in (2.30) has an eigenvalue 0 of algebraic multiplicity one if $\boldsymbol{\pi}^{\mathrm{T}}\boldsymbol{\mu} \neq 0$, of algebraic multiplicity two and geometric multiplicity one if $\boldsymbol{\pi}^{\mathrm{T}}\boldsymbol{\mu} = 0$.* □

The splitting property of Theorem 2.11 is characteristic of the Markov-modulated processes that have been analysed with Matrix-analytic methods. We prove it in Theorem 2.20 for Fluid flows, and a similar property has been shown for QBDs, where the splitting is with respect to the unit circle C instead of the imaginary axis, see [36, Theorem 5.20].

The fact that λ_m and λ_{m+1} are two real numbers, strictly separated from the other roots of $a(z)$, plays an important role in the design of efficient numerical procedures, but this is beyond our purpose in the present book. The splitting property is important here because it defines a very clean separation of the eigenvalues of U and $-\widehat{U}$ and translates into unique properties of U and $\widehat{U}$ among the solutions of (2.23), an equation which may, therefore, be seen as a defining characteristic of the two matrices. This is easily expressed in Theorem 2.13 in case $\boldsymbol{\pi}^{\mathrm{T}}\boldsymbol{\mu} \neq 0$. When $\boldsymbol{\pi}^{\mathrm{T}}\boldsymbol{\mu} = 0$, however, we need to deal with the eigenvalue 0 as there are two of them, with only one associated eigenvector. To give a precise distinction between U and $\widehat{U}$ in that case, one defines as follows two *canonical* invariant subspaces.

If $\boldsymbol{\pi}^{\mathrm{T}}\boldsymbol{\mu} = 0$,

- the *stable canonical invariant subspace* has for basis the union of the vector $\mathbf{1}$ and the Jordan chains associated with the roots λ_1 to λ_{m-1},
- the *antistable canonical invariant subspace* has for basis the union of the vector $\mathbf{1}$ and the Jordan chains associated with the roots λ_{m+2} to λ_{2m}.

The next statement is a direct consequence of [40, Theorem 2.5 and 2.6].

Theorem 2.13. *If $\boldsymbol{\pi}^{\mathrm{T}}\boldsymbol{\mu} > 0$, then U is the unique solution of (2.23) with eigenvalues in $\mathbb{C}_<$ and $-\widehat{U}$ is the unique solution with eigenvalues in $\mathbb{C}_\geq$.*

If $\boldsymbol{\pi}^{\mathrm{T}}\boldsymbol{\mu} < 0$, then U is the unique solution with eigenvalues in $\mathbb{C}_\leq$ and $-\widehat{U}$ is the unique solution with eigenvalues in $\mathbb{C}_>$.

If $\pi^{\mathsf{T}}\mu = 0$, then U is the unique solution of (2.23) associated with the stable canonical invariant subspace of $a(z)$, while $-\widehat{U}$ is the unique solution associated with the antistable canonical invariant subspace. □

Remark 2.14. One also writes that the roots $\lambda_1, \ldots, \lambda_{2m}$ of $a(z)$ have an (m, m) splitting with respect to the imaginary axis, the splitting is *proper* if $\pi^{\mathsf{T}}\mu \neq 0$ and *weak* if $\pi^{\mathsf{T}}\mu = 0$. For details, see [40, Section 1.4.5].

Let us return to (2.34) and pay attention to the submatrix

$$Y(X) = -X - 2\Delta_\sigma^{-2}\Delta_\mu \tag{2.36}$$

in the right-hand side. Clearly, $Y(X)$ is an actor in the decomposition of the matrix D, just like X, and it would be natural to guess that U being a solution of (2.28), $Y(U)$ might be one as well. This is nearly the case, for $Y(U)$ is a solution of

$$X^2 + 2X\Delta_\sigma^{-2}\Delta_\mu + 2\Delta_\sigma^{-2}Q = 0, \tag{2.37}$$

an equation which is not exactly the same as (2.28), since the unknown matrix X is here a *left factor* of the linear term. To prove that $Y(U)$ is a solution of (2.37), we write

$$\begin{aligned} Y(U)^2 + & 2Y(U)\Delta_\sigma^{-2}\Delta_\mu + 2\Delta_\sigma^{-2}Q \\ &= (U + 2\Delta_\sigma^{-2}\Delta_\mu)^2 - 2(U + 2\Delta_\sigma^{-2}\Delta_\mu)\Delta_\sigma^{-2}\Delta_\mu + 2\Delta_\sigma^{-2}Q \\ &= (U + 2\Delta_\sigma^{-2}\Delta_\mu)U + 2\Delta_\sigma^{-2}Q = 0. \end{aligned}$$

A similar calculation shows that $Y(-\widehat{U})$ is also a solution of (2.37). Thus, both $Y(U)$ and $Y(-\widehat{U})$ are *left solutions* of $\mathcal{F}(\cdot) = 0$; they are, however, of less interest than the two matrices

$$R = -\Delta_\sigma^2\, Y(U)\, \Delta_\sigma^{-2} \qquad \text{and} \qquad \widehat{R} = \Delta_\sigma^2\, Y(-\widehat{U})\, \Delta_\sigma^{-2} \tag{2.38}$$

which we shall meet in Chaps. 4 and 5.

Remark 2.15. The matrix polynomial $A(z)$ may be written as

$$A(z) = (zI - Y(U))(zI - U). \tag{2.39}$$

The roots of $\det(zI - Y(U))$ and those of $\det(zI - U)$ are split with respect to the imaginary axis, and we write that the matrices U and $Y(U)$ determine a *Wiener-Hopf factorisation* of $A(z)$. More comments about this expression are given in Sect. 4.7.

To conclude this section, we mention that when $\pi^{\mathsf{T}}\mu \neq 0$, there is, in addition to U and $\widehat{U}$, a third solution of (2.28) that has the characteristics of a generator.

Lemma 2.16. *If $\pi^{\mathsf{T}}\mu > 0$, the matrix $U^* = U - U\mathbf{1} \cdot \mathbf{u}^{\mathsf{T}}$ is a right solution of $\mathcal{F}(\cdot)$, where $\mathbf{u}$ is the left eigenvector of U associated with the eigenvalue λ_m, normalised by $\mathbf{u}^{\mathsf{T}}\mathbf{1} = 1$. The eigenvalues of U^* are 0 and $\lambda_1, \ldots, \lambda_{m-1}$ in $\mathbb{C}_<$.*

If $\boldsymbol{\pi}^{\mathrm{T}}\boldsymbol{\mu} < 0$, the matrix $-U^*$, with $U^* = \widehat{U} - \widehat{U}\mathbf{1}\cdot\widehat{\boldsymbol{u}}^{\mathrm{T}}$, is a right solution of $\mathcal{F}(\cdot) = 0$, where $\widehat{\boldsymbol{u}}$ is the left eigenvector of $\widehat{U}$ associated with the eigenvalue $-\lambda_{m+1}$, normalised by $\widehat{\boldsymbol{u}}^{\mathrm{T}}\mathbf{1} = 1$. The eigenvalues of U^* are 0 and $-\lambda_{m+1}, \ldots, -\lambda_{2m}$ in $\mathbb{C}_<$.

In both cases, U^* is a generator.

Proof. Assume that $\boldsymbol{\pi}^{\mathrm{T}}\boldsymbol{\mu} > 0$. We replace X with U^* in (2.28) and use the fact that $(U^*)^2 = U^2 - U^2\mathbf{1}\cdot\boldsymbol{u}^{\mathrm{T}}$ to obtain

$$
\begin{aligned}
&(U^*)^2 + 2\Delta_\sigma^{-2}\Delta_\mu U^* + 2\Delta_\sigma^{-2}Q \\
&\quad = U^2 - U^2\mathbf{1}\cdot\boldsymbol{u}^{\mathrm{T}} + 2\Delta_\sigma^{-2}\Delta_\mu(U - U\mathbf{1}\cdot\boldsymbol{u}^{\mathrm{T}}) + 2\Delta_\sigma^{-2}Q \\
&\quad = -U^2\mathbf{1}\cdot\boldsymbol{u}^{\mathrm{T}} - 2\Delta_\sigma^{-2}\Delta_\mu U\mathbf{1}\cdot\boldsymbol{u}^{\mathrm{T}} \\
&\quad = -2\Delta_\sigma^{-2}Q\mathbf{1}\cdot\boldsymbol{u}^{\mathrm{T}} = \mathbf{0},
\end{aligned}
$$

the second and third equalities are justified by U being a solution of (2.28), and the fourth by Q being a generator.

Since $-U\mathbf{1}\cdot\boldsymbol{u}^{\mathrm{T}} > \mathbf{0}$, the off-diagonal elements of U^* are positive; moreover, $U^*\mathbf{1} = \mathbf{0}$ by construction. This shows that U^* is a generator.

We use a similar argument if $\boldsymbol{\pi}^{\mathrm{T}}\boldsymbol{\mu} < 0$. □

2.4 Linear Fluid Flows

For the second special case, we assume that $\sigma_i = 0$ for all i, so that $\mathcal{E}_{\mathrm{d}} = \mathcal{E}_-$ and $\mathcal{E}_{\mathrm{u}} = \mathcal{E}_+$. The generator of the phase process and the vector of fluid rates are

$$
Q = \begin{bmatrix} Q_{++} & Q_{+-} & Q_{+0} \\ Q_{-+} & Q_{--} & Q_{-0} \\ Q_{0+} & Q_{0-} & Q_{00} \end{bmatrix} \qquad \text{and} \qquad \boldsymbol{\mu} = \begin{bmatrix} \boldsymbol{\mu}_+ \\ \boldsymbol{\mu}_- \\ \mathbf{0} \end{bmatrix}.
$$

Instead of tracking the evolution of the process in *time* during the interval $(0, \tau_{-x})$, it is more convenient to follow the changes in *level*. This is what we did in the proof of Theorem 2.8 when we wrote (2.21), but this approach is made explicit here. To that end, we use $|\mu_i|$, i in $\mathcal{E}_+ \cup \mathcal{E}_-$, as a conversion rate between *time* and *level*, and we interpret $|\mu_i|^{-1}Q_{ij}h$, for small h, as the elementary probability that the phase jumps from i to j during an interval of time when the level increases or decreases by h units. If i is in $\mathcal{E}_0$, of course, the fluid remains constant until the phase jumps to a state in $\mathcal{E}_-$ or $\mathcal{E}_+$.

Like in the proof of the fixed point Theorem 2.8, we condition on the level v reached at the first jump time. If $i \in \mathcal{E}_-$, the level is continuously decreasing during that interval, and there is no need to refer to Lemma 2.6. Beyond this difference, the justification of the next equation is the same as that of (2.21). Recall that U is indexed here by the phases in $\mathcal{E}_-$ and if, having reached level $-v$, the phase jumps to $\mathcal{E}_+$ or $\mathcal{E}_0$, the process must return to level $-v$ in a phase of $\mathcal{E}_-$ before it may again reach lower levels. For i, j in $\mathcal{E}_-$,

$$(e^{Ux})_{ij} = e^{|\mu_i|^{-1}Q_{ii}x}\delta_{ij} + \int_0^x e^{|\mu_i|^{-1}Q_{ii}(x-v)}|\mu_i|^{-1}|Q_{ii}| \times \Big(\sum_{k\in\mathcal{E}_-}\widetilde{P}_{ik}(e^{Uv})_{kj} + \sum_{k\in\mathcal{E}_+\cup\mathcal{E}_0}\widetilde{P}_{ik}(\Psi e^{Uv})_{kj}\Big)\,\mathrm{d}v,$$

where $\widetilde{P}$ defined in (1.30) is the transition matrix of the jump chain, and Ψ defined in (2.8) is the transition matrix at the first return to the initial level.

We take the derivative with respect to x, evaluate it at 0, and obtain

$$U_{ij} = |\mu_i|^{-1}Q_{ii}\delta_{ij} + |\mu_i|^{-1}|Q_{ii}|(\widetilde{P}_{ij} + \sum_{k\in\mathcal{E}_+\cup\mathcal{E}_0}\widetilde{P}_{ik}\Psi_{kj}),$$

or

$$U = |C_-|^{-1}Q_{--} + |C_-|^{-1}\begin{bmatrix}Q_{-+} & Q_{-0}\end{bmatrix}\begin{bmatrix}\Psi_{+-}\\ \Psi_{0-}\end{bmatrix},$$

in matrix notation, where the matrix C indexed with the phases in $\mathcal{E}_+ \cup \mathcal{E}_-$ is defined as

$$C = \mathrm{diag}(\mu_i : i \in \mathcal{E}_+ \cup \mathcal{E}_-). \tag{2.40}$$

We replace Ψ_{0-} with its expression in (2.11), and the expression for U becomes

$$U = |C_-|^{-1}T_{--} + |C_-|^{-1}T_{-+}\Psi_{+-}, \tag{2.41}$$

with

$$T = \begin{bmatrix}Q_{++} & Q_{+-}\\ Q_{-+} & Q_{--}\end{bmatrix} + \begin{bmatrix}Q_{+0}\\ Q_{-0}\end{bmatrix}(-Q_{00})^{-1}\begin{bmatrix}Q_{0+} & Q_{0-}\end{bmatrix}. \tag{2.42}$$

The matrix T is the generator of the process $\{\varphi\}$ when the states in $\mathcal{E}_0$ are censored out: These phases do not contribute to changes of level and we may remove them from view in our analysis of U and Ψ.

The physical interpretation of (2.41) is straightforward:

- the first term in the right-hand side is the rate at which the phase changes from one value in $\mathcal{E}_-$ to another, possibly after a visit to $\mathcal{E}_0$,
- the second term is the rate at which the phase jumps from $\mathcal{E}_-$ to $\mathcal{E}_+$, possibly after a visit to $\mathcal{E}_0$, and is eventually in $\mathcal{E}_-$ upon return to the initial level.

This is illustrated in Fig. 2.3: During the intervals of time $(0, t_1)$, (t_2, t_3), (t_4, t_5), and (t_6, τ_0^*), changes of phase are induced by the transition rates T_{--} and the level keeps going down. At times t_1, t_3, and t_5, the phase changes from $\mathcal{E}_-$ to $\mathcal{E}_+$, the level begins to increase, and eventually returns to the value of the running minimum. Details of the trajectory during (t_1, t_2), (t_3, t_4), and (t_5, t_6) are irrelevant here, the values $\varphi(t_2)$, $\varphi(t_4)$, and $\varphi(t_6)$ are the only ones to matter, and their distribution is given by the matrix Ψ_{+-}.

To determine Ψ_{+-}, we split the interval $(0, \tau_0^*)$ into two subintervals: $(0, \tau_+)$ and $[\tau_+, \tau_0^*)$, where τ_+ is the epoch when the phase, starting from a state in $\mathcal{E}_+$, enters $\mathcal{E}_-$ for the first time. At time τ_+, the downward ladder process takes over, see Fig. 2.4.

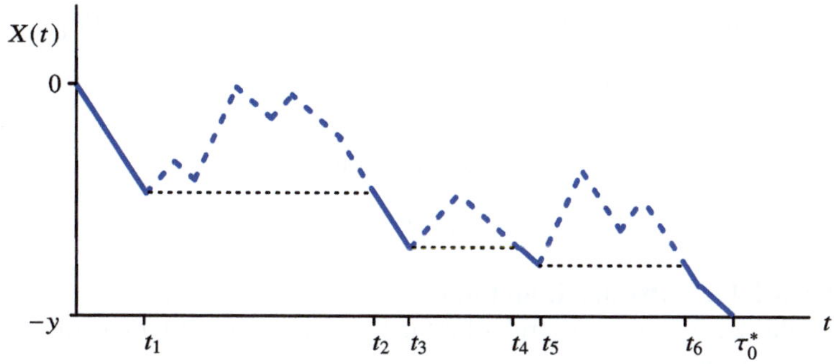

Fig. 2.3 Trajectory from $(0, \mathcal{E}_-)$ to $(-y, \mathcal{E}_-)$ for the first time

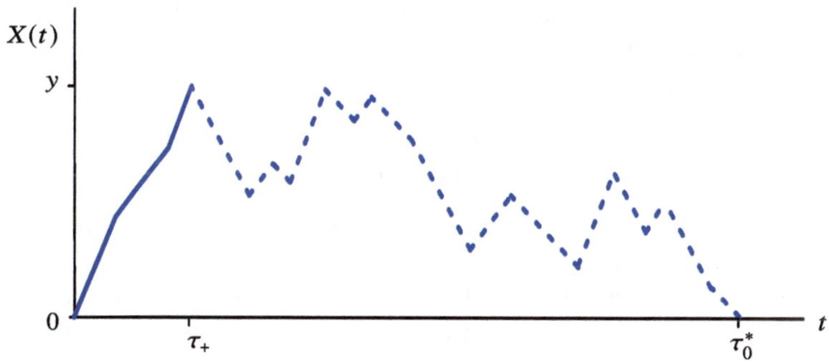

Fig. 2.4 Trajectory from $(0, \mathcal{E}_+)$ to $(0, \mathcal{E}_-)$ for the first time

The epoch τ_+ itself is irrelevant, the level y reached only matters, and we write

$$\Psi_{+-} = \int_0^\infty e^{C_+^{-1}T_{++}y} C_+^{-1} T_{+-} e^{Uy} \, \mathrm{d}y. \tag{2.43}$$

The argument is similar to the one for Φ in (2.26):

- for i and k in $\mathcal{E}_+$, $(e^{C_+^{-1}T_{++}y})_{ik}$ is the probability that, starting from i, the phase constantly remains in $\mathcal{E}_+ \cup \mathcal{E}_0$, the level increases by y units, and the process is in phase k at that level,
- $(C_+^{-1}T_{+-})_{k\ell} \, \mathrm{d}y$ is the elementary probability that the process $\{\varphi\}$ jumps from k in $\mathcal{E}_+$ to ℓ in $\mathcal{E}_-$ when the level is in the interval $(y, y + \mathrm{d}y)$, and
- $(e^{Uy})_{\ell j}$ is the probability that, from level y, the process eventually returns to 0 in phase j in $\mathcal{E}_-$.

Note the similarity, but also the differences, between the pair of Eqs. (2.41, 2.43) for U and Ψ_{+-}, and the pair (2.25, 2.26) for U and Φ.

We pre-multiply both sides of (2.43) by $C_+^{-1}T_{++}$ and integrate by part to obtain

$$C_+^{-1}T_{++}\Psi_{+-} = \left[e^{C_+^{-1}T_{++}y}C_+^{-1}T_{+-}e^{Uy}\right]_0^\infty - \int_0^\infty e^{C_+^{-1}T_{++}y}C_+^{-1}T_{+-}e^{Uy}U\,\mathrm{d}y$$
$$= -C_+^{-1}T_{+-} - \Psi_{+-}U,$$

so that Ψ_{+-} is a solution of the Sylvester equation

$$C_+^{-1}T_{+-} + C_+^{-1}T_{++}\Psi_{+-} + \Psi_{+-}U = 0. \tag{2.44}$$

Parenthesis 2.17. Sylvester Equations

This is the section where we refer for the first time to Sylvester equations, that is, linear equations of the form

$$AX + XB = C, \tag{2.45}$$

where A, B and C are given matrices and X is the unknown. The parameters A and B are square matrices of order n and m, respectively, and C and X both have n rows and m columns, see Lancaster and Tismenetsky [99, Section 12.5].

Lemma 2.18. *Equation (2.45) has a unique solution if and only if $\lambda_a + \lambda_b \neq 0$ for any pair (λ_a, λ_b) of an eigenvalue of A and an eigenvalue of B.* □

In this book, it will usually be obvious that Sylvester equations are nonsingular because one matrix coefficient will have all its eigenvalues in $\mathbb{C}_<$ and the other all its eigenvalues in $\mathbb{C}_\leq$ so that there will be no risk of eigenvalue cancellation.

The proof of Lemma 2.18 is straightforward if we rewrite (2.45) in the equivalent, more traditional form

$$(I_m \otimes A + B^{\mathrm{T}} \otimes I_n)\mathrm{vec}(X) = \mathrm{vec}(C), \tag{2.46}$$

where vec(·) is the operator which stacks the columns of a matrix on top of one another:

$$\mathrm{vec}(X) = \begin{bmatrix} x_{11} & \cdots & x_{n1} & \cdots & x_{1m} & \cdots & x_{nm} \end{bmatrix}^{\mathrm{T}},$$

when X is an $n \times m$ matrix.

To compute the solution of (2.45), we might transform it into the form (2.46) and use any general algorithm for linear systems, such as Gaussian elimination for instance, at a cost of $O(m^3n^3)$ ops.[4] Much more efficient algorithms exist, however, with numerical complexity $O(m^3)$ ops overall when $m = n$. See Bartels and Stewart [23], and Golub et al. [69].

△

Getting back to (2.44): As T is an irreducible generator, its submatrix T_{++} has all its eigenvalues in $\mathbb{C}_<$, furthermore, U is by definition the generator of a Markov process, and its eigenvalues are all in $\mathbb{C}_\leq$. Therefore, (2.44) is nonsingular by Lemma 2.18,

[4] We occasionally use the $o(\cdot)$ and $O(\cdot)$ notation. Their meanings here are that

$$\lim_{x\to 0} o(x^k)/x^k = 0 \qquad \text{and} \qquad \lim_{x\to 0} O(x^k)/x^k < \infty.$$

An "op" denotes any of the four elementary arithmetic operations with real numbers.

and Ψ_{+-} is known if U is known. Conversely, U is explicitly given in (2.41) as a function of Ψ_{+-}. Thus, it appears that the pair of matrices (U, Ψ_{+-}) might have to be determined simultaneously from the pair of equations (2.41, 2.44). However, if we replace U in (2.44) with the right-hand side of (2.41), we obtain the Eq. (2.47) below, a fixed point equation with Ψ_{+-} as the only unknown. The same argument may be repeated for the matrices $\widehat{U}$ and $\widehat{\Psi}_{-+}$, and we summarise the results in the next theorem.

Theorem 2.19. *If $\sigma_i = 0$ for all i, then the first return probability matrices Ψ_{+-} and $\widehat{\Psi}_{-+}$ are solutions respectively of the Riccati equations*

$$C_+^{-1}T_{+-} + C_+^{-1}T_{++}X + X|C_-|^{-1}T_{--} + X|C_-|^{-1}T_{-+}X = 0, \tag{2.47}$$

where X is indexed by $\mathcal{E}_+ \times \mathcal{E}_-$, and

$$|C_-|^{-1}T_{-+} + |C_-|^{-1}T_{--}Y + YC_+^{-1}T_{++} + YC_+^{-1}T_{+-}Y = 0, \tag{2.48}$$

where Y is indexed by $\mathcal{E}_- \times \mathcal{E}_+$. The generators U and $\widehat{U}$ are given by

$$U = |C_-|^{-1}T_{--} + |C_-|^{-1}T_{-+}\Psi_{+-} \tag{2.49}$$

and

$$\widehat{U} = C_+^{-1}T_{++} + C_+^{-1}T_{+-}\widehat{\Psi}_{-+}, \tag{2.50}$$

respectively. □

Equation (2.47) and its dual (2.48) have received much attention since their first manifestation in Rogers [133]. A good reference for algebraic properties and numerical resolution methods is Bini et al. [40].

Wiener-Hopf Factorisation

Our intention here is to obtain a property which uniquely identifies U and $\widehat{U}$, like we did in Sect. 2.3. This, however, comes indirectly: We start from (2.47, 2.48) and obtain a unique characterisation of Ψ_{+-} and $\widehat{\Psi}_{-+}$, which translates into a characterisation of U and $\widehat{U}$ through (2.49, 2.50).

We show by direct verification that X is a solution of (2.47) if and only if

$$C^{-1}T \begin{bmatrix} I & X \\ 0 & I \end{bmatrix} = \begin{bmatrix} I & X \\ 0 & I \end{bmatrix} \begin{bmatrix} K(X) & 0 \\ C_-^{-1}T_{-+} & -V(X) \end{bmatrix}, \tag{2.51}$$

where

$$C = \begin{bmatrix} C_+ & \\ & C_- \end{bmatrix},$$

$$K(X) = C_+^{-1}T_{++} + X|C_-|^{-1}T_{-+}, \tag{2.52}$$

and

$$V(X) = |C_-|^{-1}T_{--} + |C_-|^{-1}T_{-+}X, \tag{2.53}$$

and that Y is a solution of (2.48) if and only if

$$C^{-1}T\begin{bmatrix} I & 0 \\ Y & I \end{bmatrix} = \begin{bmatrix} I & 0 \\ Y & I \end{bmatrix}\begin{bmatrix} \widehat{V}(Y) & C_+^{-1}T_{+-} \\ 0 & -\widehat{K}(Y) \end{bmatrix}, \tag{2.54}$$

where

$$\widehat{V}(Y) = C_+^{-1}T_{++} + |C_-|^{-1}T_{-+}Y$$

and

$$\widehat{K}(Y) = |C_-|^{-1}T_{--} + YC_+^{-1}T_{+-}. \tag{2.55}$$

If we put together the first block of columns of (2.54) and the second block from (2.51), and replace X with Ψ_{+-} and Y with $\widehat{\Psi}_{-+}$, we obtain the Wiener-Hopf factorisation of the Markov chain with generator $C^{-1}T$ (Rogers [133, Theorem 1]):

$$C^{-1}T\begin{bmatrix} I & \Psi_{+-} \\ \widehat{\Psi}_{-+} & I \end{bmatrix} = \begin{bmatrix} I & \Psi_{+-} \\ \widehat{\Psi}_{-+} & I \end{bmatrix}\begin{bmatrix} \widehat{U} & 0 \\ 0 & -U \end{bmatrix}, \tag{2.56}$$

a nice expression which brings together the four fundamental matrices U, Ψ_{+-}, $\widehat{U}$ and $\widehat{\Psi}_{-+}$.

The matrix $C^{-1}T$ plays here a role similar to D in (2.29) and, as we state next, its eigenvalues are split in a manner similar to those of D in Theorem 2.11. We omit the proof as it closely follows the steps in the argument for that theorem. Alternately, one may recognise in Theorem 2.20 a direct consequence of [40, Theorem 2.7].

Theorem 2.20. *Denote by n and n' the number of phases in $\mathcal{E}_+$ and $\mathcal{E}_-$, respectively. The eigenvalues $\lambda_1, \lambda_2, \ldots, \lambda_{n+n'}$ of $C^{-1}T$, indexed in increasing order of their real parts, satisfy the inequalities*

$$\mathrm{Re}(\lambda_1) \le \cdots \le \mathrm{Re}(\lambda_{n-1}) < \lambda_n \le \lambda_{n+1} < \mathrm{Re}(\lambda_{n+2}) \le \cdots \le \mathrm{Re}(\lambda_{n+n'}).$$

Furthermore:

(a) If $\boldsymbol{\pi}^{\mathrm{T}}\boldsymbol{\mu} > 0$, then $\lambda_n = 0 < \lambda_{n+1}$.
(b) If $\boldsymbol{\pi}^{\mathrm{T}}\boldsymbol{\mu} < 0$, then $\lambda_n < 0 = \lambda_{n+1}$.
(c) If $\boldsymbol{\pi}^{\mathrm{T}}\boldsymbol{\mu} = 0$, then $\lambda_n = 0 = \lambda_{n+1}$. □

The similarity between $C^{-1}T$ and D extends to the fact that, when $\boldsymbol{\pi}^{\mathrm{T}}\boldsymbol{\mu} = 0$, $C^{-1}T$ has only one eigenvector for the two eigenvalues equal to 0. Obviously, $C^{-1}T\boldsymbol{v} = \mathbf{0}$ if and only if $\boldsymbol{v}$ is proportional to $\mathbf{1}$. With two eigenvalues 0, and $\mathbf{1}$ as the only eigenvector, we ought to find a vector $\boldsymbol{x}$ such that $C^{-1}T\boldsymbol{x} = \mathbf{1}$, or $T\boldsymbol{x} = C\mathbf{1}$. For this system to have a solution, it is necessary that $\boldsymbol{\gamma}^{\mathrm{T}}C\mathbf{1} = 0$, where $\boldsymbol{\gamma}$ is the stationary probability vector of T. By construction of T, $\boldsymbol{\gamma}^{\mathrm{T}}$ is proportional to $\begin{bmatrix}\boldsymbol{\pi}_+^{\mathrm{T}} & \boldsymbol{\pi}_-^{\mathrm{T}}\end{bmatrix}$, thus, $\boldsymbol{\gamma}^{\mathrm{T}}C\mathbf{1}$ is proportional to $\boldsymbol{\pi}_+^{\mathrm{T}}\boldsymbol{\mu}_+ + \boldsymbol{\pi}_-^{\mathrm{T}}\boldsymbol{\mu}_- = \boldsymbol{\pi}^{\mathrm{T}}\boldsymbol{\mu}$ and is equal to 0. By (1.32), the system $T\boldsymbol{x} = C\mathbf{1}$ has the solutions

$$\boldsymbol{x} = T^{\#}C\mathbf{1} + c\mathbf{1},$$

where c is an arbitrary constant. In summary, the vectors $\mathbf{1}$ and $T^{\#}C\mathbf{1}$ do form a Jordan chain for the eigenvalue 0 of $C^{-1}T$ when $\boldsymbol{\pi}^{\mathrm{T}}\boldsymbol{\mu} = 0$. The similarity with the Jordan chain (2.35) of D is clear.

The next theorem establishes the uniqueness of Ψ_{+-} and $\widehat{\Psi}_{-+}$ within a class of solutions of (2.47, 2.48) and its importance is similar to that of Theorem 2.13.

Theorem 2.21. *Denote as K and $\widehat{K}$ the matrices $K(U)$ and $\widehat{K}(\widehat{U})$, with $K(\cdot)$ and $\widehat{K}(\cdot)$ defined in (2.52, 2.55).*

The matrices Ψ_{+-} and $\widehat{\Psi}_{-+}$ are strictly positive. They are the smallest nonnegative solutions of (2.47) and (2.48). They are the unique solutions such that

$$\rho(U) = \rho(\widehat{K}) = \{-\lambda_{n+1}, \ldots, -\lambda_{n+n'}\}$$

and

$$\rho(\widehat{U}) = \rho(K) = \{\lambda_1, \ldots, \lambda_n\}.$$

Furthermore, U, K, $\widehat{U}$ and $\widehat{K}$ are irreducible.

Proof. We refer to Bini et al. [40] for the detailed proof. In summary, it follows from [40, Theorem 2.9] that

- there exists a minimal nonnegative solution to (2.47), that is, a matrix $X_{\min}$ such that $X_{\min} \geq 0$ and $X_{\min} \leq X$ for any nonnegative solution X,
- in addition, $X_{\min} > 0$ and the matrices $U_{\min} = V(X_{\min})$ and $K_{\min} = K(X_{\min})$ are irreducible, with $V(\cdot)$ and $K(\cdot)$ defined in (2.52, 2.53).

Next, $\rho(U_{\min}) = \{-\lambda_{n+1}, \ldots, -\lambda_{n+n'}\}$ and $\rho(K_{\min}) = \{\lambda_1, \ldots, \lambda_n\}$ by [40, Theorem 2.11]. Finally, by [40, Corollary 2.8], $X_{\min}$ is the unique such matrix. This implies that $X_{\min} = \Psi_{+-}$ and the rest of the statements follow. The same argument applies to $\widehat{\Psi}_{-+}$, $\widehat{U}$ and $\widehat{K}$. □

We see from (2.41, 2.43) that U is a solution of the fixed point equation

$$X = |C_{-}|^{-1}T_{--} + |C_{-}|^{-1}T_{-+}\int_0^{\infty} e^{C_{+}^{-1}T_{++}y}C_{+}^{-1}T_{+-}e^{Xy}\,\mathrm{d}y.$$

From an algorithmic point of view, this equation for U is not as convenient as the matrix-quadratic equation (2.47) for Ψ_{+-} and that explains why the calculation of Ψ_{+-} and its twin $\widehat{\Psi}_{-+}$ has drawn most of the attention so far.

We briefly conclude this section by noting that Theorem 2.21 does not tell the whole story about the solutions of (2.47). For instance, [40, Theorem 2.13] gives a property that resembles our Lemma 2.16:

- If $\boldsymbol{\pi}^{\mathrm{T}}\boldsymbol{\mu} = 0$, then Ψ_{+-} is the unique positive solution.
- If $\boldsymbol{\pi}^{\mathrm{T}}\boldsymbol{\mu} \neq 0$, then there exist two strictly positive solutions to (2.47), Ψ_{+-} and $\Psi_{\max}$, with

$$\Psi_{\max} = \Psi_{+-} + \gamma \boldsymbol{r}\boldsymbol{u}^{\mathrm{T}},$$

  where $\boldsymbol{r}$ is the right eigenvector of K for the eigenvalue λ_n, $\boldsymbol{u}$ is the left eigenvector of U for the eigenvalue $-\lambda_{n+1}$, and $\gamma = (\lambda_{n+1} - \lambda_n)/(\boldsymbol{u}^{\mathrm{T}}T_{-+}\boldsymbol{r})$.

A similar property holds for (2.48), naturally.

2.5 General MMBM

We turn our attention now to the fully general Markov-modulated Brownian motion, without any restriction on the various subsets of phases in $\mathcal{E}$. The characteristic equations for U and Ψ on the one hand, $\widehat{U}$ and $\widehat{\Psi}$ on the other hand, have a structure that is reminiscent, but more involved, than the two special cases analysed in Sects. 2.3 and 2.4.

Like in the last section, we censor out from the Markov chain the phases in $\mathcal{E}_0$, and we mostly use the transition matrix

$$T = \begin{bmatrix} Q_{\sigma\sigma} & Q_{\sigma+} & Q_{\sigma-} \\ Q_{+\sigma} & Q_{++} & Q_{+-} \\ Q_{-\sigma} & Q_{-+} & Q_{--} \end{bmatrix} + \begin{bmatrix} Q_{\sigma 0} \\ Q_{+0} \\ Q_{-0} \end{bmatrix} (-Q_{00})^{-1} \begin{bmatrix} Q_{0\sigma} & Q_{0+} & Q_{0-} \end{bmatrix}, \tag{2.57}$$

the difference with (2.42) being that we need to take $\mathcal{E}_\sigma$ into account here. The matrix U is indexed with the phases in $\mathcal{E}_{\mathrm{d}} = \mathcal{E}_\sigma \cup \mathcal{E}_-$ and the matrix $\widehat{U}$ with the phases in $\mathcal{E}_{\mathrm{u}} = \mathcal{E}_\sigma \cup \mathcal{E}_+$. To facilitate the comparison with equations in Sects. 2.3 and 2.4, we keep the notation

$$\begin{aligned} \Delta_\mu &= \operatorname{diag}(\mu_i : i \in \mathcal{E}_\sigma), \\ C_- &= \operatorname{diag}(\mu_i : i \in \mathcal{E}_-), \\ \text{and} \qquad C_+ &= \operatorname{diag}(\mu_i : i \in \mathcal{E}_+). \end{aligned}$$

Theorem 2.22. *The matrices U and Ψ are solutions of the system*

$$\begin{bmatrix} \frac{1}{2}\Delta_\sigma^2 & \\ & 0 \end{bmatrix} U^2 + \begin{bmatrix} \Delta_\mu & \\ & C_- \end{bmatrix} U + T_{dd} + T_{d+}\Psi_{+d} = 0, \tag{2.58}$$

$$C_+^{-1} T_{+d} + C_+^{-1} T_{++}\Psi_{+d} + \Psi_{+d} U = 0, \tag{2.59}$$

and

$$\Psi_{0d} = (-Q_{0,0})^{-1}(Q_{0,d} + Q_{0,+}\Psi_{+d}). \tag{2.60}$$

The matrices $\widehat{U}$ and $\widehat{\Psi}$ are solutions of the system

$$\begin{bmatrix} \frac{1}{2}\Delta_\sigma^2 & \\ & 0 \end{bmatrix} \widehat{U}^2 - \begin{bmatrix} \Delta_\mu & \\ & C_+ \end{bmatrix} \widehat{U} + T_{uu} + T_{u-}\widehat{\Psi}_{-u} = 0, \tag{2.61}$$

$$|C_-|^{-1} T_{-u} + |C_-|^{-1} T_{--}\widehat{\Psi}_{-u} + \widehat{\Psi}_{-u}\widehat{U} = 0, \tag{2.62}$$

and

$$\widehat{\Psi}_{0u} = (-Q_{0,0})^{-1}(Q_{0,u} + Q_{0,-}\widehat{\Psi}_{-u}). \tag{2.63}$$

Proof. Equations (2.59) and (2.60) cover the cases where the initial phase $\varphi(0)$ is in $\mathcal{E}_+$ or in $\mathcal{E}_0$, their proofs are identical to those of (2.44) and (2.11), respectively, and we do not repeat the details.

If $\varphi(0)$ is in $\mathcal{E}_-$, we condition on the first jump to a phase in either of the subsets $\mathcal{E}_\sigma$ and $\mathcal{E}_+$ and write that

$$(e^{Ux})_{-d} = e^{|C_-|^{-1}T_{--}x}\begin{bmatrix}0 & I_-\end{bmatrix} + \int_0^x e^{|C_-|^{-1}T_{--}y} \\ \times \{|C_-|^{-1}T_{-\sigma}(e^{U(x-y)})_{\sigma d} + |C_-|^{-1}T_{-+}\Psi_{+d}e^{U(x-y)}\}\, dy;$$

the first term on the right means that the phase remains in $\mathcal{E}_- \cup \mathcal{E}_0$ during the whole trip from level 0 to level $-x$, the second term means that the phase leaves $\mathcal{E}_- \cup \mathcal{E}_0$ when the level is in $(y, y + dy)$, and jumps to $\mathcal{E}_\sigma$ or $\mathcal{E}_+$.

Next, we take the derivative with respect to x on both sides, evaluate it at $x = 0$, and obtain

$$U_{-d} = |C_-|^{-1}T_{-d} + |C_-|^{-1}T_{-+}\Psi_{+d} \tag{2.64}$$

(note that $(e^{U(x-y)})_{\sigma -}\big|_{y-x} = \begin{bmatrix}I_\sigma & 0\end{bmatrix}$.) The interpretation of (2.64) is the same as the one we gave for (2.41).

Finally, if $\varphi(0)$ is in $\mathcal{E}_\sigma$, we condition on the first jump at time τ^* to any other phase and this leads to an equation similar to (2.22):

$$(e^{Ux})_{\sigma d} = \begin{bmatrix}e^{-\Delta_b x} & 0\end{bmatrix} + \int_0^x \int_0^\infty V_{ab}(y, v)\, P^*_{\sigma d}\, e^{U(x+y-v)}\, dv\, dy. \tag{2.65}$$

We need the new transition matrix $P^*_{\sigma d}$ here, where P^*_{ij}, i in $\mathcal{E}_\sigma$ and j in $\mathcal{E}_d$, is the conditional probability, given that there is a jump at time 0 from phase i to some other phase, that the first return to a state in $(X(0), \mathcal{E}_d)$ occurs in phase j. We have

$$P^*_{\sigma d} = \widetilde{P}_{\sigma d} + \begin{bmatrix}\widetilde{P}_{\sigma +} & \widetilde{P}_{\sigma 0}\end{bmatrix}\begin{bmatrix}\Psi_{+d} \\ \Psi_{0d},\end{bmatrix}$$

where $\widetilde{P}$ is the transition matrix (1.30) of the jump chain, or

$$P^*_{\sigma -} = \begin{bmatrix}I_\sigma & 0\end{bmatrix} + \widetilde{\Delta}^{-1}\begin{bmatrix}Q_{\sigma d} + Q_{\sigma +}\Psi_{+d} + Q_{\sigma 0}\Psi_{0d}\end{bmatrix} \tag{2.66}$$

with $\widetilde{\Delta} = \text{diag}(|q_{ii}| : i \in \mathcal{E}_\sigma)$.

Take the derivative with respect to x on both sides of (2.65), evaluate it at $x = 0$, and obtain

$$\begin{aligned} U_{\sigma d} &= -\Delta_b \begin{bmatrix}I_\sigma & 0\end{bmatrix} + \Delta_b \int_0^\infty \Delta_a e^{-\Delta_a y}\, P^*_{\sigma d}\, e^{Uy}\, dy \\ &= -\Delta_b \begin{bmatrix}I_\sigma & 0\end{bmatrix} + \Delta_b P^*_{\sigma d} + \Delta_a^{-1}(U_{\sigma d} + \Delta_b \begin{bmatrix}I_\sigma & 0\end{bmatrix})U, \end{aligned}$$

an equation that we now take through a sequence of transformations. We pre-multiply both sides by Δ_a and, using $\begin{bmatrix}I_\sigma & 0\end{bmatrix} U = U_{\sigma d}$, obtain that

$$U_{\sigma d}U + (\Delta_b - \Delta_a)U_{\sigma d} + \Delta_a\Delta_b(P^*_{\sigma d} - \begin{bmatrix}I_\sigma & 0\end{bmatrix}) = 0.$$

Next, we refer to (2.15, 2.16), we replace $\Delta_b - \Delta_a$ with $2\Delta_\mu\Delta_\sigma^{-2}$ and $\Delta_a\Delta_b$ with $2\widetilde{\Delta}\Delta_\sigma^{-2}$, and we replace $P^*_{\sigma\mathrm{d}}$ with the right-hand side of (2.66). This yields

$$\frac{1}{2}\Delta_\sigma^2 U_{\sigma\mathrm{d}} U + \Delta_\mu U_{\sigma\mathrm{d}} + Q_{\sigma\mathrm{d}} + Q_{\sigma+}\Psi_{+\mathrm{d}} + Q_{\sigma 0}\Psi_{0\mathrm{d}} = 0.$$

Then, we replace $\Psi_{0\mathrm{d}}$ with the right-hand side of (2.11) and we use (2.57), to write that

$$\frac{1}{2}\Delta_\sigma^2 U_{\sigma\mathrm{d}} U + \Delta_\mu U_{\sigma\mathrm{d}} + T_{\sigma\mathrm{d}} + T_{\sigma+}\Psi_{+\mathrm{d}} = 0. \tag{2.67}$$

Finally, this may be written as

$$\left[\tfrac{1}{2}\Delta_\sigma^2 \ 0\right] U^2 + \Delta_\mu U_{\sigma\mathrm{d}} + T_{\sigma\mathrm{d}} + T_{\sigma+}\Psi_{+\mathrm{d}} = 0. \tag{2.68}$$

The proof of (2.58) is now complete: (2.68) corresponds to the first block of rows and (2.64) to the second.

The argument for (2.61–2.63) is the same and is omitted. We mention for future reference that

$$\widehat{U}_{+\mathrm{u}} = C_+^{-1} T_{+\mathrm{u}} + C_+^{-1} T_{+-} \widehat{\Psi}_{-\mathrm{u}}, \tag{2.69}$$

this follows in a manner analogous to (2.64). □

Wiener-Hopf Factorisation The matrices $\Psi_{0\mathrm{d}}$ and $U_{-\mathrm{d}}$ are explicitly given by the right-hand sides of (2.60) and (2.64) respectively, once $\Psi_{+\mathrm{d}}$ is known; the key components in U and Ψ, therefore, are the submatrices $U_{\sigma\mathrm{d}}$ and $\Psi_{+\mathrm{d}}$. We combine them as the matrix M_{ud} indexed with $\mathcal{E}_+ \times \mathcal{E}_\mathrm{d}$:

$$M_{\mathrm{ud}} = \begin{bmatrix} U_{\sigma\mathrm{d}} \\ \Psi_{+\mathrm{d}} \end{bmatrix}. \tag{2.70}$$

Similarly, $\widehat{\Psi}_{0\mathrm{u}}$ and $\widehat{U}_{+\mathrm{u}}$ are determined by $\widehat{\Psi}_{-\mathrm{u}}$ and we combine $\widehat{U}_{\sigma\mathrm{d}}$ and $\widehat{\Psi}_{-\mathrm{u}}$ as

$$\widehat{M}_{\mathrm{du}} = \begin{bmatrix} \widehat{U}_{\sigma\mathrm{u}} \\ \widehat{\Psi}_{-\mathrm{u}} \end{bmatrix}.$$

We shall now characterise M_{ud} and $\widehat{M}_{\mathrm{du}}$ as unique solutions, in a sense to be made precise, of two Riccati equations. The final result, Theorem 2.25, has a simple formulation, but the argument itself is a bit involved.

We start from (2.59, 2.67) which we write as the $(m_\sigma + m_+) \times (m_\sigma + m_-)$ system

$$\begin{bmatrix} \frac{1}{2}\Delta_\sigma^2 & \cdot \\ \cdot & C_+ \end{bmatrix} M_{\mathrm{ud}} U + \begin{bmatrix} \Delta_\mu & T_{\sigma+} \\ \cdot & T_{++} \end{bmatrix} M_{\mathrm{ud}} + \begin{bmatrix} T_{\sigma\mathrm{d}} \\ T_{+\mathrm{d}} \end{bmatrix} = 0 \tag{2.71}$$

and we replace the matrix U with

$$\begin{aligned} U &= \begin{bmatrix} U_{\sigma\mathrm{d}} \\ U_{-\mathrm{d}} \end{bmatrix} \\ &= \begin{bmatrix} U_{\sigma\mathrm{d}} \\ |C_-|^{-1}(T_{-\mathrm{d}} + T_{-+}\Psi_{+\mathrm{d}}) \end{bmatrix} \qquad \text{by (2.64)} \\ &= A_{\mathrm{dd}} + B_{\mathrm{du}} M_{\mathrm{ud}}, \end{aligned} \tag{2.72}$$

where

$$A_{\mathrm{dd}} = \begin{bmatrix} I_\sigma & \cdot \\ \cdot & |C_-|^{-1} \end{bmatrix} \begin{bmatrix} 0 \\ T_{-\mathrm{d}} \end{bmatrix} \quad \text{and} \quad B_{\mathrm{du}} = \begin{bmatrix} I_\sigma & \cdot \\ \cdot & |C_-|^{-1} \end{bmatrix} \begin{bmatrix} I_\sigma & 0 \\ 0 & T_{-+} \end{bmatrix}.$$

The system (2.71) becomes

$$\begin{bmatrix} \frac{1}{2}\Delta_\sigma^2 & \cdot \\ \cdot & C_+ \end{bmatrix} M_{\mathrm{ud}}(A_{\mathrm{dd}} + B_{\mathrm{du}} M_{\mathrm{ud}}) + \begin{bmatrix} \Delta_\mu & T_{\sigma+} \\ \cdot & T_{++} \end{bmatrix} M_{\mathrm{ud}} + T_{\mathrm{ud}} = 0$$

and M_{ud} is a solution of the Riccati equation

$$Z B_{\mathrm{du}} Z + Z A_{\mathrm{dd}} + D_{\mathrm{uu}} Z + C_{\mathrm{ud}} = 0, \tag{2.73}$$

with

$$C_{\mathrm{ud}} = \begin{bmatrix} 2\Delta_\sigma^{-2} & \cdot \\ \cdot & C_+^{-1} \end{bmatrix} T_{\mathrm{ud}} \quad \text{and} \quad D_{\mathrm{uu}} = \begin{bmatrix} 2\Delta_\sigma^{-2} & \cdot \\ \cdot & C_+^{-1} \end{bmatrix} \begin{bmatrix} \Delta_\mu & T_{\sigma+} \\ 0 & T_{++} \end{bmatrix}.$$

By [40, Theorem 2.1], Z is a solution of (2.73) if and only if

$$H \begin{bmatrix} I_{\mathrm{d}} & 0 \\ Z & I_{\mathrm{u}} \end{bmatrix} = \begin{bmatrix} I_{\mathrm{d}} & 0 \\ Z & I_{\mathrm{u}} \end{bmatrix} \begin{bmatrix} A_{\mathrm{dd}} + B_{\mathrm{du}} Z & B_{\mathrm{du}} \\ 0 & -(D_{\mathrm{uu}} + Z B_{\mathrm{du}}) \end{bmatrix}$$

where

$$H = \begin{bmatrix} A_{\mathrm{dd}} & B_{\mathrm{du}} \\ -C_{\mathrm{ud}} & -D_{\mathrm{uu}} \end{bmatrix}. \tag{2.74}$$

In particular,

$$H \begin{bmatrix} I_{\mathrm{d}} & 0 \\ M_{\mathrm{ud}} & I_{\mathrm{u}} \end{bmatrix} = \begin{bmatrix} I_{\mathrm{d}} & 0 \\ M_{\mathrm{ud}} & I_{\mathrm{u}} \end{bmatrix} \begin{bmatrix} U & B_{\mathrm{du}} \\ 0 & -(D_{\mathrm{uu}} + M_{\mathrm{ud}} B_{\mathrm{du}}) \end{bmatrix}. \tag{2.75}$$

We show through a similar sequence of steps that $\widehat{M}_{\mathrm{du}}$ is a solution of the Riccati equation

$$Z \widehat{B}_{\mathrm{ud}} Z + Z \widehat{A}_{\mathrm{uu}} + \widehat{D}_{\mathrm{dd}} Z + \widehat{C}_{\mathrm{du}} = 0, \tag{2.76}$$

with

$$\widehat{A}_{\mathrm{uu}} = \begin{bmatrix} I_\sigma & \cdot \\ \cdot & C_+^{-1} \end{bmatrix} \begin{bmatrix} 0 \\ T_{+\mathrm{u}} \end{bmatrix}, \qquad \widehat{B}_{\mathrm{ud}} = \begin{bmatrix} I_\sigma & \cdot \\ \cdot & C_+^{-1} \end{bmatrix} \begin{bmatrix} I_\sigma & 0 \\ 0 & T_{+-} \end{bmatrix},$$

$$\widehat{C}_{\mathrm{du}} = \begin{bmatrix} 2\Delta_\sigma^{-2} & \cdot \\ \cdot & |C_-|^{-1} \end{bmatrix} T_{\mathrm{du}}, \qquad \widehat{D}_{\mathrm{dd}} = \begin{bmatrix} 2\Delta_\sigma^{-2} & \cdot \\ \cdot & |C_-|^{-1} \end{bmatrix} \begin{bmatrix} -\Delta_\mu & T_{\sigma-} \\ 0 & T_{--} \end{bmatrix}$$

and, in consequence, that

$$\widehat{H} \begin{bmatrix} I_{\mathrm{u}} & 0 \\ \widehat{M}_{\mathrm{du}} & I_{\mathrm{d}} \end{bmatrix} = \begin{bmatrix} I_{\mathrm{u}} & 0 \\ \widehat{M}_{\mathrm{du}} & I_{\mathrm{d}} \end{bmatrix} \begin{bmatrix} \widehat{U} & \widehat{B}_{\mathrm{ud}} \\ 0 & -(\widehat{D}_{\mathrm{dd}} + \widehat{M}_{\mathrm{du}} \widehat{B}_{\mathrm{ud}}) \end{bmatrix} \tag{2.77}$$

with

$$\widehat{H} = \begin{bmatrix} \widehat{A}_{\mathrm{uu}} & \widehat{B}_{\mathrm{ud}} \\ -\widehat{C}_{\mathrm{du}} & -\widehat{D}_{\mathrm{dd}} \end{bmatrix} \tag{2.78}$$

and

$$\widehat{U} = \widehat{A}_{\mathrm{uu}} + \widehat{B}_{\mathrm{ud}} \widehat{M}_{\mathrm{du}}. \tag{2.79}$$

It is tedious but not difficult to verify that $\widehat{H} = -S^{-1}HS$, where S is the permutation matrix

$$S = \begin{bmatrix} I_\sigma & \cdot & \cdot & \cdot \\ \cdot & \cdot & \cdot & I_- \\ \cdot & \cdot & -I_\sigma & \cdot \\ \cdot & I_+ & \cdot & \cdot \end{bmatrix},$$

and so (2.77) may be rewritten as

$$HS \begin{bmatrix} I_{\mathrm{u}} & 0 \\ \widehat{M}_{\mathrm{du}} & I_{\mathrm{d}} \end{bmatrix} = S \begin{bmatrix} I_{\mathrm{u}} & 0 \\ \widehat{M}_{\mathrm{du}} & I_{\mathrm{d}} \end{bmatrix} \begin{bmatrix} -\widehat{U} & -\widehat{B}_{\mathrm{ud}} \\ 0 & \widehat{D}_{\mathrm{dd}} + \widehat{M}_{\mathrm{du}} \widehat{B}_{\mathrm{ud}} \end{bmatrix} \tag{2.80}$$

with the same matrix H in the left-most position as in (2.75).

We are now in a position to prove the following splitting property for the eigenvalues of the matrix H. It generalises both Theorems 2.11 and 2.20.

Theorem 2.23. *Denote by m_u and m_d the number of phases in $\mathcal{E}_\sigma \cup \mathcal{E}_+$ and $\mathcal{E}_\sigma \cup \mathcal{E}_-$, respectively. The eigenvalues $\lambda_1, \lambda_2, \ldots, \lambda_{m_d+m_u}$ of H, indexed in increasing order of their real parts, satisfy the inequalities*

$$\mathrm{Re}(\lambda_1) \cdots \mathrm{Re}(\lambda_{m_d-1}) < \lambda_{m_d} \le \lambda_{m_d+1} < \mathrm{Re}(\lambda_{m_d+2}) \cdots \mathrm{Re}(\lambda_{m_d+m_u}). \tag{2.81}$$

Furthermore:

(a) If $\boldsymbol{\pi}^{\mathrm{T}}\boldsymbol{\mu} > 0$, then $\lambda_{m_d} < 0 = \lambda_{m_d+1}$.
(b) If $\boldsymbol{\pi}^{\mathrm{T}}\boldsymbol{\mu} < 0$, then $\lambda_{m_d} = 0 < \lambda_{m_d+1}$.
(c) If $\boldsymbol{\pi}^{\mathrm{T}}\boldsymbol{\mu} = 0$, then $\lambda_{m_d} = 0 = \lambda_{m_d+1}$; and there is only one eigenvector of H for the eigenvalue 0.

Proof. It follows from (2.75, 2.80) that the eigenvalues of U and those of $-\widehat{U}$ are eigenvalues of H, and we repeat the argument of Theorem 2.11: The m_{d} eigenvalues of U are in $\mathbb{C}_{\leq}$ and the m_u eigenvalues of $-\widehat{U}$ are in $\mathbb{C}_{\geq}$. For both matrices, the eigenvalue closest to 0 is real. This justifies the splitting (2.81).

In addition, if $\boldsymbol{\pi}^{\mathrm{T}}\boldsymbol{\mu} > 0$, then the eigenvalues of U are in $\mathbb{C}_{<}$ and the eigenvalues of $-\widehat{U}$ include 0, while the eigenvalues of U include 0 and the eigenvalues of $-\widehat{U}$ are in $\mathbb{C}_{>}$ when $\boldsymbol{\pi}^{\mathrm{T}}\boldsymbol{\mu} < 0$. This justifies the additional statements in Cases (a) and (b).

To verify in Case (c) that H has an eigenvalue 0 of algebraic multiplicity two with one eigenvector only, it helps to reorganise the blocks in H so that the transition matrix T is made apparent, and to write H as the product

$$H = -\begin{bmatrix} \cdot & \cdot & \cdot & I_\sigma \\ \cdot & \cdot & I_- & \cdot \\ I_\sigma & \cdot & \cdot & \cdot \\ \cdot & I_+ & \cdot & \cdot \end{bmatrix} \begin{bmatrix} 2\Delta_\sigma^{-2} & & & \\ & C_+^{-1} & & \\ & & C_-^{-1} & \\ & & & -I_\sigma \end{bmatrix} \left[\begin{array}{ccc|c} & & & \Delta_\mu \\ & T & & \cdot \\ & & & \cdot \\ \hline \cdot & \cdot & \cdot & I_\sigma \end{array}\right] \begin{bmatrix} I_\sigma & \cdot & \cdot & \cdot \\ \cdot & \cdot & \cdot & I_+ \\ \cdot & I_- & \cdot & \cdot \\ \cdot & \cdot & I_\sigma & \cdot \end{bmatrix},$$

the proof being by direct verification. This expression makes it easy to verify that a right eigenvector of H for the eigenvalue 0 is

$$\boldsymbol{v}_1 = \begin{bmatrix} \mathbf{1}_\sigma \\ \mathbf{1}_- \\ \mathbf{0}_\sigma \\ \mathbf{1}_+, \end{bmatrix}$$

and, like in the proof of Theorem 2.11, we need to show that there is a vector

$$\boldsymbol{v}_2 = \begin{bmatrix} \boldsymbol{x}_\sigma \\ \boldsymbol{x}_- \\ \boldsymbol{y}_\sigma \\ \boldsymbol{x}_+ \end{bmatrix}$$

such that $H\boldsymbol{v}_2 = \boldsymbol{v}_1$. After a few simple manipulations, we find that $H\boldsymbol{v}_2 = \boldsymbol{v}_1$ is equivalent to the system

$$T \begin{bmatrix} \boldsymbol{x}_\sigma \\ \boldsymbol{x}_- \\ \boldsymbol{x}_+ \end{bmatrix} + \begin{bmatrix} \Delta_\mu \boldsymbol{y}_\sigma \\ 0 \\ 0 \end{bmatrix} = \begin{bmatrix} \mathbf{0}_\sigma \\ -\boldsymbol{\mu}_+ \\ -\boldsymbol{\mu}_- \end{bmatrix}, \qquad \boldsymbol{y}_\sigma = \mathbf{1}_\sigma.$$

The subvector $\boldsymbol{y}_\sigma$ is known, and the remaining components of $\boldsymbol{v}_2$ have to be a solution of

$$T \begin{bmatrix} \boldsymbol{x}_\sigma \\ \boldsymbol{x}_- \\ \boldsymbol{x}_+ \end{bmatrix} = -\begin{bmatrix} \boldsymbol{\mu}_\sigma \\ \boldsymbol{\mu}_+ \\ \boldsymbol{\mu}_- \end{bmatrix}. \tag{2.82}$$

The matrix T is singular, and the system (2.82) has a solution if and only if $\boldsymbol{\gamma}_\sigma^{\mathrm{T}}\boldsymbol{\mu}_\sigma + \boldsymbol{\gamma}_+^{\mathrm{T}}\boldsymbol{\mu}_+ + \boldsymbol{\gamma}_-^{\mathrm{T}}\boldsymbol{\mu}_- = 0$, where $\boldsymbol{\gamma}$ is the stationary probability vector of T. This vector is proportional to $\begin{bmatrix}\boldsymbol{\pi}_\sigma^{\mathrm{T}} & \boldsymbol{\pi}_+^{\mathrm{T}} & \boldsymbol{\pi}_-^{\mathrm{T}}\end{bmatrix}^{\mathrm{T}}$, so that the matrix H does have two eigenvalues equal to 0, with one eigenvector, when $\boldsymbol{\pi}^{\mathrm{T}}\boldsymbol{\mu} = 0$. This concludes our proof. □

Corollary 2.24. *The eigenvalues of H are split into two subsets as follows:*

$$\rho(U) = \rho(\widehat{D}_{dd} + \widehat{M}_{du}\widehat{B}_{ud}) = \{\lambda_1, \ldots, \lambda_{m_d}\},$$

and

$$\rho(\widehat{U}) = \rho(D_{uu} + M_{ud}B_{du}) = \{-\lambda_{m_d+1}, \ldots, -\lambda_{m_d+m_u}\}.$$

□

Just as in Sect. 2.3, the eigenvalues of H have an $(m_\mathrm{d}, m_\mathrm{u})$-splitting with respect to the imaginary axis, the splitting is proper when $\boldsymbol{\pi}^\mathrm{T}\boldsymbol{\mu} \neq 0$ and weak if $\boldsymbol{\pi}^\mathrm{T}\boldsymbol{\mu} = 0$. This helps us to uniquely identify M_ud and $\widehat{M}_\mathrm{du}$ as solutions of their respective Riccati equations. The next theorem fully characterises U and $\widehat{U}$. Like Theorem 2.13, it follows from [40, Theorems 2.5 and 2.6].

Theorem 2.25. *If $\boldsymbol{\pi}^\mathrm{T}\boldsymbol{\mu} > 0$, then the matrix M_{ud} is the unique solution of the Riccati equation (2.73) such that $\rho(U) \subset \mathbb{C}_<$, where U is given by (2.72), and the matrix $\widehat{M}_{du}$ is the unique solution of (2.76) such that $\rho(\widehat{U}) \subset \mathbb{C}_\leq$, where $\widehat{U}$ is given by (2.79).*

If $\boldsymbol{\pi}^\mathrm{T}\boldsymbol{\mu} < 0$, then M_{ud} and $\widehat{M}_{du}$ are the unique solutions of (2.73) and (2.76), respectively, such that $\rho(U) \subset \mathbb{C}_\leq$ and $\rho(\widehat{U}) \subset \mathbb{C}_<$.

If $\boldsymbol{\pi}^\mathrm{T}\boldsymbol{\mu} = 0$, then M_{ud} is the unique solution of (2.73) associated with the stable canonical invariant subspace of H, while $\widehat{M}_{du}$ is the unique solution of (2.76) associated with the antistable canonical invariant subspace. □

Remark 2.26. To conclude this section, we mention that Ivanovs [87, Eq. (2.2)] gives the elegant equation

$$\begin{bmatrix} \frac{1}{2}\Delta_\sigma^2 & & & \\ & 0 & & \\ & & 0 & \\ & & & 0 \end{bmatrix} \begin{bmatrix} I_\mathrm{u} \\ \widehat{\Psi} \end{bmatrix} \widehat{U}^2 - \begin{bmatrix} \Delta_\mu & & & \\ & C_+ & & \\ & & C_- & \\ & & & 0 \end{bmatrix} \begin{bmatrix} I_\mathrm{u} \\ \widehat{\Psi} \end{bmatrix} \widehat{U} + Q \begin{bmatrix} I_\mathrm{u} \\ \widehat{\Psi} \end{bmatrix} = 0, \qquad (2.83)$$

equivalent to the system (2.61–2.63). A similar equation holds for the pair of matrices (U, Ψ), but we need to reorganise the matrix Q so that its rows and columns are given in the order of the subsets $\mathcal{E}_\sigma, \mathcal{E}_-, \mathcal{E}_+, \mathcal{E}_0$. To that end, we exchange the positions of the phases in $\mathcal{E}_+$ and $\mathcal{E}_-$, and the phase generator becomes

$$Q' = \begin{bmatrix} I_\sigma & \cdot & \cdot & \cdot \\ \cdot & \cdot & I_+ & \cdot \\ \cdot & I_- & \cdot & \cdot \\ \cdot & \cdot & \cdot & I_0 \end{bmatrix}^{-1} Q \begin{bmatrix} I_\sigma & \cdot & \cdot & \cdot \\ \cdot & \cdot & I_+ & \cdot \\ \cdot & I_- & \cdot & \cdot \\ \cdot & \cdot & \cdot & I_0 \end{bmatrix}$$

after permutation. The system (2.58–2.60) is equivalent to the equation

$$\begin{bmatrix} \frac{1}{2}\Delta_\sigma^2 & & & \\ & 0 & & \\ & & 0 & \\ & & & 0 \end{bmatrix} \begin{bmatrix} I_\mathrm{d} \\ \Psi \end{bmatrix} U^2 + \begin{bmatrix} \Delta_\mu & & & \\ & C_- & & \\ & & C_+ & \\ & & & 0 \end{bmatrix} \begin{bmatrix} I_\mathrm{d} \\ \Psi \end{bmatrix} U + Q' \begin{bmatrix} I_\mathrm{d} \\ \Psi \end{bmatrix} = 0. \qquad (2.84)$$

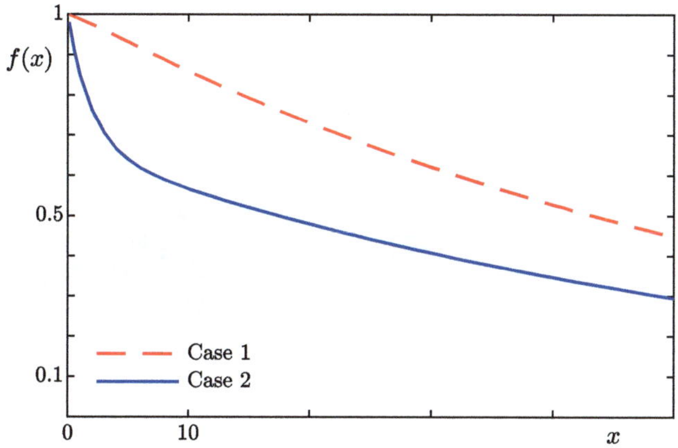

Fig. 2.5 Example 2.27. Probability $f_\alpha(x)$ of reaching level x in a finite time, starting from level 0, for two initial distributions of $\varphi(0)$

As illustration, we give two examples of distribution of $\varphi(\tau_x)$, the phase upon reaching level x, starting from level 0.

Example 2.27. This is the singularly perturbed system from Example 1.22, with $\varepsilon = 0.01$. The stationary drift is $\boldsymbol{\pi}^{\mathrm{T}}\boldsymbol{\mu} = -0.0666$, and thus $\lim_{t\to\infty} X(t) = -\infty$ with probability 1, and $\mathrm{P}[\tau_x < \infty] < 1$ for all strictly positive x. Actually, the eigenvalue of $\widehat{U}$ with maximal real part is -0.0163 and $\lim_{x\to\infty} e^{\widehat{U}x} = 0$. From (2.10), we know that

$$\mathrm{P}[\tau_x < \infty] = \boldsymbol{\alpha}^{\mathrm{T}} e^{\widehat{U}x}\mathbf{1},$$

where $\boldsymbol{\alpha}$ is the distribution of $\varphi(0)$, and we denote that probability as $f_\alpha(x)$. We show its value in Fig. 2.5 for $0 \le x \le 50$ and for two different initial distributions:

Case 1: $\boldsymbol{\alpha}(1) = \begin{bmatrix}0.5 & 0.5 & 0 & 0\end{bmatrix}^{\mathrm{T}}$.
Case 2: $\boldsymbol{\alpha}(2) = \begin{bmatrix}0 & 0 & 0.5 & 0.5\end{bmatrix}^{\mathrm{T}}$.

The dashed (red) curve is for f_1, the function f_α in Case 1, the continuous (blue) curve is for f_2 in Case 2. As expected, both are decreasing functions of x.

In Case 1, the MMBM starts in the set $\{1, 2\}$ for which the overall tendency is for the level to go up. As a matter of fact, $f_1(x)$ is very close to $e^{-0.0163x}$, the asymptotically dominant component of $e^{\widehat{U}x}$; the maximum difference over the interval $(0, 50)$ is less than 0.01, and if we had plotted this function in addition to $f_1(x)$, the two curves would have been hardly distinguishable at the scale of Fig. 2.5.

The shape of $f_2(x)$, however, is significantly different. In Case 2, the phase starts in the set $\{3, 4\}$ and remains there for an expected 100 units of time. The tendency in this set is for the level to go down, and, by the time the phase jumps to the set $\{1, 2\}$, the level is likely to be negative and might never return to positive levels again. This explains why $f_2(x)$ is much smaller than $f_1(x)$. However, the behaviour of Brownian motion as described in Sect. 1.4 ensures that, with probability 1, the level takes strictly positive values straightaway, that is, before time η for any $\eta > 0$.

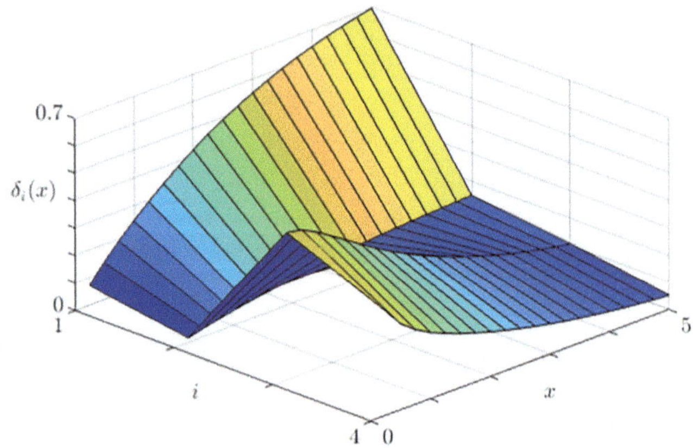

Fig. 2.6 Example 2.27. Conditional distribution $\boldsymbol{\delta}^{\mathrm{T}}(x) = (\boldsymbol{\alpha}^{\mathrm{T}} e^{\widehat{U}x}\mathbf{1})^{-1}\boldsymbol{\alpha}^{\mathrm{T}} e^{\widehat{U}x}$ of the phase at time τ_x, given that τ_x is finite, for Case 2

This explains why $f_2(x)$ is close to 1 for small values of x, before dropping sharply. After the first drop, the curve decreases at a slower rate, asymptotically the same as $e^{\widehat{U}x}$, and the probability of reaching level x is still slightly above 0.5 for $x = 17$.

Next, we turn our attention to the conditional distribution $\boldsymbol{\delta}^{\mathrm{T}}(x) = f_\alpha(x)^{-1}\boldsymbol{\alpha}^{\mathrm{T}} e^{\widehat{U}x}$ of $\varphi(\tau_x)$, given that the process does reach level x. We give in Fig. 2.6 the values of $\boldsymbol{\delta}_i(x)$ in Case 2 for $i = 1$ to 4 and $0 \le x \le 5$. As x increases from 0 to 5, we see that the probability mass moves from the initial set $\{3, 4\}$ to $\{1, 2\}$. Before leaving the set $\{3, 4\}$, the system has the opportunity of reaching strictly positive but relatively small levels, and $\boldsymbol{\delta}_3(x)$ is the bigger component. For higher values of x, the system is more likely to be in the set $\{1, 2\}$ at time τ_x and $\boldsymbol{\delta}_1(x)$ becomes dominant. In this example, $\boldsymbol{\delta}_3(x) > \boldsymbol{\delta}_1(x)$ for $0 \le x < 1.6558$, approximately.

For x increasing further, one shows that $\lim_{x\to\infty} \boldsymbol{\delta}(x) = \widehat{\boldsymbol{v}}$, the normalised left eigenvector of $\widehat{U}$ associated to the eigenvalue closest to 0, independently of the initial distribution $\boldsymbol{\alpha}$. In this example,

$$\widehat{\boldsymbol{v}} = \begin{bmatrix}0.7931 & 0.1688 & 0.0255 & 0.0127\end{bmatrix}^{\mathrm{T}}.$$

In Fig. 2.6, the conditional distribution for $x = 5$ is

$$\boldsymbol{\delta}(5) = \begin{bmatrix}0.6940 & 0.1477 & 0.1053 & 0.0529\end{bmatrix}^{\mathrm{T}},$$

already showing the same profile as its asymptotic value.

Example 2.28. This example is inspired from Latouche and Ramaswami [106, Example 9.2.2], we use it to illustrate the influence of the volatility parameter. The phase generator is

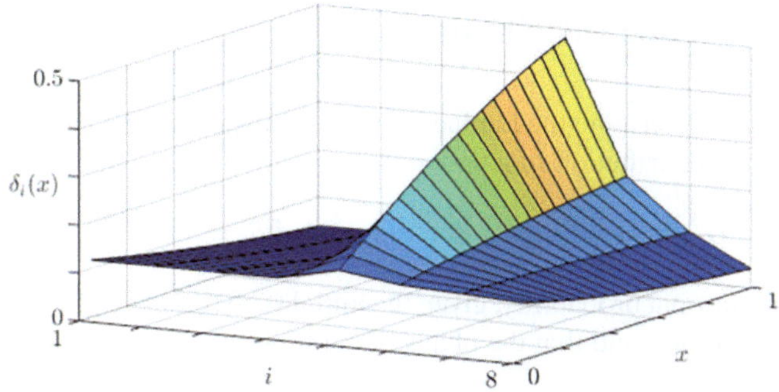

Fig. 2.7 Example 2.28. Conditional distribution $\delta^{\mathrm{T}}(x) = (\boldsymbol{\pi}^{\mathrm{T}} e^{\widehat{U}x}\mathbf{1})^{-1}\boldsymbol{\pi}^{\mathrm{T}} e^{\widehat{U}x}$ of the phase at time τ_x

$$Q = \begin{bmatrix} -1 & 1 & \cdot & \cdot & \cdot & \cdot & \cdot & \cdot \\ \cdot & -1 & 1 & \cdot & \cdot & \cdot & \cdot & \cdot \\ \cdot & \cdot & -1 & 1 & \cdot & \cdot & \cdot & \cdot \\ \cdot & \cdot & \cdot & -1 & 1 & \cdot & \cdot & \cdot \\ \cdot & \cdot & \cdot & \cdot & -1 & 1 & \cdot & \cdot \\ \cdot & \cdot & \cdot & \cdot & \cdot & -1 & 1 & \cdot \\ \cdot & \cdot & \cdot & \cdot & \cdot & \cdot & -1 & 1 \\ 1 & \cdot & \cdot & \cdot & \cdot & \cdot & \cdot & -1 \end{bmatrix}$$

so that the phase cycles through the states 1–8 at a constant rate. The Brownian parameters are

$$\boldsymbol{\mu} = -\begin{bmatrix}1 \; 1 \; 1 \; 1 \; 1 \; 1 \; 1 \; 1\end{bmatrix}^{\mathrm{T}},$$
$$\boldsymbol{\sigma} = \begin{bmatrix}1 \; 1 \; 1 \; 1 \; 3 \; 2 \; \sqrt{2} \; 1\end{bmatrix}^{\mathrm{T}}.$$

The drift is negative, equal to -1 for all phases, and the differences between the phases occur only through the variances, which are equal to 1 for most of the cycle, before jumping to 9 in phase 5 and getting progressively smaller until returning to 1.

We give in Fig. 2.7 the conditional distribution $\boldsymbol{\delta}(x)$ for $0 \le x \le 1$, with $\varphi(0)$ having the stationary distribution $\boldsymbol{\pi}$ of Q. We see that the high variance of phase 5, and to a lesser extent of phases 6 and 7, makes it much more likely that the system is in these phases when the level first reaches high values. For this example, the limiting vector $\widehat{\boldsymbol{v}}$ is

$$\widehat{\boldsymbol{v}} = \begin{bmatrix}0.0065 \; 0.0035 \; 0.0019 \; 0.0011 \; 0.6515 \; 0.2529 \; 0.0696 \; 0.0131\end{bmatrix}^{\mathrm{T}},$$

and the conditional distribution at time τ_1 is

$$\boldsymbol{\delta}(1) = \begin{bmatrix}0.0407 \; 0.0421 \; 0.0429 \; 0.0432 \; 0.4827 \; 0.2059 \; 0.1035 \; 0.0390\end{bmatrix}^{\mathrm{T}},$$

not close to $\widehat{\boldsymbol{v}}$ but showing a similar shape already.

2.6 Crossing Time Distribution

It is quite easy to determine the Laplace-Stieltjes transform (LST) of the distribution of τ_x. Define the distribution function

$$G_{ij}(t;x) = \mathrm{P}_{0,i}[\tau_{-x} \le t, \varphi(\tau_{-x}) = j], \qquad t \ge 0,$$

that is, the conditional probability, given the initial state $(0, i)$, that the process reaches level $-x$ before time t and that the phase is j at time τ_{-x}, and denote its LST as

$$\Gamma_{ij}(s;x) = \int_0^\infty e^{-st}\, \mathrm{d}G_{ij}(t;x)$$

for $s \ge 0$. We may interpret the transform as

$$\Gamma_{ij}(s;x) = \mathrm{P}_{0,i}[\tau_{-x} \le e_s, \varphi(\tau_{-x}) = j],$$

where e_s is an exponential random variable with parameter s, independent of the MMBM. Like in Sect. 1.8, we may think of e_s as the epoch when the MMBM is killed and enters a cemetery state, and $\Gamma_{ij}(s;x)$ as the probability that the MMBM reaches level $-x$ before being killed. To determine $\Gamma(s;x)$ and, in consequence, to determine the distributions of τ_{-x} and of $\varphi(\tau_{-x})$, it suffices therefore to replace the generator Q with $Q_s = Q - sI$ in the analysis of the matrix $G(x)$. Similarly, we use Q_s in the analysis of $\widehat{G}(x)$ and so determine $\widehat{\Gamma}(s;x) = \mathrm{P}[\tau_x < e_s; \varphi(\tau_x)]$.

From there, we obtain as usual moments of ladder times by taking derivatives of their LST with respect to s, evaluated at 0, and we obtain the whole distribution by numerical inversion (Abate and Whitt [1], Horváth et al. [86]). The expectation, however, may be determined with a direct argument, as we show in the next lemma.

Lemma 2.29. *Assume that $X(0) = 0$ and that $\varphi(0)$ has the distribution $\boldsymbol{\alpha}$. Take $x > 0$ arbitrary.*

- *If $\boldsymbol{\pi}^{\mathrm{T}}\boldsymbol{\mu} > 0$, then*

$$\mathrm{E}[\tau_x] = \{x + \boldsymbol{\alpha}^{\mathrm{T}}(e^{\widehat{U}x} - I)(-Q^{\#})\boldsymbol{\mu}\}/(\boldsymbol{\pi}^{\mathrm{T}}\boldsymbol{\mu}), \tag{2.85}$$

- *if $\boldsymbol{\pi}^{\mathrm{T}}\boldsymbol{\mu} < 0$, then*

$$\mathrm{E}[\tau_{-x}] = \{-x + \boldsymbol{\alpha}^{\mathrm{T}}(e^{Ux} - I)(-Q^{\#})\boldsymbol{\mu}\}/(\boldsymbol{\pi}^{\mathrm{T}}\boldsymbol{\mu}), \tag{2.86}$$

where U and $\widehat{U}$ are analysed in Sect. 2.3.

Proof. Assume $\boldsymbol{\pi}^{\mathrm{T}}\boldsymbol{\mu} < 0$. In that case, $\mathrm{E}[\tau_{-x}] < \infty$ for $x > 0$ and it results from (1.40) that

$$\mathrm{E}[X(\tau_{-x})] = \mathrm{E}[\tau_{-x}]\boldsymbol{\pi}^{\mathrm{T}}\boldsymbol{\mu} + \boldsymbol{\alpha}^{\mathrm{T}}(\mathrm{E}[e^{Q\tau_{-x}}] - I)Q^{\#}\boldsymbol{\mu}. \tag{2.87}$$

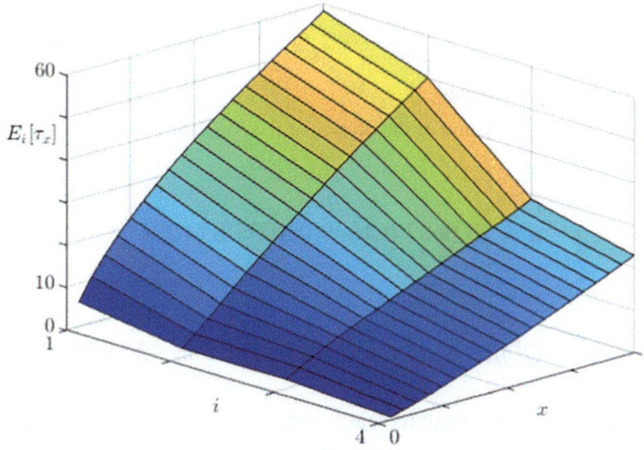

Fig. 2.8 Example 2.31. Expectation of τ_{-x} given the initial phase $\varphi(0) = i, 0 \le x \le 2, i = 1, \ldots, 4$

Now, $X(\tau_{-x}) = -x$ by definition, and

$$(\mathrm{E}[e^{Q\tau_{-x}}])_{ij} = \mathrm{P}_i[\varphi(\tau_{-x}) = j] = (e^{Ux})_{ij}.$$

Therefore, (2.87) may be written as

$$-x = \mathrm{E}[\tau_{-x}]\boldsymbol{\pi}^{\mathrm{T}}\boldsymbol{\mu} + \boldsymbol{\alpha}^{\mathrm{T}}(e^{Ux} - I)Q^{\#}\boldsymbol{\mu},$$

which proves (2.86). The proof of (2.85) is similar. □

Remark 2.30. If all μ_is are equal, then $\boldsymbol{\mu}$ is proportional to $\mathbf{1}$ and $Q^{\#}\boldsymbol{\mu} = \mathbf{0}$, so that $\mathrm{E}_i[\tau_x] = x/(\boldsymbol{\pi}^{\mathrm{T}}\boldsymbol{\mu})$ for any x of the same sign as $\boldsymbol{\pi}^{\mathrm{T}}\boldsymbol{\mu}$, independently of the starting phase i and independently of the volatility parameters $\boldsymbol{\sigma}$.

Example 2.31. This is the singularly perturbed system from Examples 1.22 and 2.27, with $\varepsilon = 0.1$, so that transitions from one set of phases to the other occur 10 times more rapidly than in Example 2.27. The stationary drift is –0.0661.

We plot in Fig. 2.8 the expected time $\mathrm{E}_i[\tau_{-x}]$ to reach negative levels, for each initial phase i in $\{1 \cdots 4\}$ and for $0 \le x \le 2$. In the set $\{1, 2\}$, the tendency is for the level to increase, and it takes more time to reach negative levels than if the process starts in $\{3, 4\}$, where the tendency is to move down.

The linear term $-x/(\boldsymbol{\pi}^{\mathrm{T}}\boldsymbol{\mu})$ in (2.86) is independent of the distribution of $\varphi(0)$, and it is equal to $15.129\,x$ in this example. The differences between the expected time and the linear term are due to the component

$$\mathrm{E}[\tau_{-x}] + x/(\boldsymbol{\pi}^{\mathrm{T}}\boldsymbol{\mu}) = (\boldsymbol{\alpha}^{\mathrm{T}}(e^{Ux} - I)(-Q^{\#})\boldsymbol{\mu})/(\boldsymbol{\pi}^{\mathrm{T}}\boldsymbol{\mu}),$$

with the vector $\boldsymbol{\alpha}$ placing all the probability mass on each phase in turn. The exponential of Ux in the right-hand side is bounded, and $\lim_{x\to\infty} e^{Ux} = \mathbf{1} \cdot \boldsymbol{u}^{\mathrm{T}}$, where $\boldsymbol{u}$ is the stationary probability vector of e^U. In this example,

Network	Video		
	Buffering	*Playing*	*Finishing*
Congested	1	2	5
Normal	3	4	5

Table 2.1 Example 2.32, association of the phases of the MMBM with the states of the model

$$\boldsymbol{u} = \left[0.0687\ 0.1649\ 0.1037\ 0.6627\right]^{\mathrm{T}},$$

and

$$\lim_{x\to\infty}(e^{Ux} - I)(-Q^{\#})\boldsymbol{\mu}/(\boldsymbol{\pi}^{\mathrm{T}}\boldsymbol{\mu}) = \left[36.1029\ 30.0383\ -3.8971\ -10.6068\right]^{\mathrm{T}}$$

so that, indeed, it takes more time to reach very low levels if the MMBM starts from phases 1 or 2, less time if it starts from phase 3 and least if it starts from phase 4. Furthermore, Fig. 2.8 shows that the expected value of τ_{-x} is nearly linear if $\varphi(0)$ is in $\{3, 4\}$, but it is not yet linear for $x = 2$ if $\varphi(0)$ is in $\{1, 2\}$.

Example 2.32. This example is based on Gribaudo et al. [73, Section 5] where the authors analyse a model for the receiving buffer in a video streaming application. There are two components: the network and the video. The network may be in either of two states, *normal* and *congested*. The video may be in three states, *buffering*, *playing* and *finishing*. The MMBM phases represent a combination of the states of both components as indicated in Table 2.1.

The transition matrix is

$$Q = \begin{bmatrix} * & \beta_B & \alpha_{LH} & \cdot & \cdot \\ \cdot & * & \cdot & \alpha_{LH} & \beta_P \\ \alpha_{HL} & \cdot & * & \beta_B & \cdot \\ \cdot & \alpha_{HL} & \cdot & * & \beta_P \\ p_1\beta_F & \cdot & p_3\beta_F & \cdot & * \end{bmatrix}$$

with $\beta_B = 0.1, \beta_F = 0.1, \beta_P = 0.03, \alpha_{HL} = \alpha_{LH} = 1/60$ (the time unit is 1 second). The parameters p_1 and p_3 are the stationary fraction of time spent by the network in each of its two states, and they are chosen to be equal, so that $p_1 = p_3 = 0.5$. The diagonal elements of Q, represented by the symbol "$*$", are such that the row sums are equal to 0.

The drift parameters are

$$\boldsymbol{\mu} = \left[0.250\ -0.750\ 0.625\ -0.375\ -1.000\right]^{\mathrm{T}}.$$

In phases 1 and 3, the video application is buffering data and the drift components are equal to the input rate from the network, in phases 2, 4 and 5, the application is processing data from the buffer at the rate of 1 Mb per second.

The variance components are the sum of a contribution from the network input and a smaller contribution from the video application. They are given as $\sigma_i^2 = r_i$, $1 \le i \le 5$, with

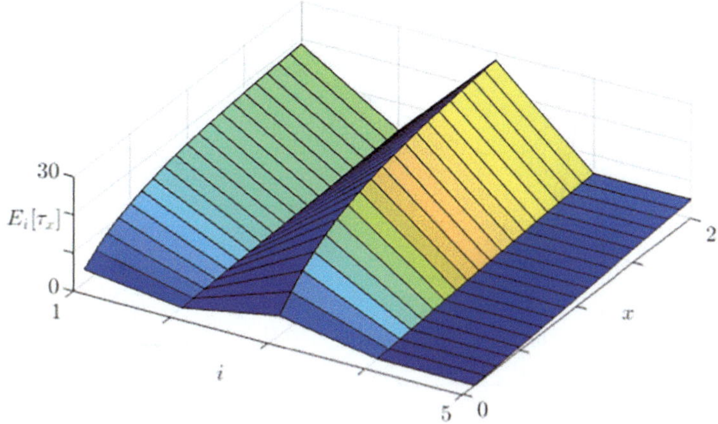

Fig. 2.9 Example 2.32. Expected time $E_i[\tau_{-x}]$to reduce the buffer content given the initial phase i of the system, $0 \le x \le 2$

$$r = \begin{bmatrix}0.250 & 0.350 & 0.625 & 0.725 & 0.100\end{bmatrix}^{\mathrm{T}}.$$

We give in Fig. 2.9 the expected time τ_{-x} needed to reduce the buffer content by x units, conditionally given the initial phase i. We also give in the table below the expected time to reduce the buffer content by 0.1 Mb and by 2 Mb.

	Phases				
x	1	2	3	4	5
0.1	3.3466	0.1346	4.8040	0.2598	0.1381
2	18.6175	2.8281	27.9553	5.1256	5.0469

The effect of the video application is clear: When it is processing data in phases 2, 4 or 5, the net rate μ is negative, reduction of the buffer content occurs nearly instantaneously, and it takes only a few seconds for the buffer to drop by 2 Mb. If the system is in phases 1 or 3, when the application is buffering, the net rate is positive and it remains so for 10 units of time, on average. During that time, the level has the opportunity to go up quite a bit, and it will take a significant amount of time before the system drops below the initial buffer level.

Example 2.33. We illustrate here the influence of the volatility parameters and of the initial phase distribution on the expected time to reach a given level. This example is a two-phase system with transition matrix

$$Q = \begin{bmatrix}-1 & 1 \\ 1 & -1\end{bmatrix},$$

and parameters

$$\mu = \begin{bmatrix}1 & -0.5\end{bmatrix}^{\mathrm{T}} \qquad \sigma = c\begin{bmatrix}1 & 1\end{bmatrix}^{\mathrm{T}}.$$

The volatility is the same for both phases, and we compare four cases, with c_i, $1 \le i \le 4$ taking the values in

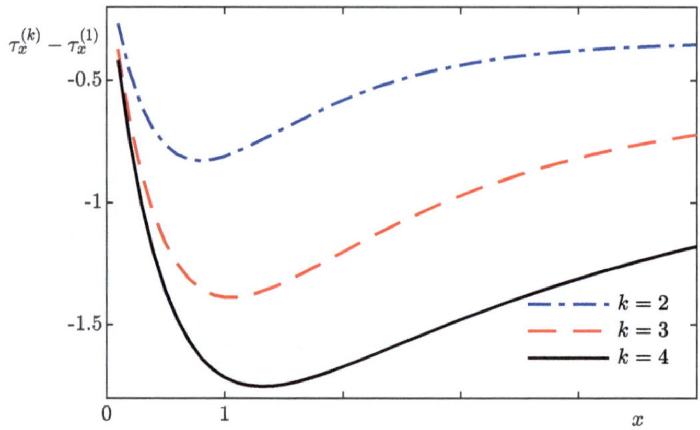

Fig. 2.10 Example 2.33. Difference $\mathrm{E}[\tau_x^{(k)}] - \mathrm{E}[\tau_x^{(1)}]$, for $0 \le x \le 5$, and $k = 2$ (blue dash-dotted curve), $k = 3$ (red dashed curve), and $k = 4$ (black plain curve). The initial phase is $\varphi(0) = 2$

$$\boldsymbol{c} = \begin{bmatrix} 1 & 2 & 4 & 8 \end{bmatrix}^{\mathrm{T}}.$$

In the discussion of this example, we denote as $\tau_x^{(k)}$ the first passage time to level x when the volatility parameter is c_k. We use c_1 as the reference value and we analyse the difference $\mathrm{E}[\tau_x^{(k)}] - \mathrm{E}[\tau_x^{(1)}]$ as the volatility increases from c_2 to c_4. By (2.85), we have

$$\mathrm{E}[\tau_x^{(k)}] - \mathrm{E}[\tau_x^{(1)}] = \boldsymbol{\alpha}^{\mathrm{T}}(e^{\widehat{U}_k x} - e^{\widehat{U}_1 x})(-Q^{\#})\boldsymbol{\mu}/(\boldsymbol{\pi}^{\mathrm{T}}\boldsymbol{\mu}),$$

where we denote by $\widehat{U}_k$ the matrix $\widehat{U}$ when the volatility is c_k. Since $\boldsymbol{\pi}^{\mathrm{T}}\boldsymbol{\mu} > 0$, the matrices $e^{\widehat{U}_k}$ are stochastic and

$$\lim_{x\to\infty}(\mathrm{E}[\tau_x^{(k)}] - \mathrm{E}[\tau_x^{(1)}]) = (\boldsymbol{u}_k^{\mathrm{T}} - \boldsymbol{u}_1^{\mathrm{T}})(-Q^{\#})\boldsymbol{\mu}/(\boldsymbol{\pi}^{\mathrm{T}}\boldsymbol{\mu}),$$

where $\boldsymbol{u}_k$ is the stationary probability vector of $e^{\widehat{U}_k}$. In this example,

$$\lim_{x\to\infty}(\mathrm{E}[\tau_x^{(k)}] - \mathrm{E}[\tau_x^{(1)}])_{k=2,3,4} = \begin{bmatrix} -0.3404 & -0.5720 & -0.7025 \end{bmatrix}$$

and

$$\mathrm{E}[\tau_x^{(4)}] < \mathrm{E}[\tau_x^{(3)}] < \mathrm{E}[\tau_x^{(2)}] < \mathrm{E}[\tau_x^{(1)}] \tag{2.88}$$

asymptotically as x goes to infinity: High levels are reached faster, on average, when the volatility is larger.

We see from Fig. 2.10 that the same inequalities hold for all x when the process starts from phase 2: For all MMBMs, the drift at time 0 has the same value $\mu_2 = -0.5$ and the average level decreases by same amount until the phase jumps to 1. The system with a higher volatility reaches positive levels faster, even for small values of x.

The story is different if $\varphi(0) = 1$, as we show in Fig. 2.11. For small x, the expected times to reach x follow the order

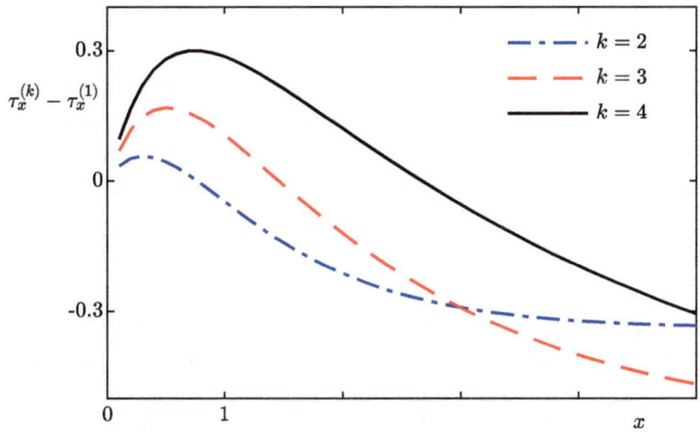

Fig. 2.11 Example 2.33. Difference $\mathrm{E}[\tau_x^{(k)}] - \mathrm{E}[\tau_x^{(1)}]$, for $0 \le x \le 5$, and $k = 2$ (blue dash-dotted curve), $k = 3$ (red dashed curve), and $k = 4$ (black plain curve). The initial phase is $\varphi(0) = 1$

$$\mathrm{E}[\tau_x^{(1)}] < \mathrm{E}[\tau_x^{(2)}] < \mathrm{E}[\tau_x^{(3)}] < \mathrm{E}[\tau_x^{(4)}],$$

the opposite of (2.88). Here, μ_1 is positive at time 0, so that the expected level starts immediately to increase. However, the process with a bigger volatility has a higher probability of moving to negative levels, from where it would take more time to reach positive values. As x grows to ∞, naturally, the influence of the initial phase vanishes and the ordering becomes (2.88) again.

2.7 Notes

Different authors have used different arguments to obtain characteristic equations for the four fundamental matrices U, $\widehat{U}$, Ψ, and $\widehat{\Psi}$. The approach I follow in Sects. 2.3 (purely Brownian MMBM) and 2.4 (linear Fluid flows) is similar to the one in Asmussen [13] and, in the case of Fluid flows, in Rogers [133]. A proof of (2.83) in Sect. 2.5 (general MMBMs) is given in Breuer [44, Eqn (6)], its derivation is based on the infinitesimal generator of the MMBM. Asmussen and Kella [16] and Asmussen [15] follow an approach based on martingales.

Ramaswami [130] builds a connection between Fluid flow systems and discrete-time, discrete state space QBDs, his approach is further developed in da Silva Soares and Latouche [53]. Ramaswami's main objective is to start from efficient algorithms available for QBDs and adapt these to the computation of the matrix Ψ. In [53, 130], the discrete level of the QBD is not directly related to the level of the Fluid queue; Bean and O'Reilly [25], to the contrary, give a spatially coherent uniformisation which preserves a natural mapping between the level of the two processes. Ahn and Ramaswami [5] construct a QBD model for a queueing process such that the workload in the queue has the same distribution as the Fluid flow at epochs of an independent Poisson process.

D'Auria et al. [55] and Ivanovs [88] show that the Jordan normal form of U is directly related to the Jordan normal form of $\mathcal{F}(z)$. This gives a connection between the matrices U and $\widehat{U}$ on the one hand and the Laplace exponent $\mathcal{F}(z)$ on the other hand, a connection which generalises Corollary 2.9. Actually, the authors prove their results for the more general Markov-modulated Lévy processes, provided that jumps are restricted to one direction only.

As there is scant computationally effective results available for arbitrary Markov-modulated Lévy processes, some authors suggest to use models for which jumps have a phase-type (PH) distribution, that is, a distribution of the time until absorption in a finite, absorbing Markov chain (Neuts [116, Chapter 2], Latouche and Ramaswami [106, Chapter 2], He [82, Chapter 1]). Examples are Breuer [44] (general negative jumps, PH positive jumps), Ahn [3, 4] (bilateral PH jumps), Pistorius [127] (law of the up-crossing ladder processes), Jiang and Pistorius [93] and Deelstra et al. [58] (valuation of financial options). The advantages are twofold: Firstly, if one adds to the set $\mathcal{E}$ the phases from the PH jumps, and replaces the jumps with linear intervals of slope 1 or -1, one replaces the Lévy process with a general MMBM, for which numerical procedures are readily available (see Sect. 6.4). Secondly, a general Lévy process may be approximated arbitrarily closely with a PH jumps process (Asmussen et al. [20]) and so, not much generality is lost in applications through this restriction.

I have given in this chapter several equations for the matrices U (and for its twin the matrix $\widehat{U}$). Each of these has its own strong points. The fixed point equation (2.19) is used in Sect. 6.1 as the starting point for an algorithm to compute U by successive iteration. At first sight, it does not seem to be very appealing to repeatedly evaluate the integral from 0 to ∞ but, as I shall show, the computation at each step reduces to the resolution of a simple linear system. In addition, I shall give a nice physical justification to prove the convergence of the sequence of approximations. The matrix-quadratic equation (2.23) is found in much of the literature on MMBMs (Asmussen [13], Ivanovs [87], D'auria et al. [57]). Furthermore, it has long been connected to (4.45), a characteristic equation for the important matrix R, which I shall introduce in Chap. 4. I should also mention that matrix-quadratic equations are the object of much attention in the numerical linear algebra literature (Bini et al. [36]). Equation (2.29) and the concomitant Wiener-Hopf analysis are at the root of the identification in Theorem 2.13 of U and $\widehat{U}$ as uniquely characterised solutions of (2.19). Finally, the system (2.25, 2.27) is used in Sect. 6.2 to design a quadratically convergent algorithm which is easily implemented.

Chapter 3
Exit from an Interval

◇ *First passage outside an interval* ◇ *Exit probability* ◇ *Scale matrix function* ◇ *Zero drift* ◇ *Time to exit* ◇ *Regulated process with a barrier* ◇

We illustrate in this short chapter an argument which has been used in different contexts, to analyse the behaviour of a Markov-modulated process until it leaves a finite interval for the first time. Among others, Hajek [79] and Latouche and Ramaswami [106, Chapter 10] use the same approach for QBDs evolving over a finite interval, da Silva Soares and Latouche [54] do so for Fluid flows, and Ivanovs [87] follows it for MMBMs. In Sect. 3.1, the stationary drift is assumed to be different from zero, an assumption that is often made and which considerably simplifies the analysis. We show in Sect. 3.2 how to overcome the technical difficulty when $\boldsymbol{\pi}^{\mathsf{T}}\boldsymbol{\mu} = 0$. Section 3.3 is about how much time is necessary for the MMBM to exit from the interval, and in the last sections, we assume that the process is regulated at one boundary, so that there is only one way to escape.

Assumption 3.1. *As announced at the end of Sect. 2.2, we assume for the remainder of the book that $\sigma_i > 0$ for all $i \in \mathcal{E}$, unless there is an explicit mention to the contrary.*

3.1 First Passage Outside an Interval

We assume that $X(0) = 0$, and we look for the distribution of the phase when the process visits for the first time the boundary of some interval $(-a, b)$, with $a, b \geq 0$.

Theorem 3.2. *The exit probability matrices $\Psi^{(a,b)}$ and $\widehat{\Psi}^{(a,b)}$ with*

$$\Psi_{ij}^{(a,b)} = \mathrm{P}_{0,i}[\tau_{-a} < \tau_b, \varphi(\tau_{-a}) = j],$$

© The Author(s), under exclusive license to Springer Nature Switzerland AG 2026
G. Latouche, *Markov-Modulated Brownian Motion*, Springer Series in Operations Research and Financial Engineering, https://doi.org/10.1007/978-3-031-98576-8_3

and

$$\widehat{\Psi}_{ij}^{(a,b)} = \mathrm{P}_{0,i}[\tau_b < \tau_{-a}, \varphi(\tau_b) = j],$$

for $i, j \in \mathcal{E}$, $a, b \geq 0$, $a + b > 0$, *are solutions of the linear system*

$$\left[\Psi^{(a,b)}\ \widehat{\Psi}^{(a,b)}\right](I + \mathcal{P}^{(a+b)}) = \mathcal{U}^{(a,b)}, \tag{3.1}$$

where

$$\mathcal{U}^{(a,b)} = \left[e^{Ua}\ e^{\widehat{U}b}\right] \qquad \text{and} \qquad \mathcal{P}^{(x)} = \begin{bmatrix} 0 & e^{\widehat{U}x} \\ e^{Ux} & 0 \end{bmatrix}, \tag{3.2}$$

for $x \geq 0$.

Proof. We decompose the trajectories from 0 to b in two sets: Either the MMBM reaches level b without having gone down to $-a$, or it goes down to $-a$ before eventually going up to b. Thus,

$$e^{\widehat{U}b} = \widehat{\Psi}^{(a,b)} + \Psi^{(a,b)} e^{\widehat{U}(a+b)}. \tag{3.3}$$

Observe that (3.3) holds if $a = 0$, $b > 0$: In that case, $\Psi^{(0,b)} = I$ and $\widehat{\Psi}^{(0,b)} = 0$ since, starting from $X(0) = 0$, the Brownian motion instantaneously visits negative values as we saw in Sect. 1.4. For the same reason, (3.3) also holds if $a > 0$ and $b = 0$. That equation does not hold, however, if $a = b = 0$.

Similarly,

$$e^{Ua} = \Psi^{(a,b)} + \widehat{\Psi}^{(a,b)} e^{U(a+b)},$$

and the two equations together form the system (3.1). □

We may write

$$I + \mathcal{P}^{(x)} = \begin{bmatrix} I & 0 \\ 0 & -I \end{bmatrix} (I - \mathcal{P}^{(x)}) \begin{bmatrix} I & 0 \\ 0 & -I \end{bmatrix}. \tag{3.4}$$

The matrix $\begin{bmatrix} I & 0 \\ 0 & -I \end{bmatrix}$ being its own inverse, the two matrices $I + \mathcal{P}^{(x)}$ and $I - \mathcal{P}^{(x)}$ are similar, they have the same set of eigenvalues and, in particular, they are singular or nonsingular together.

Assume for the time being that $\boldsymbol{\pi}^{\mathrm{T}}\boldsymbol{\mu} \neq 0$. In that case, either e^{Ux} or $e^{\widehat{U}x}$ is substochastic by Lemma 2.4. In consequence, $\mathcal{P}^{(x)}$ is substochastic, $I - \mathcal{P}^{(x)}$ is nonsingular, and so is $I + \mathcal{P}^{(x)}$. Moreover, the products $e^{Ux}e^{\widehat{U}x}$ and $e^{\widehat{U}x}e^{Ux}$ themselves are substochastic and

$$(I + \mathcal{P}^{(x)})^{-1} = \begin{bmatrix} I & -e^{\widehat{U}x} \\ -e^{Ux} & I \end{bmatrix} \begin{bmatrix} (I - e^{\widehat{U}x}e^{Ux})^{-1} & 0 \\ 0 & (I - e^{Ux}e^{\widehat{U}x})^{-1} \end{bmatrix} \tag{3.5}$$

for $x \geq 0$, as we verify by pre-multiplying both sides by $I + \mathcal{P}^{(x)}$. The system (3.1), therefore, is nonsingular, and the exit probability matrices are fully determined. We give their explicit expressions in Theorem 3.3.

If $\boldsymbol{\pi}^{\mathrm{T}}\boldsymbol{\mu} = 0$, however, $I + \mathcal{P}^{(x)}$ is singular, the system (3.1) does not completely characterise the matrices $\Psi^{(a,b)}$ and $\widehat{\Psi}^{(a,b)}$, and we need one additional equation. We attend to this question in Sect. 3.2.

Theorem 3.3. *If $\boldsymbol{\pi}^{\mathrm{T}}\boldsymbol{\mu} \neq 0$, then*

$$\Psi^{(a,b)} = W_b W_{a+b}^{-1}, \qquad \widehat{\Psi}^{(a,b)} = \widehat{W}_a \widehat{W}_{a+b}^{-1}, \tag{3.6}$$

where

$$W_x = e^{-Ux} - e^{\widehat{U}x} \qquad \text{and} \qquad \widehat{W}_x = e^{-\widehat{U}x} - e^{Ux}, \tag{3.7}$$

for $x \geq 0$. Furthermore, $\widehat{W}_x = -W_{-x}$ and $\widehat{\Psi}^{(a,b)} = W_{-a}W_{-(a+b)}^{-1}$.

Proof. We have seen that $I + \mathcal{P}^{(a+b)}$ is nonsingular if $\boldsymbol{\pi}^{\mathrm{T}}\boldsymbol{\mu} \neq 0$. In that case,

$$\left[\Psi^{(a,b)}\ \widehat{\Psi}^{(a,b)}\right] = \left[e^{Ua}\ e^{\widehat{U}b}\right](I + \mathcal{P}^{(a+b)})^{-1}$$

and we find from (3.5) that

$$\begin{aligned}
\Psi^{(a,b)} &= (e^{Ua} - e^{\widehat{U}b}e^{U(a+b)})(I - e^{\widehat{U}(a+b)}e^{U(a+b)})^{-1} \\
&= (e^{-Ub} - e^{\widehat{U}b})e^{U(a+b)}(I - e^{\widehat{U}(a+b)}e^{U(a+b)})^{-1} \\
&= W_b W_{a+b}^{-1}.
\end{aligned}$$

Similarly,

$$\begin{aligned}
\widehat{\Psi}^{(a,b)} &= (e^{\widehat{U}b} - e^{Ua}e^{\widehat{U}(a+b)})(I - e^{U(a+b)}e^{\widehat{U}(a+b)})^{-1} \\
&= \widehat{W}_a \widehat{W}_{a+b}^{-1}.
\end{aligned}$$

The proof of (3.6) is complete, the last statement is obvious. □

The equations in (3.6) have a nice physical interpretation. Take $x > 0$ arbitrary but fixed. Define as follows the two sequences $\{\theta_n : n \geq 1\}$ and $\{\theta'_n : n \geq 1\}$ of alternate visits to level 0 and level x:

$$\begin{aligned}
\theta_0 &= 0, \\
\theta'_n &= \inf\{t > \theta_n : X(t) = x\}, \\
\theta_n &= \inf\{t > \theta'_{n-1} : X(t) = 0\},
\end{aligned}$$

for $n \geq 1$. As usual, we set θ_n or θ'_n to be infinite if there is no finite t to satisfy their definition. Define $N_0 = \max\{n : \theta_n < \infty\}$, that is, N_0 is the number of visits to 0 that are separated from the previous visit to 0 by at least one visit to level x. As $\boldsymbol{\pi}^{\mathrm{T}}\boldsymbol{\mu} \neq 0$, the MMBM process is transient, $X(t)$ tends to $+\infty$ or to $-\infty$, N_0 is finite a.s., and $\mathrm{E}[N_0]$ is finite.

The inverse of W_x may be written as

$$
\begin{aligned}
W_x^{-1} &= e^{Ux}(I - e^{\widehat{U}x} e^{Ux})^{-1} \\
&= \sum_{\nu \geq 0} e^{Ux}(e^{\widehat{U}x} e^{Ux})^{\nu},
\end{aligned}
$$

we recognise that

$$
\begin{aligned}
(e^{Ux}(e^{\widehat{U}x} e^{Ux})^{\nu})_{ij} &= \mathrm{P}_{x,i}[\theta_{\nu+1} < \infty, \varphi(\theta_{\nu+1}) = j] \\
&= \mathrm{P}_{x,i}[N_0 \geq \nu + 1, \varphi(\theta_{\nu+1}) = j],
\end{aligned}
$$

and so

$$
(W_x^{-1})_{ij} = \sum_{\nu \geq 0} \mathrm{P}_{x,i}[N_0 > \nu; \varphi(\theta_{\nu+1}) = j].
$$

In short, the entries of W_x^{-1} are the expected total number of visits to level 0 that are separated by visits to level x, given that the process starts in level x.

Similarly, W_{a+b}^{-1} is the matrix of the expected number of visits to level 0 that are separated by a visit to level $a + b$, starting from $a + b$. As level 0 is a purely conventional mark, W_{a+b}^{-1} is also the number of visits to $-a$ separated by visits to b, starting from b.

Now, let us write the first equation in (3.6) as

$$
W_b^{-1} \Psi^{(a,b)} = W_{a+b}^{-1}.
$$

The first factor on the left counts the number of visits to level 0 separated by visits to b, the second factor is the probability that the process will visit level $-a$ before the possible next visit to b, and their product is indeed the expected number of visits to $-a$ separated by visits to b, starting from b.

A similar interpretation holds for the second equation in (3.6): Both sides of

$$
\widehat{W}_a^{-1} \widehat{\Psi}^{(a,b)} = \widehat{W}_{a+b}^{-1}
$$

are the expected number of visits to level b separated by visits to $-a$, given that the process starts from level $-a$.

It must be noted that W_x and $\widehat{W}_x$ are related to the so-called *scale matrix functions*, details are given in Remark 4.6.

Remark 3.4. From a computational standpoint, the expressions (3.6) are not very useful and it is better to numerically solve the linear system (3.1). The reason is that all eigenvalues of U and of $\widehat{U}$ have a negative real part, and the matrix exponentials e^{-Ux} and $e^{-\widehat{U}x}$ grow very fast with $x > 0$. On the contrary, the coefficient matrix $I + \mathcal{P}^{(a+b)}$ in (3.1) is numerically stable as long as $\boldsymbol{\pi}^{\mathrm{T}} \boldsymbol{\mu}$ is not too close to 0.

Example 3.5. This is a model for the workload of a *noisy M/E$_k$/1 queue*, it is inspired from d'Auria et al. [57]. Customers arrive to a server according to a Poisson process with parameter λ and they request an Erlang-distributed service time with k stages, each of rate k so that the expected length of a service is 1 unit of time. The workload is represented as a Lévy process whereby work to be processed arrives in Erlang

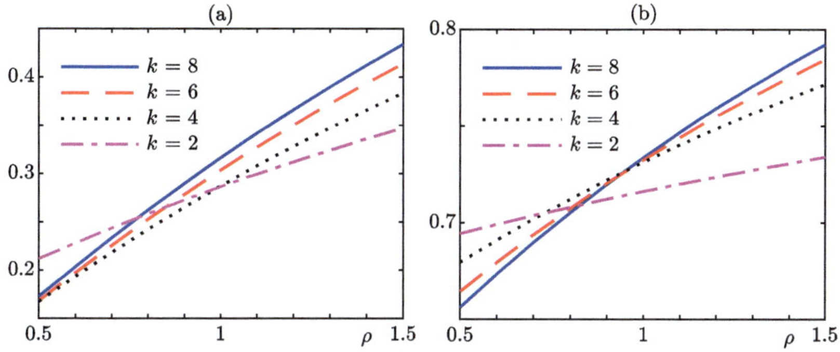

Fig. 3.1 Example 3.5. Probability $\boldsymbol{\alpha}^{\mathrm{T}}\Psi^{(a,b)}\mathbf{1}$ of exit through the upper boundary, as a function of the traffic coefficient ρ, for different distributions of $\varphi(0)$ and different values of k. In (**a**) $\varphi(0)$ is equal to 1, in (**b**) it is uniformly distributed in $\{2, \ldots, m\}$

jumps at the epochs marked by the Poisson process and decreases at rate 1 between arrivals.

To represent the process as an MMBM, the jumps are replaced with linear intervals with slope 1, as described at the end of Sect. 2.7. The number of phases in $\mathcal{E}$ is $m = k + 1$, the generator is

$$Q = \begin{bmatrix} -\lambda & \lambda & & & \\ & -k & k & & \\ & & \ddots & \ddots & \\ & & & -k & k \\ k & & & & -k \end{bmatrix}$$

(we only display the non-zero elements), and the vector of drift parameters is

$$\boldsymbol{\mu} = \begin{bmatrix} -1 & 1 & \ldots & 1 \end{bmatrix}^{\mathrm{T}}.$$

Sojourns in Phase 1 correspond to the intervals between two arrivals, during which the workload gets depleted at rate $\mu_1 = -1$. The phases 2 to m correspond to the k exponential stages of the Erlang distribution. It is assumed in [57] that phase 1 has a strictly positive volatility, which is why the system is called a *noisy* queue.

The stationary drift $\boldsymbol{\pi}^{\mathrm{T}}\boldsymbol{\mu}$ is equal to $(1 - 1/\lambda)/(1 + 1/\lambda)$, that is, it is proportional to the expected increase of workload between two successive visits to Phase 1, and it is strictly negative if and only if the traditional traffic coefficient $\rho = \lambda \mathrm{E}[S]$ is less than 1. As $\mathrm{E}[S] = 1$ in this example, λ is equal to ρ here.

The authors in [57] determine for $k = 2$ the Jordan decomposition of the Laplace exponent $\mathcal{F}(z)$, thereby solving for the matrices U and Ψ. We use the same model, with the difference that $\sigma_i > 0$ for all i, in accordance with our Assumption 3.1.

We show in Fig. 3.1 the probability of exit through the upper level, as a function of ρ. The parameters are $b = 0.5$, $a = 0.5$, $\sigma_i = r = \sqrt{1/k}$ for all i, ρ ranges from 0.5

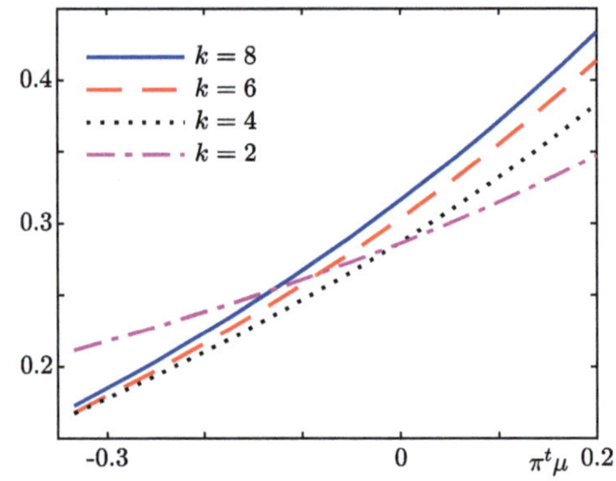

Fig. 3.2 Example 3.5. Probability $\boldsymbol{\alpha}^{\mathrm{T}}\Psi^{(a,b)}\mathbf{1}$ of escape through the upper boundary given as a function of the stationary drift, for $\varphi(0) = 0$

to 1.5 and k takes the values 2, 4, 6, and 8. When ρ is equal to 1, $\boldsymbol{\pi}^{\mathrm{T}}\boldsymbol{\mu}$ is equal to 0 and we use (3.20, 3.21) instead of (3.1) to compute the exit probabilities.

In Fig. 3.1a on the left, the initial distribution of $\varphi(0)$ is $\boldsymbol{\alpha} = \begin{bmatrix}1\ 0 \ldots 0\end{bmatrix}^{\mathrm{T}}$ so that the system is initially in an interval between arrivals, the workload is depleted at rate -1, and the level is moving, on average, to the lower bound. In consequence, the probability $\boldsymbol{\alpha}^{\mathrm{T}}\Psi^{(a,b)}\mathbf{1}$ is less than 0.5, even when $\rho = 1.5$.

In Fig. 3.1b on the right, $\boldsymbol{\alpha} = \begin{bmatrix}0\ 1 \ldots 1\end{bmatrix}^{\mathrm{T}}/k$ and $\varphi(0)$ is uniformly distributed on the Erlang stages. This means that an arrival has occurred and the workload is being increased. The probability is greater than 0.6 that the level will cross the upper boundary first, even for $\rho = 0.5$. The volatility parameter r has a minor influence, except for $k = 2$, when r takes its largest value.

In Fig. 3.2, $\varphi(0)$ is equal to 1, like in Fig. 3.1a, but here we plot $\boldsymbol{\alpha}^{\mathrm{T}}\Psi^{(a,b)}\mathbf{1}$ as a function of the stationary drift $\boldsymbol{\pi}^{\mathrm{T}}\boldsymbol{\mu}$ instead of the traffic coefficient ρ. Comparison of the two plots shows, quite naturally, that the probability increases in both, and ranges in the same interval of values. The impressions given in each plot are, however, different: The curves in Fig. 3.2 are concave functions of $\boldsymbol{\pi}^{\mathrm{T}}\boldsymbol{\mu}$, meaning that the *rate of increase of* $\boldsymbol{\alpha}^{\mathrm{T}}\Psi^{(a,b)}\mathbf{1}$ is itself increasing with $\boldsymbol{\pi}^{\mathrm{T}}\boldsymbol{\mu}$, while the rate of increase in Fig. 3.1a diminishes as ρ increases.

3.2 Zero Drift

If $\boldsymbol{\pi}^{\mathrm{T}}\boldsymbol{\mu} = 0$, both e^{Ux} and $e^{\widehat{U}x}$ are stochastic matrices, both W_x and $\widehat{W}_x$ are singular matrices, and we need to adapt Theorem 3.3. With this in mind, we get back to the system (3.1), where the coefficient matrix $I + \mathcal{P}^{(x)}$ is singular, and we need one additional equation to completely characterise the two transition matrices. One such equation is (3.9) in Lemma 3.7, our choice being motivated by Eq. (1.33) given in our discussion of the group inverse. To that end we shall need the stationary probability vector $\boldsymbol{\eta}$ of $\mathcal{P}^{(a+b)}$.

We easily see that $\boldsymbol{\eta}^{\mathrm{T}} = \left[\boldsymbol{\eta}_1^{\mathrm{T}}\ \boldsymbol{\eta}_2^{\mathrm{T}}\right]$, with

$$\boldsymbol{\eta}_1^{\mathrm{T}} = \boldsymbol{\eta}_1^{\mathrm{T}} e^{\widehat{U}(a+b)} e^{U(a+b)} \qquad \text{and} \qquad \boldsymbol{\eta}_2^{\mathrm{T}} = \boldsymbol{\eta}_2^{\mathrm{T}} e^{U(a+b)} e^{\widehat{U}(a+b)}, \tag{3.8}$$

normalised so that $\boldsymbol{\eta}_1^{\mathrm{T}}\mathbf{1} = \boldsymbol{\eta}_2^{\mathrm{T}}\mathbf{1} = 0.5$. The eigenvector of $I + \mathcal{P}^{(a+b)}$ associated to the eigenvalue 0 is proportional to $\left[\boldsymbol{\eta}_1^{\mathrm{T}}\ -\boldsymbol{\eta}_2^{\mathrm{T}}\right]$.

Remark 3.6. We note for future reference that the *stationary probability vectors* of $e^{\widehat{U}(a+b)} e^{U(a+b)}$ and of $e^{U(a+b)} e^{\widehat{U}(a+b)}$, being normalised to 1, are $2\boldsymbol{\eta}_1$ and $2\boldsymbol{\eta}_2$, respectively.

Lemma 3.7. *If $\boldsymbol{\pi}^{\mathrm{T}}\boldsymbol{\mu} = 0$, then the exit probability matrix is such that*

$$\left[\Psi^{(a,b)}\ \widehat{\Psi}^{(a,b)}\right] \boldsymbol{\beta} = \boldsymbol{h}, \tag{3.9}$$

where $\boldsymbol{h} = -Q^{\#}\boldsymbol{\mu}$ and

$$\boldsymbol{\beta} = \begin{bmatrix} -a\mathbf{1} + \boldsymbol{h} \\ b\mathbf{1} + \boldsymbol{h} \end{bmatrix}. \tag{3.10}$$

The solution of (3.1, 3.9) is

$$\left[\Psi^{(a,b)}\ \widehat{\Psi}^{(a,b)}\right] = \mathcal{U}^{(a,b)}(I + \mathcal{P}^{(a+b)})^{\#} + \boldsymbol{w} \cdot \widetilde{\boldsymbol{\eta}}^{\mathrm{T}}, \tag{3.11}$$

where

$\mathcal{U}^{(a,b)}$ and $\mathcal{P}^{(a+b)}$ are defined in (3.2),

$\boldsymbol{\eta} = \begin{bmatrix} \boldsymbol{\eta}_1 \\ \boldsymbol{\eta}_2 \end{bmatrix}$ is the stationary probability vector of $\mathcal{P}^{(a+b)}$,

$\widetilde{\boldsymbol{\eta}} = \begin{bmatrix} \boldsymbol{\eta}_1 \\ -\boldsymbol{\eta}_2 \end{bmatrix}$ is a left eigenvector of $I + \mathcal{P}^{(a+b)}$ for the eigenvalue 0,

and

$$\boldsymbol{w} = (\widetilde{\boldsymbol{\eta}}^{\mathrm{T}}\boldsymbol{\beta})^{-1}(\boldsymbol{h} - \mathcal{U}^{(a,b)}(I + \mathcal{P}^{(a+b)})^{\#}\boldsymbol{\beta}). \tag{3.12}$$

Proof. We know from Proposition 2.3 that $h_i = \lim_{t\to\infty} \mathrm{E}_{0,i}[X(t)]$, for i in $\mathcal{E}$. We condition on the first visit to either level b or level $-a$ and verify that

$$\boldsymbol{h} = \widehat{\Psi}^{(a,b)}\boldsymbol{h}^{(b)} + \Psi^{(a,b)}\boldsymbol{h}^{(-a)},$$

where $h_i^{(x)} = \lim_{t\to\infty} \mathrm{E}_{x,i}[X(t)]$, or

$$\boldsymbol{h} = \widehat{\Psi}^{(a,b)}(b\mathbf{1} + \boldsymbol{h}) + \Psi^{(a,b)}(-a\mathbf{1} + \boldsymbol{h})$$

and this concludes the proof of (3.9).

The matrix $\mathcal{P}^{(a+b)}$ is stochastic and irreducible, therefore $I + \mathcal{P}^{(a+b)}$, just like $I - \mathcal{P}^{(a+b)}$, has a unique eigenvalue equal to 0, and we may refer to its group inverse: By (1.33), the system (3.1) has the solution

$$\left[\Psi^{(a,b)}\ \widehat{\Psi}^{(a,b)}\right] = \mathcal{U}^{(a,b)}(I + \mathcal{P}^{(a+b)})^{\#} + \boldsymbol{w} \cdot \widetilde{\boldsymbol{\eta}}^{\mathrm{T}} \tag{3.13}$$

for some vector $\boldsymbol{w}$. We post-multiply both sides of (3.13) by $\boldsymbol{\beta}$, the left-hand side becomes equal to $\boldsymbol{h}$ by (3.9) and this proves (3.12), provided that $\widetilde{\boldsymbol{\eta}}^{\mathrm{T}}\boldsymbol{\beta} \neq 0$. Incidentally, this will show that when $\boldsymbol{\pi}^{\mathrm{T}}\boldsymbol{\mu} = 0$, the system (3.1, 3.9) is nonsingular indeed.

By (3.8), $\boldsymbol{\eta}_1^{\mathrm{T}} = \boldsymbol{\eta}_2^{\mathrm{T}} e^{U(a+b)}$ and we may write the product $\widetilde{\boldsymbol{\eta}}^{\mathrm{T}}\boldsymbol{\beta}$ as

$$\widetilde{\boldsymbol{\eta}}^{\mathrm{T}}\boldsymbol{\beta} = -(a+b)/2 + \boldsymbol{\eta}_2^{\mathrm{T}}(e^{U(a+b)} - I)\boldsymbol{h}. \tag{3.14}$$

To complete the proof, it suffices to show that $\boldsymbol{\eta}_2^{\mathrm{T}}(e^{U(a+b)} - I)\boldsymbol{h} \leq 0$. Now,

$$\begin{aligned} \boldsymbol{h} &= \lim_{t\to\infty} \mathrm{E}[X(t) - (a+b)] + (a+b)\mathbf{1} \\ &= \lim_{t\to\infty} (\mathrm{E}[(X(t) - (a+b))\mathbb{1}\{\tau_{a+b} < t\}] \\ &\qquad + \mathrm{E}[(X(t) - (a+b))\mathbb{1}\{t \leq \tau_{a+b}\}]) + (a+b)\mathbf{1} \\ &\leq \lim_{t\to\infty} \mathrm{E}[(X(t) - (a+b))\mathbb{1}\{\tau_{a+b} < t\}] + (a+b)\mathbf{1} \end{aligned}$$

since $X(t) - (a+b) \leq 0$ for $t \leq \tau_{a+b}$,

$$= e^{\widehat{U}(a+b)}\boldsymbol{h}$$

because τ_{a+b} is finite a.s. In consequence, (3.14) becomes

$$\begin{aligned} \widetilde{\boldsymbol{\eta}}^{\mathrm{T}}\boldsymbol{\beta} &\leq -(a+b)/2 + \boldsymbol{\eta}_2^{\mathrm{T}}(e^{U(a+b)}e^{\widehat{U}(a+b)} - I)\boldsymbol{h} \\ &= -(a+b)/2 \end{aligned}$$

by (3.8). This completes the proof. □

Remark 3.8. Our choice of vectors $\boldsymbol{\beta}$ and $\boldsymbol{h}$ in (3.9) is somewhat arbitrary: As the MMBM escapes the interval $(-a, b)$ in finite time almost surely, one has $(\Psi^{(a,b)} + \widehat{\Psi}^{(a,b)})\mathbf{1} = \mathbf{1}$ and we might have written the additional equation as

$$(\Psi^{(a,b)} + \widehat{\Psi}^{(a,b)})(\boldsymbol{\beta} + c\mathbf{1}) = \boldsymbol{h} + c\mathbf{1},$$

where c would be any real number.

We used in Theorem 3.3 the expression (3.5) for the inverse of $I + \mathcal{P}^{(x)}$ and so gave explicit expressions for the exit probability matrices when $\boldsymbol{\pi}^{\mathrm{T}}\boldsymbol{\mu} \neq 0$. We shall need for the same purpose an expression for $(I + \mathcal{P}^{(x)})^{\#}$ if $\boldsymbol{\pi}^{\mathrm{T}}\boldsymbol{\mu} = 0$, and this is the object of the next lemma.

Lemma 3.9. *Assume that A and $\widehat{A}$ are two irreducible stochastic matrices of the same order. Define*

$$M = \begin{bmatrix} I & \widehat{A} \\ A & I \end{bmatrix}. \tag{3.15}$$

The group inverse of M is

$$M^{\#} = \begin{bmatrix} I & -\widehat{A} \\ -A & I \end{bmatrix} \begin{bmatrix} (I-\widehat{A}A)^{\#} & 0 \\ 0 & (I-A\widehat{A})^{\#} \end{bmatrix} + \begin{bmatrix} 0 & \mathbf{1}\cdot\boldsymbol{\gamma}_2^{\mathrm{T}} \\ \mathbf{1}\cdot\boldsymbol{\gamma}_1^{\mathrm{T}} & 0 \end{bmatrix}, \tag{3.16}$$

where $\boldsymbol{\gamma}_1$ and $\boldsymbol{\gamma}_2$ are respectively the left eigenvectors of $\widehat{A}A$ and of $A\widehat{A}$ for the eigenvalue 1, normalised by $\boldsymbol{\gamma}_1^{\mathrm{T}}\mathbf{1} = \boldsymbol{\gamma}_2^{\mathrm{T}}\mathbf{1} = 0.5$.

Proof. We use the fact that the group inverses of our matrices are the unique solutions of linear systems like (1.34). Here, we need to verify that

$$MM^{\#} = I - \boldsymbol{j}\cdot\widetilde{\boldsymbol{\gamma}}^{\mathrm{T}}, \tag{3.17}$$

where $M^{\#}$ is given in (3.16), and

$$\boldsymbol{j} = \begin{bmatrix} \mathbf{1} \\ -\mathbf{1} \end{bmatrix} \qquad \text{and} \qquad \widetilde{\boldsymbol{\gamma}}^{\mathrm{T}} = \begin{bmatrix} \boldsymbol{\gamma}_1^{\mathrm{T}} & -\boldsymbol{\gamma}_2^{\mathrm{T}} \end{bmatrix}$$

are the right and left eigenvectors of the matrix M for the eigenvalue 0, normalised by $\widetilde{\boldsymbol{\gamma}}^{\mathrm{T}}\boldsymbol{j} = 1$. We replace M and $M^{\#}$ with the right-hand sides of (3.15, 3.16) and obtain the following. Note that the stationary probability vectors associated with $\widehat{A}A$ and $A\widehat{A}$ are $2\boldsymbol{\gamma}_1$ and $2\boldsymbol{\gamma}_2$, respectively.

$$MM^{\#} = \begin{bmatrix} (I-\widehat{A}A)(I-\widehat{A}A)^{\#} & 0 \\ 0 & (I-A\widehat{A})(I-A\widehat{A})^{\#} \end{bmatrix} + \mathbf{1}\cdot\boldsymbol{\gamma}^{\mathrm{T}} \tag{3.18}$$

$$= \begin{bmatrix} I - 2\mathbf{1}\cdot\boldsymbol{\gamma}_1^{\mathrm{T}} & 0 \\ 0 & I - 2\mathbf{1}\cdot\boldsymbol{\gamma}_2^{\mathrm{T}} \end{bmatrix} + \mathbf{1}\cdot\boldsymbol{\gamma}^{\mathrm{T}} \qquad \text{by (1.34)}$$

$$= I + \begin{bmatrix} -\mathbf{1}\cdot\boldsymbol{\gamma}_1^{\mathrm{T}} & \mathbf{1}\cdot\boldsymbol{\gamma}_2^{\mathrm{T}} \\ \mathbf{1}\cdot\boldsymbol{\gamma}_1^{\mathrm{T}} & -\mathbf{1}\cdot\boldsymbol{\gamma}_2^{\mathrm{T}} \end{bmatrix}, \tag{3.19}$$

which is the right-hand side of (3.17). □

Theorem 3.10. *If $\boldsymbol{\pi}^{\mathrm{T}}\boldsymbol{\mu} = 0$, then*

$$\Psi^{(a,b)} = W_b W_{a+b}^{\#} + (\mathbf{1}+\boldsymbol{w})\cdot\boldsymbol{\eta}_1^{\mathrm{T}}, \tag{3.20}$$

$$\widehat{\Psi}^{(a,b)} = \widehat{W}_a \widehat{W}_{a+b}^{\#} + (\mathbf{1}-\boldsymbol{w})\cdot\boldsymbol{\eta}_2^{\mathrm{T}}, \tag{3.21}$$

where $\begin{bmatrix} \boldsymbol{\eta}_1^{\mathrm{T}} & \boldsymbol{\eta}_2^{\mathrm{T}} \end{bmatrix}$ is the stationary probability vector of $\mathcal{P}^{(a+b)}$ and $\boldsymbol{w}$ is given in Lemma 3.7.

Proof. To simplify the presentation, we use the temporary notation

$$\Lambda = e^{U(a+b)} \qquad \text{and} \qquad \widehat{\Lambda} = e^{\widehat{U}(a+b)},$$

and we use (3.2) to rewrite (3.11) as

$$\begin{bmatrix} \Psi^{(a,b)} & \widehat{\Psi}^{(a,b)} \end{bmatrix} = \begin{bmatrix} e^{Ua} & e^{\widehat{U}b} \end{bmatrix} \begin{bmatrix} I & \widehat{\Lambda} \\ \Lambda & I \end{bmatrix}^{\#} + \boldsymbol{w}\cdot\widetilde{\boldsymbol{\eta}}^{\mathrm{T}}.$$

By Lemma 3.9, this becomes

$$\left[\Psi^{(a,b)}\ \widehat{\Psi}^{(a,b)}\right] = \left[e^{Ua} - e^{\widehat{U}b}\Lambda\ \ e^{\widehat{U}b} - e^{Ua}\widehat{\Lambda}\right] \begin{bmatrix} (I-\widehat{\Lambda}\Lambda)^{\#} & 0 \\ 0 & (I-\Lambda\widehat{\Lambda})^{\#} \end{bmatrix} + \left[\mathbf{1}\cdot\boldsymbol{\eta}_1^{\mathrm{T}}\ \ \mathbf{1}\cdot\boldsymbol{\eta}_2^{\mathrm{T}}\right] + \boldsymbol{w}\cdot\widetilde{\boldsymbol{\eta}}^{\mathrm{T}},$$

or

$$\Psi^{(a,b)} = W_b\Lambda(I-\widehat{\Lambda}\Lambda)^{\#} + (\mathbf{1}+\boldsymbol{w})\boldsymbol{\eta}_1^{\mathrm{T}}, \tag{3.22}$$

$$\widehat{\Psi}^{(a,b)} = \widehat{W}_a\widehat{\Lambda}(I-\Lambda\widehat{\Lambda})^{\#} + (\mathbf{1}-\boldsymbol{w})\boldsymbol{\eta}_2^{\mathrm{T}}, \tag{3.23}$$

and it remains for us to prove that (3.20) follows from (3.22), Eq. (3.21) following from (3.23) by a similar argument.

For any $y > 0$, $e^{Uy}e^{\widehat{U}y}$ is an irreducible stochastic matrix when $\boldsymbol{\pi}^{\mathrm{T}}\boldsymbol{\mu} = 0$. Therefore, W_y, which is equal to $e^{-Uy}(I - e^{Uy}e^{\widehat{U}y})$, has a unique eigenvalue 0 with right eigenvector $\mathbf{1}$ and left eigenvector $\boldsymbol{w}_y$. By (1.34),

$$I = W_y^{\#}W_y + \mathbf{1}\cdot\boldsymbol{w}_y^{\mathrm{T}}$$

and we may replace the first term in the right-hand side of (3.22) with

$$\begin{aligned} W_b\Lambda(I-\widehat{\Lambda}\Lambda)^{\#} &= W_bW_{a+b}^{\#}W_{a+b}\Lambda(I-\widehat{\Lambda}\Lambda)^{\#} \\ &= W_bW_{a+b}^{\#}(I-\widehat{\Lambda}\Lambda)(I-\widehat{\Lambda}\Lambda)^{\#} \\ &= W_bW_{a+b}^{\#}(I-2\mathbf{1}\cdot\boldsymbol{\eta}_1^{\mathrm{T}}) \qquad \text{by (1.34) again,} \\ &= W_bW_{a+b}^{\#}. \end{aligned}$$

This concludes our proof. □

Before moving to the next section, we give some limiting properties of the exit probability matrices.

Case 1: $a > 0$ is fixed, and $b \to \infty$. In this case, $\lim_{b\to\infty}\Psi^{(a,b)} = e^{Ua}$, as one would expect: The influence of the upper boundary vanishes, and we are reduced to the first passage probabilities to level $-a$ that are analysed in Chap. 2. The limit of $\widehat{\Psi}^{(a,b)}$ depends on the sign of the stationary drift.

If $\boldsymbol{\pi}^{\mathrm{T}}\boldsymbol{\mu} < 0$, then e^{U} is stochastic and $e^{U(a+b)} \to \mathbf{1}\cdot\boldsymbol{u}^{\mathrm{T}}$ as $b \to \infty$, where $\boldsymbol{u}$ is the stationary probability vector of U. The matrix $e^{\widehat{U}}$ is substochastic and $e^{\widehat{U}b} \to 0$ as $b \to \infty$. In the limit, (3.1) becomes

$$\lim_{b\to\infty}\left[\Psi^{(a,b)}\ \widehat{\Psi}^{(a,b)}\right]\begin{bmatrix} I & 0 \\ \mathbf{1}\cdot\boldsymbol{u}^{\mathrm{T}} & I \end{bmatrix} = \left[e^{Ua}\ 0\right],$$

so that

$$\lim_{b\to\infty}\left[\Psi^{(a,b)}\ \widehat{\Psi}^{(a,b)}\right] = \left[e^{Ua}\ 0\right].$$

If $\boldsymbol{\pi}^{\mathrm{T}}\boldsymbol{\mu} > 0$, then $e^{\widehat{U}}$ is stochastic, e^{U} is substochastic and (3.1) becomes

$$\lim_{b\to\infty} \left[\Psi^{(a,b)}\ \widehat{\Psi}^{(a,b)}\right] \begin{bmatrix} I & \mathbf{1}\cdot\widehat{\boldsymbol{u}}^{\mathrm{T}} \\ 0 & I \end{bmatrix} = \left[e^{Ua}\ \mathbf{1}\cdot\widehat{\boldsymbol{u}}^{\mathrm{T}}\right]$$

where $\widehat{\boldsymbol{u}}$ is the stationary probability vector of $\widehat{U}$. In consequence,

$$\lim_{b\to\infty} \left[\Psi^{(a,b)}\ \widehat{\Psi}^{(a,b)}\right] = \left[e^{Ua}\ (I - e^{Ua})\mathbf{1}\cdot\widehat{\boldsymbol{u}}^{\mathrm{T}}\right]. \tag{3.24}$$

This makes sense from a heuristic point of view: The sum $\Psi^{(a+b)} + \widehat{\Psi}^{(a+b)}$ is stochastic for all b, e^{Ua} is substochastic, and the limit $(I - e^{Ua})\mathbf{1}\cdot\widehat{\boldsymbol{u}}^{\mathrm{T}}$ in (3.24) is there to compensate for the probability mass missing from e^{Ua}.

Finally, if $\boldsymbol{\pi}^{\mathrm{T}}\boldsymbol{\mu} = 0$, then both e^{U} and $e^{\widehat{U}}$ are stochastic. In the limit, (3.1) becomes the equation

$$\lim_{b\to\infty} \left[\Psi^{(a,b)}\ \widehat{\Psi}^{(a,b)}\right] \begin{bmatrix} I & \mathbf{1}\cdot\widehat{\boldsymbol{u}}^{\mathrm{T}} \\ \mathbf{1}\cdot\boldsymbol{u}^{\mathrm{T}} & I \end{bmatrix} = \left[e^{Ua}\ \mathbf{1}\cdot\widehat{\boldsymbol{u}}^{\mathrm{T}}\right]$$

and one easily verifies that $\left[e^{Ua}\ 0\right]$ is one solution. To see that it is the solution we are looking for, we develop the left-hand side of the additional equation (3.9). Every term is obviously a bounded function of b, with the possible exception of $b\widehat{\Psi}^{(a,b)}\mathbf{1}$; therefore, that term is bounded also, which implies that $\widehat{\Psi}^{(a,b)} \to 0$ and

$$\lim_{b\to\infty} \left[\Psi^{(a,b)}\ \widehat{\Psi}^{(a,b)}\right] = \left[e^{Ua}\ 0\right].$$

Case 2: $b > 0$ **is fixed and** $a \to \infty$**.** We adapt the argument above, and we find that

$$\begin{aligned} \lim_{a\to\infty} \left[\Psi^{(a,b)}\ \widehat{\Psi}^{(a,b)}\right] &= \left[0\ \ e^{\widehat{U}b}\right] && \text{if } \boldsymbol{\pi}^{\mathrm{T}}\mu \ge 0, \\ &= \left[(I - e^{\widehat{U}b})\mathbf{1}\cdot\boldsymbol{u}\ \ e^{\widehat{U}b}\right] && \text{if } \boldsymbol{\pi}^{\mathrm{T}}\mu < 0, \end{aligned}$$

where $\boldsymbol{u}$ is the stationary probability vector of U.

Case 3: $a = 0$ **and** $b > 0$ **is fixed.** We would guess that $\Psi^{(0,b)} = I$, as we know from Sect. 1.4 that Brownian motion instantly visits negative as well as positive values. It is easy to verify that

$$\left[\Psi^{(0,b)}\ \widehat{\Psi}^{(0,b)}\right] = \left[I\ 0\right] \tag{3.25}$$

is a solution of (3.1) and, when $\boldsymbol{\pi}^{\mathrm{T}}\boldsymbol{\mu} = 0$, that it is a solution of (3.9) as well.

Case 4: $a > 0$ **is fixed and** $b = 0$**.** By the same argument as used in Case 3, we verify that

$$\left[\Psi^{(a,0)}\ \widehat{\Psi}^{(a,0)}\right] = \left[0\ I\right].$$

3.3 Expected Time to Exit

In addition to the probability of leaving the interval $(-a, b)$ through the upper or the lower boundary, and the distribution of the phase occupied then, one would be interested in determining how much time the MMBM needs to get there. One approach, like the one mentioned in Sect. 2.6, is to introduce a killing mechanism and to define the LS transforms

$$\Psi^{(a,b)}(s) = \mathrm{P}[\tau_{-a} < \tau_b, \tau_{-a} < e_s; \varphi(\tau_{-a})]$$

and

$$\widehat{\Psi}^{(a,b)}(s) = \mathrm{P}[\tau_b < \tau_{-a}, \tau_b < e_s; \varphi(\tau_b)],$$

where e_s is an exponential random variable with parameter s, independent of the MMBM. The LS transform of the total time spent in $(-a, b)$ until exit is

$$\mathrm{P}[\tau_{-a} \wedge \tau_b < e_s] = \boldsymbol{\alpha}(\Psi^{(a,b)}(s) + \widehat{\Psi}^{(a,b)}(s))\mathbf{1},$$

where $\boldsymbol{\alpha}$ is the distribution of the initial phase $\varphi(0)$. We do not pursue this investigation any more than we did in Sect. 2.6, and we only determine below the *expected time to exit.*

Theorem 3.11. *Assume that* $X(0) = 0$ *and denote as* $\boldsymbol{\alpha}$ *the distribution of* $\varphi(0)$*; take* $a, b \geq 0$, $a + b > 0$.

- *If* $\boldsymbol{\pi}^{\mathrm{T}}\boldsymbol{\mu} > 0$, *then*

$$\mathrm{E}[\tau_{-a} \wedge \tau_b] = \mathrm{E}[\tau_b] - \boldsymbol{\alpha}\Psi^{(a,b)}\mathrm{E}[\tau_{a+b}; \varphi(\tau_{a+b})]\mathbf{1},$$

- *if* $\boldsymbol{\pi}^{\mathrm{T}}\boldsymbol{\mu} < 0$, *then*

$$\mathrm{E}[\tau_{-a} \wedge \tau_b] = \mathrm{E}[\tau_{-a}] - \boldsymbol{\alpha}\widehat{\Psi}^{(a,b)}\mathrm{E}[\tau_{-(a+b)}; \varphi(\tau_{-(a+b)})]\mathbf{1},$$

where the matrix notation $\mathrm{E}[\cdot; \varphi(t)]$ *is defined in (1.26).*

Proof. Assume that $\boldsymbol{\pi}^{\mathrm{T}}\boldsymbol{\mu} > 0$. In that case, τ_b is finite, and it is equal to the time spent in $(-a, b)$ plus, in case the exit is through the lower boundary, the time to reach level b from level $-a$. Therefore,

$$\begin{aligned}\mathrm{E}[\tau_b | X(0) = 0, \varphi(0) = i] &= \mathrm{E}[\tau_{-a} \wedge \tau_b | X(0) = 0, \varphi(0) = i] \\ &\quad + \sum_{k \in \mathcal{E}} \Psi_{ik}^{(a,b)} \mathrm{E}[\tau_b | X(0) = -a, \varphi(0) = k], \quad i \in \mathcal{E}.\end{aligned}$$

As

$$\mathrm{E}[\tau_b | X(0) = -a, \varphi(0) = k] = \mathrm{E}[\tau_{a+b} | X(0) = 0, \varphi(0) = k],$$

we may write

$$\mathrm{E}[\tau_b] = \mathrm{E}[\tau_{-a} \wedge \tau_b] + \boldsymbol{\alpha}\Psi^{(a,b)}\mathrm{E}[\tau_{a+b}; \varphi(\tau_{a+b})]\mathbf{1}$$

and the first statement is proved.

If $\boldsymbol{\pi}^{\mathrm{T}}\boldsymbol{\mu} < 0$, the MMBM might never reach level b at all and $\mathrm{E}[\tau_b]$ is infinite, the level $-a$ is reached in finite time, however, and we adapt the argument above to justify the second statement. □

With Lemma 2.29, this theorem fully determines the expectation of $\tau_{-a} \wedge \tau_b$ when $\boldsymbol{\pi}^{\mathrm{T}}\boldsymbol{\mu} \neq 0$. In case $\boldsymbol{\pi}^{\mathrm{T}}\boldsymbol{\mu} > 0$,

$$\mathrm{E}[\tau_{a+b}; \varphi(\tau_{a+b})]\mathbf{1} = \{(a+b)\mathbf{1} + (e^{\widehat{U}(a+b)} - I)(-Q^{\#})\boldsymbol{\mu}\}/(\boldsymbol{\pi}^{\mathrm{T}}\boldsymbol{\mu})$$

by (2.85). To see this, we take $\boldsymbol{\alpha} = \boldsymbol{e}_i$, successively for all i in $\mathcal{E}$. If $\boldsymbol{\pi}^{\mathrm{T}}\boldsymbol{\mu} < 0$, we use (2.86) to obtain

$$\mathrm{E}[\tau_{-(a+b)}; \varphi(\tau_{-(a+b)})]\mathbf{1} = \{-(a+b)\mathbf{1} + (e^{-U(a+b)} - I)(-Q^{\#})\boldsymbol{\mu}\}/(\boldsymbol{\pi}^{\mathrm{T}}\boldsymbol{\mu}).$$

We shall deal in Lemma 4.25 with the case when $\boldsymbol{\pi}^{\mathrm{T}}\boldsymbol{\mu} = 0$.

3.4 Lower Control Barrier at 0

Regulated Brownian motions were introduced in Sect. 1.3, and their definition may be extended to MMBMs in a natural manner.

Definition 3.12. *Assume that* $\{X, \varphi\}$ *is a Markov-modulated Brownian motion with* $X(0) = 0$, *and define its running minimum* $m_X(t) = \inf\{X(s) : 0 \le s \le t\}$, $t \ge 0$.

The process $\{Z(t), \varphi(t) : t \ge 0\}$ *with*

$$Z(t) = X(t) - m_X(t)$$

is called the regulated MMBM with a lower control barrier at zero.

Clearly, $Z(t) \ge 0$ for all t, and we recognise that $-m_X(t)$ is the Skorokhod regulator of Parenthesis 1.6. We analyse in this section the ascending ladder time process $\{\delta_x : x \ge 0\}$, where

$$\delta_x = \inf\{t \ge 0 : Z(t) = x\}$$

and its associated ladder process $\{\varphi(\delta_x) : x \ge 0\}$. We denote as $H^{(b)}$, with

$$H_{ij}^{(b)} = \mathrm{P}_{0,i}[\delta_b < \infty, \varphi(\delta_b) = j], \qquad i, j \in \mathcal{E}, \tag{3.26}$$

the transition probability matrix of eventual exit from the interval $(0, b)$.

One of the main differences with the processes analysed in Chap. 2 is that $\delta_x < \infty$ for all x. If $\boldsymbol{\pi}^{\mathrm{T}}\boldsymbol{\mu} \geq 0$, $\{X\}$ will drift to steadily higher values, and so will $\{Z\}$. If $\boldsymbol{\pi}^{\mathrm{T}}\boldsymbol{\mu} < 0$, $\{X\}$ drifts towards $-\infty$ and whenever the MMBM reaches a new minimum, $\{Z\}$ is reflected back to positive values, so that it will eventually reach x for any given x; see Fig. 1.2 for an illustration. Therefore, $H^{(b)}$ is stochastic for any value of $\boldsymbol{\pi}^{\mathrm{T}}\boldsymbol{\mu}$. This, of course, means that $H^{(b)}$ cannot be equal to $\widehat{\Psi}^{(0,b)}$, which is equal to 0 as we saw in (3.25).

In the general case where the process starts at time 0 in some other level x, with $0 < x < b$, we denote the exit probability matrix as $H^{(b)}(x)$, with

$$(H^{(b)}(x))_{ij} = \mathrm{P}_{x,i}[\delta_b < \infty, \varphi(\delta_b) = j],$$

for i, j in $\mathcal{E}$. We condition on the first visit of the process to either level 0 or level b, and it is obvious that

$$H^{(b)}(x) = \widehat{\Psi}^{(x,b-x)} + \Psi^{(x,b-x)} H^{(b)},$$

so that we need to determine $H^{(b)}$ only.

In view of the behaviour described in Sect. 1.4, the regulated process oscillates and returns to its starting level infinitely often in any positive interval of time. This complicates somewhat our analysis, and to prove Theorem 3.14, we need some information about the behaviour of a regulated Brownian motion during an exponentially distributed interval of time.

Lemma 3.13. *Assume that $\{z(t) : t \geq 0\}$ is a regulated Brownian motion with a lower control barrier at 0 and parameters μ and σ. Assume that e_ω is an exponential variable with parameter ω, independent of $\{z\}$. Define the probability*

$$h_\omega^{(b)} = \mathrm{P}[\delta_b < e_\omega | z(0) = 0]$$

that the regulated Brownian motion reaches level b before time e_ω. We have

$$h_\omega^{(b)} = 2me^{-\widehat{u}_\omega b}(u_\omega + \widehat{u}_\omega e^{-2mb})^{-1}, \tag{3.27}$$

where $m = \sqrt{\mu^2 + 2\omega\sigma^2}/\sigma^2$, $u_\omega = m + \mu/\sigma^2$, and $\widehat{u}_\omega = m - \mu/\sigma^2$. Furthermore,

$$h_\omega^{(b)} = 1 - 2\omega b^2 + o(b^2) \tag{3.28}$$

asymptotically as $b \to 0$.

Proof. We denote as $F(\cdot)$ the distribution function of δ_b and rewrite $h_\omega^{(b)}$ as

$$h_\omega^{(b)} = \int_0^\infty e^{\omega y}\, \mathrm{d}F(y) = \mathrm{E}[e^{\omega\delta_b} | z(0) = 0].$$

This allows us to see that (3.27) follows from Zhou [145, Sections 5 and 6] and Breuer [45, Example 1].

Next, we show, after simple manipulations, that

$$h_\omega^{(b)}|_{b=0} = 1, \qquad \frac{\partial}{\partial b} h_\omega^{(b)}|_{b=0} = 0, \quad \text{and} \quad \frac{\partial^2}{\partial b^2} h_\omega^{(b)}|_{b=0} = -2\omega,$$

and this proves (3.28). □

Loosely stated, (3.28) means that for b small enough, the regulated Brownian motion reaches level b so fast that the killing mechanism has almost no chance to interrupt the process before then.

Theorem 3.14. *The phase transition probability matrix $H^{(b)}$ defined in (3.26) is given by*

$$H^{(b)} = -(L^{(b)})^{-1}\widehat{L}^{(b)}, \tag{3.29}$$

where

$$L^{(b)} = \frac{\partial}{\partial x}\Psi^{(x,b-x)}|_{x=0} \qquad \text{and} \qquad \widehat{L}^{(b)} = \frac{\partial}{\partial x}\widehat{\Psi}^{(x,b-x)}|_{x=0}. \tag{3.30}$$

These two matrices are solutions of the system

$$\left[L^{(b)}\ \widehat{L}^{(b)}\right](I + \mathcal{P}^{(b)}) = \left[U\ \ -\widehat{U}e^{\widehat{U}b}\right], \tag{3.31}$$

and the matrix $L^{(b)}$ is nonsingular.

Proof. We take x fixed, $0 < x < b$, and we condition the distribution of $\varphi(\delta_b)$ upon the phase at the epoch δ_x when the regulated process reaches the intermediary level x. At time δ_x, we find ourselves in the situation of Theorem 3.2: The starting level is x, in the interior of the interval $(0, b)$, the MMBM may reach level b without returning to level 0, or it may return to level 0 before getting to level b, in which case the system is back in its initial conditions. This gives us

$$H^{(b)} = H^{(x)}(\widehat{\Psi}^{(x,b-x)} + \Psi^{(x,b-x)}H^{(b)}), \tag{3.32}$$

where we remember that the probability of hitting either of the boundaries of $(0, b)$, starting from x, is the same as the probability of hitting either of $-x$ or $b-x$ starting from 0.

The left-hand side of (3.32) being independent of x, its derivative with respect to x is identically 0, the derivative of the right-hand side is equal to 0 for any x and, in the limit for $x \to 0$, we find that

$$0 = \frac{\partial}{\partial x}H^{(x)}|_{x=0}(\widehat{\Psi}^{(0,b)} + \Psi^{(0,b)}H^{(b)}) + H^{(0)}(\widehat{L}^{(b)} + L^{(b)}H^{(b)}). \tag{3.33}$$

Assume that $\varphi(0) = i$ and denote by T the sojourn time in phase i before the first change of phase. We have

$$\begin{aligned}(H^{(x)})_{ii} &= \mathrm{P}[\varphi(\delta_x) = i | Z(0) = 0, \varphi(0) = i] \\ &\geq \mathrm{P}[\varphi(\delta_x) = i, \delta_x < T | Z(0) = 0, \varphi(0) = i] \\ &= 1 - 2|Q_{ii}|x^2 + o(x^2)\end{aligned}$$

by Lemma 3.13, for all i in $\mathcal{E}$. Since $H^{(x)}$ is a stochastic matrix, the sum of its off-diagonal terms on row i is bounded by $2|Q_{ii}|x^2 + o(x^2)$, and we may write that $H^{(x)} = I + O(x^2)$. We conclude that

$$H^{(0)} = I \qquad \frac{\partial}{\partial x} H^{(x)}|_{x=0} = 0, \tag{3.34}$$

and (3.33) becomes

$$L^{(b)} H^{(b)} + \widehat{L}^{(b)} = 0 \qquad \text{for any } b. \tag{3.35}$$

Equation (3.29) will be proved once we show that $L^{(b)}$ is a nonsingular matrix.

We define as follows a discrete-time Markov chain $\{K_n : n \geq 0\}$ on the finite state space $\{x, b - x\} \otimes \mathcal{E}$, $0 < x < b$: We observe the evolution of the MMBM in $(0, b)$, starting either from level x or from level $b - x$, until it visits level 0 or level b for the first time. If level 0 is visited first, the process is re-started at level x, if level b is reached first, the process is re-started at level $b - x$. The transition matrix of $\{K_n\}$ is

$$M(x) = \begin{bmatrix} \Psi^{(x,b-x))} & \widehat{\Psi}^{(x,b-x)} \\ \Psi^{(b-x,x)} & \widehat{\Psi}^{(b-x,x)} \end{bmatrix}.$$

By (3.25), $\Psi^{(0,b)} = I$, and $\widehat{\Psi}^{(b,0)} = I$ as well by the same argument. Therefore, $\lim_{x \to 0} M(x) = I$, and we may write

$$M(x) = I + xM'(0) + o(x), \tag{3.36}$$

with

$$M'(0) = \lim_{x \to 0} \frac{1}{x}(M(x) - I) = \begin{bmatrix} L^{(b)} & \widehat{L}^{(b)} \\ D^{(b)} & \widehat{D}^{(b)} \end{bmatrix}, \tag{3.37}$$

where, in addition to $L^{(b)}$ and $\widehat{L}^{(b)}$ from (3.30), we define

$$D^{(b)} = \frac{\partial}{\partial x} \Psi^{(b-x,x)}|_{x=0} \qquad \text{and} \qquad \widehat{D}^{(b)} = \frac{\partial}{\partial x} \widehat{\Psi}^{(b-x,x)}|_{x=0}. \tag{3.38}$$

Equation (3.36) shows that, in all likelihood, the next boundary visited is the one closest to the starting level, with $M'(0)$ giving the first-order correcting factor for x close to 0. The matrix $M(x)$ is irreducible stochastic and $M(x)\mathbf{1} = \mathbf{1}$ uniformly in x. Therefore, $M'(0)\mathbf{1} = \mathbf{0}$, the off-diagonal elements of $M'(0)$ are nonnegative, and its diagonal elements are negative.

In summary, $M'(0)$ has the algebraic features of an irreducible infinitesimal generator, and its diagonal block $L^{(b)}$ is nonsingular, $(-L^{(b)})^{-1}$ is nonnegative, and

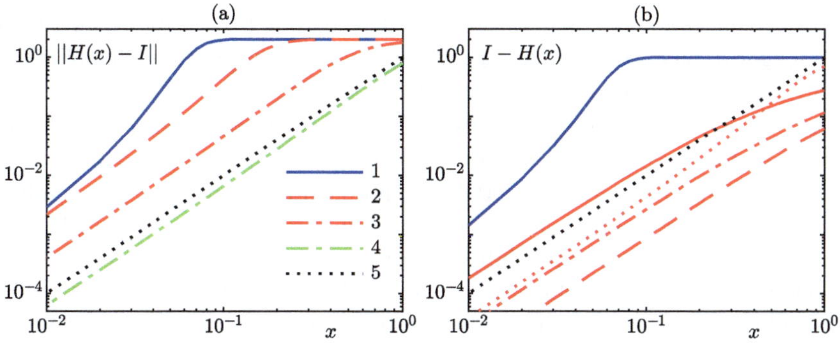

Fig. 3.3 (**a**) On the left: log-log plot of $||H^{(x)} - I||_\infty$ for four illustrative examples of MMBMs, the dotted line marked "5" is x^2 and serves as visual reference. (**b**) On the right: log-log plot of $1 - H_{ii}^{(x)}$ for the video streaming example, for $i = 1, \ldots 5$

we conclude from ((3.35) that $H^{(b)} = (-L^{(b)})^{-1}\widehat{L}^{(b)}$. We note for later reference that $\widehat{D}^{(b)}$ is nonsingular, with $(-\widehat{D}^{(b)})^{-1} \geq 0$.

Finally, we remember from (3.1) that

$$\left[\Psi^{(x,b-x)}\ \widehat{\Psi}^{(x,b-x)}\right](I + \mathcal{P}^{(b)}) = \mathcal{U}^{(x,b-x)},$$

we take on both sides the derivative with respect to x evaluated at $x = 0$ and so obtain (3.31). This concludes the proof of the theorem. □

Example 3.15. We used Lemma 3.13 in our proof of Theorem 3.14 to show that $H^{(x)} = I + O(x^2)$, and this was the key to proving (3.35). In Fig. 3.3a on the left of Fig. 3.3, we give the log-log plot of $||H^{(x)} - I||_\infty$ for a few illustrative examples, with x in the interval $0.01 \leq x \leq 1$. We include x^2 as the dotted line labelled 5, for comparison with the examples.

- The blue line, labelled 1, is for the video streaming model of Example 2.32.
- The dashed and the dot-dashed red lines, labelled 2 and 3, correspond to MMBMs with the same cyclic structure as in Example 2.28, respectively with $m = 10$ phases for Line 2, and $m = 2$ for Line 3. In both cases, the transition rates $|Q_{ii}|$ from one state to the next are equal to m, so that each cycle, from one phase back to the same phase, takes one unit of time on average. The Brownian parameters are $\mu_1 = 1.1\,m$, $\mu_2 = -1.2\,m$, and $\mu_i = 0$ otherwise, and $\sigma_i^2 = 1$ for all i.
- The green dot-dashed line, labelled 4, is for a singularly perturbed system, similar to the MMBMs of Examples 1.22 and 2.27. The parameters here are $a_1 = 0.2$, $a_2 = 0.3$, $a_3 = 0.3$, $a_4 = 0.15$, $\varepsilon = 0.01$ and are otherwise the same as in Example 1.22.

We do observe for all examples that $||H^{(x)} - I||_\infty$ evolves like x^2 for x small enough. The aspect of Line 5 is much different from the other examples and this is due to the differences among the five phases of the video streaming model, as may be seen in Fig. 3.3b on the right. There, we plot the diagonal elements of $I - H^{(x)}$:

For a given phase, $1-H_{ii}^{(x)}$ is the probability that, starting from $\varphi(0)=i$, the MMBM is no longer in phase i when the regulated process reaches level x. The blue curve is for $i=5$, and the red curves, in descending order on the left of the figure, are for phases 1–4. Selected numerical values of $1-H_{ii}^{(x)}$ are given in Table 3.1.

The line for phase 5 is significantly different and the global norm $||H^{(x)}-I||_\infty$ is dominated by that phase. Take $x=0.01$, the probability of being in a phase different from $\varphi(0)$ at time δ_x is very small for all phases in $\mathcal{E}$. For $x=0.1$, this is still the case for the phases 1–4, but that probability is nearly one for $\varphi(0)=5$. This we explain by the distinguishing features of that phase: $\mu_5=-1$, the smallest drift, and $\sigma_5^2=0.01$, the smallest volatility. With these parameters, we may envision that in phase 5 the level would go down nearly deterministically and repeatedly hit the control barrier, without reaching level 0.1, until there is a change of phase.

x	0.0100	0.1000	1.0000
$i=1$	0.0002	0.0144	0.2803
$i=2$	0.0000	0.0044	0.6979
$i=3$	0.0000	0.0027	0.1162
$i=4$	0.0000	0.0009	0.0612
$i=5$	0.0014	0.9781	1.0000

Table 3.1 Selected values of $1-H_{ii}^{(x)}$ for the video streaming example

The coefficient matrix of the system (3.31) is the same as the matrix in (3.1), and this tells us that we need to look for different expressions for the matrices $L^{(b)}$ and $\widehat{L}^{(b)}$, according to whether $\boldsymbol{\pi}^{\mathrm{T}}\boldsymbol{\mu}\neq 0$ or $\boldsymbol{\pi}^{\mathrm{T}}\boldsymbol{\mu}=0$. In the first case, we have the following lemma.

Lemma 3.16. *If $\boldsymbol{\pi}^{\mathrm{T}}\boldsymbol{\mu}\neq 0$, then*

$$L^{(b)}=(Ue^{-Ub}+\widehat{U}e^{\widehat{U}b})W_b^{-1}, \qquad \widehat{L}^{(b)}=-(U+\widehat{U})\widehat{W}_b^{-1}, \tag{3.39}$$

and

$$H^{(b)}=W_b(Ue^{-Ub}+\widehat{U}e^{\widehat{U}b})^{-1}(U+\widehat{U})\widehat{W}_b^{-1}. \tag{3.40}$$

Proof. The equations (3.39) directly follow from (3.6) and the definition (3.30) of $L^{(b)}$ and $\widehat{L}^{(b)}$, Eq. (3.40) is an obvious consequence of (3.29). □

Like in Sect. 3.2, we need an additional equation when $\boldsymbol{\pi}^{\mathrm{T}}\boldsymbol{\mu}=0$, it is given below in (3.41).

Lemma 3.17. *If $\boldsymbol{\pi}^{\mathrm{T}}\boldsymbol{\mu}=0$, then the system formed by (3.31) and*

$$\begin{bmatrix}L^{(b)} & \widehat{L}^{(b)}\end{bmatrix}\boldsymbol{\beta}^*=\mathbf{1} \tag{3.41}$$

is nonsingular, where $\boldsymbol{\beta}^*=\begin{bmatrix}\boldsymbol{h}\\ b\mathbf{1}+\boldsymbol{h}\end{bmatrix}$ and $\boldsymbol{h}$ is defined in Lemma 3.7.

The solution of (3.31, 3.41) is

$$L^{(b)} = (Ue^{-Ub} + \widehat{U}e^{\widehat{U}b})W_b^{\#} + \boldsymbol{w}^* \cdot \widehat{\boldsymbol{\eta}}_1^{\mathrm{T}} \tag{3.42}$$

$$\widehat{L}^{(b)} = (U + \widehat{U})\widehat{W}_b^{\#} - \boldsymbol{w}^* \cdot \widehat{\boldsymbol{\eta}}_2^{\mathrm{T}}, \tag{3.43}$$

where $\widehat{\boldsymbol{\eta}}$ is the stationary probability vector of $\mathcal{P}^{(b)}$, and

$$\boldsymbol{w}^* = (\boldsymbol{\eta}^{*\mathrm{T}}\boldsymbol{\beta}^*)^{-1}(\mathbf{1} - \left[U \; -\widehat{U}e^{\widehat{U}b}\right](I + \mathcal{P}^{(b)})^{\#}\boldsymbol{\beta}^*), \tag{3.44}$$

where $\boldsymbol{\eta}^{*\mathrm{T}} = \left[\widehat{\boldsymbol{\eta}}_1^{\mathrm{T}} \; -\widehat{\boldsymbol{\eta}}_2^{\mathrm{T}}\right]$.

Proof. We start from (3.9) which we write as

$$\left[\Psi^{(x,b-x)} \; \widehat{\Psi}^{(x,b-x)}\right] \begin{bmatrix} -x\mathbf{1} + \boldsymbol{h} \\ (b-x)\mathbf{1} + \boldsymbol{h} \end{bmatrix} = \boldsymbol{h},$$

then take the derivative on both sides with respect to x and set $x = 0$. This proves (3.41). One easily verifies that $\boldsymbol{\eta}^*$ is a right eigenvector of the coefficient matrix $I + \mathcal{P}^{(b)}$ of (3.31) for the eigenvalue 0, and we conclude from (1.33) that

$$\left[L^{(b)} \; \widehat{L}^{(b)}\right] = \left[U \; -\widehat{U}e^{\widehat{U}b}\right](I + \mathcal{P}^{(b)})^{\#} + \boldsymbol{w}^* \cdot \boldsymbol{\eta}^{*\mathrm{T}}$$

for some vector $\boldsymbol{w}^*$. We post-multiply this equation by $\boldsymbol{\beta}^*$ and conclude that $\boldsymbol{w}^*$ is given by (3.44). From there, we follow the steps in the proofs of Lemma 3.7 and Theorem 3.10. □

We conclude this section with an example. This is a model of independent multiplexing sources sharing a common buffer. Similar examples are found in Karandikar and Kulkarni [95] and Simon [138].

Example 3.18. We assume that m sources alternate between being active and inactive, the two states being labelled respectively as on and `off`. A source remains in its on state for an exponentially distributed amount of time, with parameter a, the parameter for the `off` state is b. The phase φ is the number of inactive sources, so that the generator Q is

$$Q = \begin{bmatrix} Q_{0,0} & ma & & & & \\ b & Q_{1,1} & (m-1)a & & & \\ & 2b & Q_{2,2} & (m-2)a & & \\ & & \ddots & \ddots & \ddots & \\ & & & (m-1)b & Q_{m-1,m-1} & a \\ & & & & mb & Q_{m,m} \end{bmatrix}$$

with $Q_{i,i} = -ib - (m-i)a$ (we show the non-zero entries only). On average, there are $mb/(a+b)$ active sources in the long run. When active, each individual source behaves like a BM(r, s_{on}^2): r is its input rate and s_{on}^2 is the volatility of its input stream.

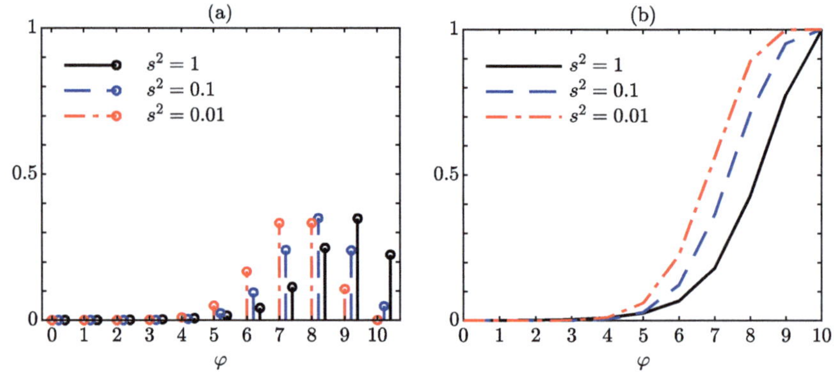

Fig. 3.4 Multiplexing sources. Distribution of the phase upon reaching level 1, for s^2 taking the values 0.01 (dot-dashed red lines), 0.1 (dashed blue lines), and 1 (continuous black lines). The probability distributions are shown in (**a**) on the left and the cumulative distribution functions in (**b**) on the right

The buffer is drained at the mean rate C, with volatility s^2, independently of the number of active sources, as long as there is data to be transmitted, and the buffer content evolves like an MMBM with lower control barrier at zero, with parameters $\mu_i = (m-i)r - C$ when there are i inactive sources, and $\sigma_i^2 = (m-i)s_{\text{on}}^2 + s^2$. The initial distribution $\boldsymbol{\alpha}$ of the phase is uniform on $\{0, 1, \ldots, m\}$.

The stationary drift $\boldsymbol{\pi}^{\mathrm{T}}\boldsymbol{\mu} = rmb/(a+b) - C$ is the difference between the mean collective input from the active sources and the multiplexer capacity, and the ratio

$$\rho = rmb/((a+b)C)$$

is the fraction of the available capacity which is being used, on average, by the multiplexed sources.

We show in Fig. 3.4 the distribution $\boldsymbol{\alpha}^{\mathrm{T}}H^{(1)}$ of the phase $\varphi(\delta_x)$ for $x = 1$. The parameters are $m = 10$, $a = 9$ and $b = 1$, so that each source is active one tenth of the time, on average. The output rate of the buffer is $C = 1$, the traffic coefficient is fixed at $\rho = 0.5$ and the input rate of each source is $r = C\rho(a+b)/(bm)$, so that the sources together use one half of the system capacity on average, and the stationary drift is $\boldsymbol{\pi}^{\mathrm{T}}\boldsymbol{\mu} = -0.5$. The volatility of each source input stream is $s_{\text{on}}^2 = 0.1\,r$. With these parameters, the individual Fluid rates are

$$\boldsymbol{\mu} = \begin{bmatrix} 4.0 & 3.5 & 3.0 & 2.5 & 2.0 & 1.5 & 1.0 & 0.5 & 0 & -0.5 & -1.0 \end{bmatrix}^{\mathrm{T}}.$$

The volatility parameter s^2 of the buffer takes three values,

A.1 the small value $s^2 = 0.01$ which corresponds to a very regular buffer mechanism,
A.2 the more realistic $s^2 = 0.1$
A.3 and the very high $s^2 = 1$.

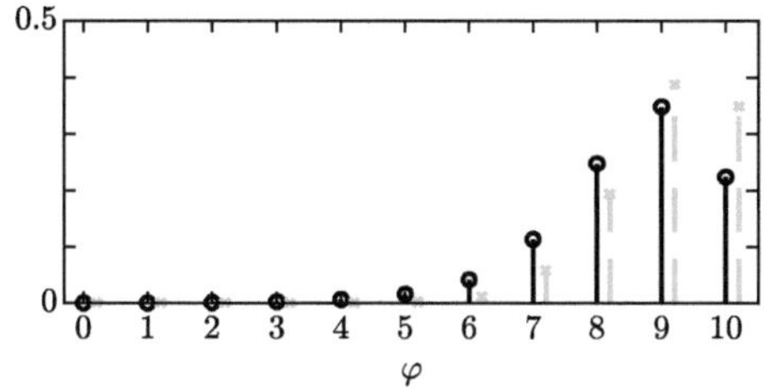

Fig. 3.5 Multiplexing sources. Distribution of the phase upon reaching level 1 for Case A.3 (continuous black lines), and stationary distribution $\boldsymbol{\pi}$ of Q (dashed green lines)

We give the distribution probabilities in Fig. 3.4a on the left and the cumulative probability distribution in Fig. 3.4b on the right. The cumulative distribution is shown as a continuous function to make it easier to read the picture. We see very clearly in both graphs that the probability mass of $\varphi(\delta_1)$ is shifted to higher values in $\mathcal{E}$ as s^2 increases, to the point where the probability is 0.8195 when $s^2 = 1$, of reaching level 1 with a phase in the set $\{8, 9, 10\}$. This is somewhat unexpected as these are the three phases for which $\mu_i \leq 0$, and one might imagine instead that the process is more likely to go up in phases for which μ_i is greatest.

An explanation may be found in the fact that the distribution $\boldsymbol{\alpha}^{\mathrm{T}} H^{(1)}$ in Case A.3 is similar to the stationary distribution $\boldsymbol{\pi}$ of Q, as we show in Fig. 3.5: The transitions in the Markov process with generator Q are fast, the phase approaches rapidly its stationary distribution before the MMBM reaches level 1, and it is most likely that the phase is in $\{8, 9, 10\}$ by then (the stationary probability of that subset is 0.9298); the high volatility helps bring the MMBM above level 1 despite the negative drift. Thus, the effect we observe in Fig. 3.4 might be the combination of fast change of phases with high variance of each Brownian motion.

To verify this interpretation, we show in Fig. 3.6 the effect of slowing down the transitions between phases. The parameters m, C, r, s_{on} and $\boldsymbol{\alpha}$ remain the same, s^2 is equal to 0.1 and the pair (a, b) of parameters takes three values,

B.1 $(a, b) = (9, 1)$ as in Case A.2,
B.2 $(a, b) = (1.8, 0.2)$ so that sojourn times in each phase are five times longer, on average, and
B.3 $(a, b) = (0.9, 0.1)$ so that sojourn times are 10 times as long as in Case B.1.

The parameters for Case B.1 are the same as for Case A.2 and we use the same line styles to represent the phase distribution. We observe that as the phase transitions become slower, the MMBM has more opportunities to reach level 1 before the phase process gets to its stationary distribution, and $\boldsymbol{\alpha}^{\mathrm{T}} H^{(1)}$ places more probability mass over small values of φ.

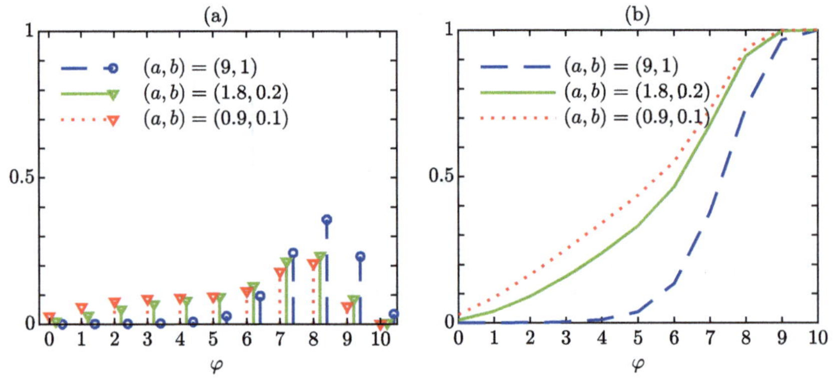

Fig. 3.6 Multiplexing sources. Distribution of the phase upon reaching level 1, for different values of the transition rates between phases: $(a, b) = (9, 1)$ (dashed blue lines), $(1.8, 0.2)$ (continuous green lines), and $(0.9, 0.1)$ (dotted red lines). The probability distributions are shown in (**a**) on the left, the cumulative distribution functions in (**b**) on the right

3.5 Upper Control Barrier at 0

We deal in Chap. 5 with processes that are forced to remain below some level $b > 0$, and we shall be interested in the distribution of the phase upon hitting level 0, starting from b. This, of course, is the same as the distribution of the phase upon hitting level $-b$, starting from level 0, for a process constrained to remain in negative levels, an observation which allows us to use much material from Sect. 3.4.

We define the process $\{\widetilde{Z}(t), \varphi(t) : t \geq 0\}$ with

$$\widetilde{Z}(t) = X(t) - M_X(t),$$

where $M_X(t) = \sup\{X(s) : 0 \leq s \leq t\}$ is the running maximum of the MMBM. The process $\{\widetilde{Z}, \varphi\}$ is called the regulated process with an upper control barrier at 0, and we analyse the descending ladder time process $\{\widetilde{\delta}_{-x} : x \geq 0\}$, where

$$\widetilde{\delta}_{-x} = \inf\{t : \widetilde{Z}(t) = -x\}$$

is the first passage time to $-x$. The transition probability matrix from $\varphi(0)$ to $\varphi(\widetilde{\delta}_{-x})$ is $\widetilde{H}^{(x)}$, with

$$\widetilde{H}^{(x)}_{ij} = \mathrm{P}_{0i}[\widetilde{\delta}_{-x} < \infty, \varphi(\widetilde{\delta}_{-x}) = j], \qquad i, j \in \mathcal{E}. \tag{3.45}$$

The next theorem brings together properties which correspond to those of δ_x in Theorem 3.14, Lemma 3.16, and Lemma 3.17. We do not give the detailed proof: Starting from the decomposition

$$\widetilde{H}^{(b)} = \widetilde{H}^{(x)}(\widehat{\Psi}^{(b-x,x)}\widetilde{H}^{(b)} + \Psi^{(b-x,x)})$$

similar to (3.32), one adapts the analysis from Sect. 3.4.

Theorem 3.19. *The transition probability matrix $\widetilde{H}^{(b)}$ from $\varphi(0)$ to $\varphi(\widetilde{\delta}_{-b})$ is given by*

$$\widetilde{H}^{(b)} = -(\widehat{D}^{(b)})^{-1} D^{(b)}, \tag{3.46}$$

where the matrices

$$D^{(b)} = \frac{\partial}{\partial x}\Psi^{(b-x,x)}|_{x=0} \quad \text{and} \quad \widehat{D}^{(b)} = \frac{\partial}{\partial x}\widehat{\Psi}^{(b-x,x)}|_{x=0} \tag{3.47}$$

are solutions of the system

$$\left[D^{(b)}\ \widehat{D}^{(b)}\right](I + \mathcal{P}^{(b)}) = \left[-Ue^{Ub}\ \widehat{U}\right], \tag{3.48}$$

and $\widehat{D}^{(b)}$ is nonsingular.

If $\boldsymbol{\pi}^{\mathrm{T}}\boldsymbol{\mu} \neq 0$, then

$$D^{(b)} = -(U + \widehat{U})W_b^{-1} \quad \text{and} \quad \widehat{D}^{(b)} = (Ue^{Ub} + \widehat{U}e^{-\widehat{U}b})\widehat{W}_b^{-1}, \tag{3.49}$$

and

$$\widetilde{H}(b) = \widehat{W}_b(\widehat{U}e^{-\widehat{U}b} + Ue^{Ub})^{-1}(U + \widehat{U})W_b^{-1}. \tag{3.50}$$

If $\boldsymbol{\pi}^{\mathrm{T}}\boldsymbol{\mu} = 0$, then the system formed by (3.48) and

$$\left[D^{(b)}\ \widehat{D}^{(b)}\right]\widetilde{\boldsymbol{\beta}} = -\mathbf{1}$$

is nonsingular, where $\widetilde{\boldsymbol{\beta}} = \begin{bmatrix} -b\mathbf{1} + \boldsymbol{h} \\ \boldsymbol{h} \end{bmatrix}$, and $\boldsymbol{h}$ is defined in Lemma 3.7. Its solution is

$$D^{(b)} = -(U + \widehat{U})W_b^{\#} + \widetilde{\boldsymbol{w}} \cdot \widehat{\boldsymbol{\eta}}_1^{\mathrm{T}} \tag{3.51}$$

$$\widehat{D}^{(b)} = -(Ue^{Ub} + \widehat{U}e^{-\widehat{U}b})\widehat{W}_b^{\#} - \widetilde{\boldsymbol{w}} \cdot \widehat{\boldsymbol{\eta}}_2^{\mathrm{T}} \tag{3.52}$$

and

$$\widetilde{\boldsymbol{w}} = -(\boldsymbol{\eta}^{*\mathrm{T}}\widetilde{\boldsymbol{\beta}})^{-1}(\mathbf{1} + \left[-Ue^{Ub}\ \widehat{U}\right](I + \mathcal{P}^{(b)})^{\#}\widetilde{\boldsymbol{\beta}}).$$

□

3.6 Notes

The probability of leaving a finite interval through the upper or lower boundary has been analysed by several authors and the proof I give for Eq. (3.1) is quite standard. The expressions given in Theorems 3.3 and 3.10, however, are new, as far as I know. Different explicit forms are given in Jiang and Pistorius [93] and in Ivanovs and Palmowski [90], the latter is similar to (3.6), but more about it in Remark 4.6.

The additional equation (3.9) to be used when $\boldsymbol{\pi}^{\mathrm{T}}\boldsymbol{\mu} = 0$ is identical to the one given in D'Auria et al. [56, Section 7], and also in Ivanovs [87], on the basis of the spectral decomposition of the matrix Laplace exponent. The proof I give is different, based as it is on the stochastic process itself, an argument that Giang T. Nguyen and I developed for Fluid flows in [104, Lemma 8.3].

In Sects. 3.4 and 3.5, I follow Harrison [80] in using the terms *regulated processes* with a lower, or an upper, control barrier at zero. These are often called processes with *reflecting boundaries* in the literature.

Theorem 3.14 is proved with a different argument in Latouche and Nguyen [102]: We start there from a fast-switching Fluid flow as described in Sect. 1.11, solve the problem in the world of Fluid flows, and take the limit as the switching rate goes to infinity. We follow the same approach in [100, 101, 103], our objective there having been to overcome the difficulties stemming from the instantaneous returns by a Brownian motion to its initial level, as described in Sect. 1.4. Breuer [45] deals with this issue by forcing the MMBM to jump to $\varepsilon > 0$ whenever it hits level 0, and then taking the limit as ε tends to 0.

I avoid using the fast-switching approximation in the present book, with one minor exception, to prove Theorem 5.2. Elsewhere, I condition on the state of the MMBM after the process has had the opportunity to evolve over a strictly positive interval of time. That was done already in the proof of Theorem 2.8, where one waits until the first time there is a change of phase. In the proof of Theorem 3.14, one conditions on the phase when the process reaches a strictly positive level x, the whole argument crucially depends on Lemma 3.13, which Zhou [145] credits to results in Avram et al. [21].

Much of the material for $\boldsymbol{\pi}^{\mathrm{T}}\boldsymbol{\mu} = 0$ is new, I believe: Either the argument in the proof is novel, or the expression of the solution is different from what has been published before or, as in the case of Theorem 3.10 and Lemma 3.17, the results themselves are new. For instance, the matrix $L^{(b)}$ in Lemma 3.16 is the same as $A(b)$ in Lemma 1 of Latouche and Simon [107], but the expressions and the methods of proof are different.

Chapter 4
Expected Local Time

⋄ Local time ⋄ Time away from 0 ⋄ Time per unit of time ⋄ Matrices U and R ⋄ Zero drift ⋄ Wiener-Hopf factorisation ⋄ Local time before crossings ⋄ Time reversal ⋄

Two families of matrices play an important role in the analysis of Markov additive processes (MAPs). The first one is related to first passage times from one level to another and was the subject of Chap. 2. The second family is about the expected time spent bythe MAP in various locations between successive visits to its initial level; it is the object of the present chapter, and we shall show in Chap. 5 that it is fundamental in expressing the stationary distribution of regulated processes. We define in Sect. 4.1 the local time spent by the process at a given level and we give some of its basic properties. In Sect. 4.2, we introduce the matrix R such that e^{Rx} is the expected local time spent at level x per unit of local time spent at level 0 during a suitably defined interval of time. We analyse the connection between R and the matrix U from Chap. 2, both when $\boldsymbol{\pi}^{\mathrm{T}}\boldsymbol{\mu} \neq 0$ and when $\boldsymbol{\pi}^{\mathrm{T}}\boldsymbol{\mu} = 0$. We show in Sect. 4.5 how the matrices R and U combine to determine the expected local time at 0 before the first visit to a given level $x \neq 0$. In closing the chapter, we prove the relation between the matrices R and U of the given MMBM and those of the time-reversed process.

4.1 Occupation Measure and Local Time

We need to adapt the definitions from Sect. 1.5 to the two-dimensional process $\{X, \varphi\}$. We follow Ivanovs and Palmowski [90] and mimic the characterisation (1.14) of the local time for Brownian motion.

The local time at level x in phase j during the time interval $[0, t]$ is defined as

© The Author(s), under exclusive license to Springer Nature Switzerland AG 2026

G. Latouche, *Markov-Modulated Brownian Motion*, Springer Series in Operations Research and Financial Engineering, https://doi.org/10.1007/978-3-031-98576-8_4

$$\ell_j(x,t) = \lim_{\epsilon \downarrow 0} \frac{1}{2\epsilon} \int_0^t \mathbb{1}\{|X(s) - x| < \epsilon, \varphi(s) = j\} \, \mathrm{d}s, \tag{4.1}$$

for $x \in \mathbb{R}$, $j \in \mathcal{E}$, $t \geq 0$. For any given x and j, $\{\ell_j(x,t), t \geq 0\}$ is a process that increases only when $(X(t), \varphi(t)) = (x, j)$. The occupation density formula (1.15) is generalised for MMBMs as

$$\int_0^t f(X(s), \varphi(s)) \, \mathrm{d}s = \sum_{j \in \mathcal{E}} \int_{\mathbb{R}} f(y, j) \ell_j(y, t) \, \mathrm{d}y \quad \text{a.s.} \tag{4.2}$$

for any measurable, nonnegative function f, and (1.16) becomes here

$$\int_0^t \mathrm{E}_{x,i}[f(X(s), \varphi(s))] \, \mathrm{d}s = \sum_{j \in \mathcal{E}} \int_{\mathbb{R}} f(y, j) \, \mathrm{E}_{x,i}[\ell_j(y, t)] \, \mathrm{d}y. \tag{4.3}$$

The local time has the following important additivity property (Ivanovs and Palmowski [90, Proposition 8]).

Proposition 4.1. *Let T denote a stopping time, with $X(T) = y$ for some y in $\mathbb{R}$ and $\varphi(T) = k$, for some k in $\mathcal{E}$.*

The shifted process $\{\ell_j(x, T + s) - \ell_j(x, T) : s \geq 0\}$ is independent of $\{X(t), \varphi(t) : 0 \leq t \leq T\}$ conditionally given $X(T) = y$, $\varphi(T) = k$; it has the same distribution as $\{\ell_j(x - y, s) : s \geq 0\}$ given $X(0) = 0$, $\varphi(0) = k$. □

We should mention that if $\sigma_j = 0$ for some phase j, with $\mu_j \neq 0$, an equivalent, more convenient definition of the local time is

$$\ell_j(x,t) = \frac{1}{|\mu_j|} N_j(x,t), \tag{4.4}$$

where

$$N_j(x,t) = \#\{s \in [0,t] : X(s) = x, \varphi(s) = j\}$$

is the number of times the process crosses level x in phase j during $[0, t]$. The expression (4.4) will be used in the proof of Theorem 5.2.

We denote by $L_x[y, T]$ the matrix function of *expected* local times over the time interval $[0, T]$, with entries

$$(L_x[y,T])_{ij} = \mathrm{E}_{x,i}[\ell_j(y,T)],$$

for x, y in $\mathbb{R}$, i, j in $\mathcal{E}$, T real, finite or infinite. Our analysis begins with the expected local time over the whole history of the process, that is, with $T = \infty$.

We know from the discussion in Sect. 2.1 that $X(t)$ drifts to $+\infty$ if $\boldsymbol{\pi}^{\mathrm{T}} \boldsymbol{\mu} > 0$, and to $-\infty$ if $\boldsymbol{\pi}^{\mathrm{T}} \boldsymbol{\mu} < 0$. If $\boldsymbol{\pi}^{\mathrm{T}} \boldsymbol{\mu} \neq 0$, therefore, there is for any given y a finite time after which the process does not visit level y, that time being equal to 0 if the process never visits level y at all. On the other hand, if $\boldsymbol{\pi}^{\mathrm{T}} \boldsymbol{\mu} = 0$, the MMBM oscillates forever without bounds, so that it visits infinitely often any given level y. Thus, the property stated in Theorem 4.3 is obvious; still, we give it a formal proof to illustrate the type of arguments we need in the context of MMBMs.

As a preliminary, we need the following property of Brownian motion, see Harrison [80, Page 51].

Lemma 4.2. *Assume that* $\{X\}$ *is a Brownian motion with parameters* μ *and* σ, *and with* $X(0) = 0$. *Assume that* T *is an exponential random variable with parameter* λ. *The expected local time at level 0 over* $[0, T]$ *is*

$$L_0[0,T] = 1/\sqrt{\mu^2 + 2\lambda\sigma^2}. \tag{4.5}$$

□

We plan to use a Markov-renewal sequence of returns to level 0 and, because of the oscillating sample path property discussed in Sect. 1.4, we need to design inter-renewal intervals with a strictly positive length. The process $\{t_n^* : n \geq 0\}$ defined here will be used in the proof of Theorem 4.3, and later in the proofs of Corollary 4.5 and Theorem 4.8, and in the definition (4.23). Briefly stated, $\{t_n^*\}$ is the sequence of visits to level 0 that are separated from the preceding visit to 0 by a change of phase:

$$t_0^* = 0,$$

$$t_n' = \inf\{t > t_{n-1}^* : \varphi(t) \neq \varphi(t_{n-1}^*)\}, \tag{4.6}$$

$$t_n^* = \inf\{t > t_n' : X(t) = 0\} \tag{4.7}$$

for $n \geq 1$, with $t_n^* = \infty$ if no finite t satisfies (4.7); at time t_n', there is a change of phase, at time t_n^* the process hits level 0 and the phase may have taken any value in $\mathcal{E}$.

Theorem 4.3. *If* $\boldsymbol{\pi}^{\mathrm{T}}\boldsymbol{\mu} \neq 0$, *then all entries of* $L_x[y,\infty]$ *are finite for any* x *and* y *in* $\mathbb{R}$. *Moreover,*

$$L_0[x,\infty] = e^{\widehat{U}x} L_0[0,\infty], \tag{4.8}$$

$$L_0[-x,\infty] = e^{Ux} L_0[0,\infty] \tag{4.9}$$

for $x \geq 0$.

If $\boldsymbol{\pi}^{\mathrm{T}}\boldsymbol{\mu} = 0$, *then all entries of* $L_x[y,\infty]$ *are infinite, for all* x *and* y.

Proof. Obviously, there is no time spent at level x until the first visit there, and so

$$\begin{aligned} L_0[x,\infty] &= \mathrm{P}_0[\tau_x < \infty] L_x[x,\infty] \\ &= \mathrm{P}_0[\tau_x < \infty] L_0[0,\infty] \end{aligned}$$

by the Markov property, and (4.8, 4.9) follow. Also,

$$L_x[y,\infty] = \begin{cases} e^{U(x-y)} L_0[0,\infty] & \text{for } x > y \\ e^{\widehat{U}(y-x)} L_0[0,\infty] & \text{for } x < y \end{cases} \tag{4.10}$$

since $\mathrm{P}_x[\tau_y < \infty] = \mathrm{P}_0[\tau_{y-x} < \infty]$, and we only need to deal with $L_0[0, \infty]$. To that end, we decompose the whole interval $[0, \infty)$ into the subintervals $[t_n^*, t_{n+1}^*)$ defined in (4.7), and we add the expected local time over each subinterval.

The sequence $\{\varphi(t_n^*) : t_n^* < \infty\}$ is a Markov chain on the state space $\mathcal{E}$, with transition probabilities

$$P_{ij}^* = \mathrm{P}_{0,i}[t_1^* < \infty, \varphi(t_1^*) = j], \qquad i, j \in \mathcal{E}.$$

Assume that $\varphi(0) = i$. By definition of t_1^*, $\ell_j(0, t_1^*)$ is equal to 0 for any $j \neq i$, and $\ell_i(0, t_1^*) = \ell_i(0, t_1')$. As t_1', the sojourn time in phase i, is exponentially distributed with parameter $|Q_{ii}|$, therefore,

$$L_0[0, t_1'] = \Delta_d \tag{4.11}$$

by Lemma 4.2, where

$$d_i = 1/\sqrt{\mu_i^2 + 2|Q_{ii}|\sigma_i^2}, \qquad i \in \mathcal{E} \tag{4.12}$$

and, by the Markov property,

$$L_0[0, \infty] = \Delta_d + P^* L_0[0, \infty]. \tag{4.13}$$

Take the case where $\boldsymbol{\pi}^{\mathrm{T}}\boldsymbol{\mu} > 0$. There is a positive probability that the level at time t_1' is strictly positive, and, by Corollary 2.2, there is a positive probability that the process would never return to level 0, so that P^* is a substochastic matrix. A similar argument holds in case $\boldsymbol{\pi}^{\mathrm{T}}\boldsymbol{\mu} < 0$. When $\boldsymbol{\pi}^{\mathrm{T}}\boldsymbol{\mu} \neq 0$, therefore, Eq. (4.13) becomes

$$\begin{aligned} L_0[0, \infty] &= \sum_{k \geq 0} (P^*)^k \Delta_d \\ &= (I - P^*)^{-1} \Delta_d \\ &< \infty \end{aligned}$$

and the first statement is proved. Note that the (i, j)th component of $(I - P^*)^{-1}$ is the expectation of the total number $\#\{k : \varphi(t_k^*) = j\}$ of returns to level 0 in phase j, given the initial phase i.

If $\boldsymbol{\pi}^{\mathrm{T}}\boldsymbol{\mu} = 0$, by Corollary 2.2 again, the process returns to 0 with probability 1, the matrix P^* is stochastic, the expected number of returns to level 0 is infinite, and $L_0[0, \infty] = \infty$. □

With the next property, we show that $L_0[0, \infty]$ (and therefore $L_x[y, \infty]$ for all x and y) is fully determined as soon as we know the matrices U and $\widehat{U}$.

Theorem 4.4. *If $\boldsymbol{\pi}^{\mathrm{T}}\boldsymbol{\mu} \neq 0$, then*

$$L_0[0, \infty] = -2(U + \widehat{U})^{-1} \Delta_\sigma^{-2}. \tag{4.14}$$

Proof. We closely follow the same argument as in the proof of Theorem 2.8. We condition on the state of the system at time t_1', when there is the first change of phase, and we obtain an equation which may be compared to (2.22):

$$L_0[0,\infty] = \Delta_d + \int_0^\infty \int_0^\infty V_{ab}(y,v)\widetilde{P} L_{y-v}[0,\infty]\,\mathrm{d}v\,\mathrm{d}y$$

with Δ_d defined in (4.11), V_{ab} defined in (2.17) and $\widetilde{P}$ in (1.30). The first term on the right is the expected local time at 0 over the interval $(0, t_1')$, in the second term we have a succession of factors,

- with density $V_{ab}(y, v)$, the process goes down to a minimum level equal to $-v$ during the interval $(0, t_1')$ and it is at level $y - v$ at time t_1',
- at that time, a new phase is chosen with the transition matrix $\widetilde{P}$,
- the third factor is the remainder of the total expected local time at zero, starting from the new level $y - v$.

We replace the second integral by the sum of two integrals, one from 0 to v and one from v to ∞, so as to use (4.10) and, after tedious but simple manipulations, we obtain that

$$\begin{aligned} L_0[0,\infty] = \Delta_d + \Big\{ &\int_0^\infty e^{-(\Delta_a+\Delta_b)y}\,\mathrm{d}y\, \Delta_a \Delta_b \int_0^\infty e^{-\Delta_b z} \widetilde{P} e^{\widehat{U} z}\,\mathrm{d}z \\ &+ \int_0^\infty e^{-(\Delta_a+\Delta_b)v}\,\mathrm{d}v\, \Delta_a \Delta_b \int_0^\infty e^{-\Delta_a z} \widetilde{P} e^{U z}\,\mathrm{d}z \Big\} L_0[0,\infty] \end{aligned}$$

with $\boldsymbol{a}$ and $\boldsymbol{b}$ defined in (2.18). Using (2.19, 2.20), this becomes

$$0 = \Delta_d + (\Delta_a + \Delta_b)^{-1}(U + \widehat{U}) L_0[0,\infty].$$

It follows from Lemma 2.4 that $U+\widehat{U}$ is a nonsingular matrix when $\boldsymbol{\pi}^{\mathrm{T}}\boldsymbol{\mu} \neq 0$: Indeed, it is irreducible, its off-diagonal elements are nonnegative, and $(U+\widehat{U})\mathbf{1} < \mathbf{0}$. Finally, we recall that $\Delta_a + \Delta_b = 2\Delta_d^{-1}\Delta_\sigma^{-2}$ and (4.14) is proved. □

From Theorem 4.3, we see that the asymptotic behaviour of $L_0[x, \infty]$, as x tends to infinity, is governed by algebraic properties of U and $\widehat{U}$ obtained in Sect. 2.2. In case $\boldsymbol{\pi}^{\mathrm{T}}\boldsymbol{\mu} > 0$, we know from Lemma 2.4 that $e^{\widehat{U}x}$ is a stochastic matrix, so that

$$\lim_{x\to\infty} L_0[x,\infty] = \mathbf{1} \cdot \widehat{\boldsymbol{u}}^{\mathrm{T}} L_0[0,\infty],$$

where $\widehat{\boldsymbol{u}}$ is the stationary probability vector associated to $\widehat{U}$: $\widehat{\boldsymbol{u}}^{\mathrm{T}}\widehat{U} = \mathbf{0}$, $\widehat{\boldsymbol{u}}^{\mathrm{T}}\mathbf{1} = 1$. The matrix U is a defective generator and, by Theorem 2.11, its eigenvalues have all strictly negative real parts, with λ_m having multiplicity 1 and being closest to 0. Therefore,

$$\lim_{x\to\infty} (e^{-\lambda_m x} L_0[-x,\infty]) = \boldsymbol{v} \cdot \boldsymbol{u}^{\mathrm{T}} L_0[0,\infty],$$

where $\boldsymbol{v}$ and $\boldsymbol{u}$ are the eigenvectors of U associated to λ_m: $\boldsymbol{u}^{\mathrm{T}} U = \lambda_m \boldsymbol{u}^{\mathrm{T}}$, $U\boldsymbol{v} = \lambda_m \boldsymbol{v}$, $\boldsymbol{u}^{\mathrm{T}}\boldsymbol{v} = 1$.

If $\boldsymbol{\pi}^{\mathrm{T}}\boldsymbol{\mu} < 0$, then e^{Ux} is stochastic, $e^{\widehat{U}x}$ is substochastic, and

$$\lim_{x\to\infty} L_0[-x,\infty] = \mathbf{1}\cdot \boldsymbol{u}^{\mathrm{T}} L_0[0,\infty],$$

where $\boldsymbol{u}$ is the stationary probability vector associated to U, while

$$\lim_{x\to\infty} (e^{\lambda_{m+1}x} L_0[x,\infty]) = \widehat{\boldsymbol{v}}\cdot \widehat{\boldsymbol{u}}^{\mathrm{T}} L_0[0,\infty],$$

where $\widehat{\boldsymbol{u}}^{\mathrm{T}}\widehat{U} = -\lambda_{m+1}\widehat{\boldsymbol{u}}^{\mathrm{T}}$, $\widehat{U}\widehat{\boldsymbol{v}} = -\lambda_{m+1}\widehat{\boldsymbol{v}}$, $\widehat{\boldsymbol{u}}^{\mathrm{T}}\widehat{\boldsymbol{v}} = 1$.

We are also interested in $L_0[0,\tau_x]$, the expected local time at level 0 until the process crosses level x for the first time. We determine in Theorem 4.22 a nice expression that holds for any value of the stationary drift. We shall, however, have to wait until after we have further developed the material in this chapter. Meanwhile, we give in Corollary 4.5 below an expression that is easily determined when $\boldsymbol{\pi}^{\mathrm{T}}\boldsymbol{\mu} \neq 0$.

Corollary 4.5. *The matrix $L_0[0,\tau_x]$ of expected local time at level 0, before the first passage to level x, is nonsingular.*

If $\boldsymbol{\pi}^{\mathrm{T}}\boldsymbol{\mu} \neq 0$, then

$$L_0[0,\tau_x] = (I - e^{\widehat{U}x}e^{Ux})L_0[0,\infty] \tag{4.15}$$

and

$$L_0[0,\tau_{-x}] = (I - e^{Ux}e^{\widehat{U}x})L_0[0,\infty] \tag{4.16}$$

for $x \geq 0$.

Proof. The matrix $e^{\widehat{U}x}e^{Ux}$ is the phase transition matrix at the epochs when the level returns to 0 *after* reaching up to level x. Thus, if $L_0[0,\infty]$ is finite, we may write that

$$L_0[0,\infty] = L_0[0,\tau_x] + e^{\widehat{U}x}e^{Ux}L_0[0,\infty],$$

for $x > 0$, and this proves (4.15). Equation (4.16) is justified by a similar argument.

If $\boldsymbol{\pi}^{\mathrm{T}}\boldsymbol{\mu} \neq 0$, we have seen in the proof of Theorem 3.2 that $I - e^{Ux}e^{\widehat{U}x}$ and $I - e^{\widehat{U}x}e^{Ux}$ are nonsingular, and $L_0[0,\infty]$ is nonsingular by Theorem 4.4. Therefore, Eqs. (4.15, 4.16) show that $L_0[0,\tau_x]$ is nonsingular for all x.

If $\boldsymbol{\pi}^{\mathrm{T}}\boldsymbol{\mu} = 0$, we need a different argument to prove that $L_0[0,\tau_x]$ is nonsingular, an argument in which we use the epochs $\{t'_n\}$ and $\{t^*_n\}$ defined in (4.6, 4.7). Define $\widetilde{\boldsymbol{\ell}}$ to be the vector of expected local time at level 0, given the initial phase, until the process either returns to level 0 at time t^*_1 or crosses level x. That is,

$$\widetilde{\ell}_i = \mathrm{E}_{0,i}[\ell_i(0, t^*_1 \wedge \tau_x)], \qquad i \in \mathcal{E}.$$

Also, define the probability $\widetilde{K}_{ij}$ of transition from $(0,i)$ at time 0 to $(0,j)$ at time t^*_1, provided this happens before the first visit to level x:

$$\widetilde{K}_{ij} = \mathrm{P}_{0,i}[t^*_1 < \tau_x, \varphi(t^*_1) = j], \qquad i,j \in \mathcal{E}.$$

As there is a positive probability that $\tau_x < t_1^*$, the matrix $\widetilde{K}$ is substochastic for any value of $\boldsymbol{\pi}^{\mathrm{T}}\boldsymbol{\mu}$ and, if we decompose the interval $(0, \tau_x)$ into successive returns to level 0 at times $t_1^*, t_2^*, \dots$, we find that

$$L_0[0, \tau_x] = \sum_{\nu \geq 0} \widetilde{K}^\nu \Delta_{\widetilde{\ell}} = (I - \widetilde{K})^{-1} \Delta_{\widetilde{\ell}}.$$

The MMBM does not visit level 0 during the interval (t_1', t_1^*), so that

$$\widetilde{\ell}_i = \mathrm{E}_{0,i}[\ell_i(0, t_1' \wedge \tau_x)].$$

There is a positive probability that $t_1' < \tau_x$, therefore $\widetilde{\ell}_i > 0$ by Lemma 4.2, and the matrix $\Delta_{\widetilde{\ell}}$ is nonsingular. This concludes the proof that $L_0[0, \tau_x]$ is nonsingular in all cases. □

Remark 4.6. Ivanovs and Palmowski [90, Theorem 1] give for $\widehat{\Psi}^{(a,b)}$ an expression that is similar to, yet different from (3.6), we write it as $\widehat{\Psi}^{(a,b)} = W_a^*(W_{a+b}^*)^{-1}$. The matrix W_x^* is called a *scale matrix* in [90] and is formally characterised by its Laplace transform

$$\int_0^\infty e^{-sx} W_x^* \, \mathrm{d}x = \mathcal{F}(s)^{-1},$$

where $\mathcal{F}(s)$ is the matrix Laplace exponent of the MMBM, defined in (1.29).

If $\boldsymbol{\pi}^{\mathrm{T}}\boldsymbol{\mu} \neq 0$, the scale matrix is related to our matrix $\widehat{W}_x$ through the relation

$$W_x^* = \widehat{W}_x \, L_0[0, \infty].$$

This follows from the expression (4.15) for $L_0[0, \tau_x]$ and the definition (3.7) of $\widehat{W}_x$, and from [90, Eq. 2] which states that $W_x^* = e^{-\widehat{U}x} L_0[0, \tau_x]$.

4.2 Time Away from 0

We analyse in this section the amount of time spent away from level 0 between two successive visits to level 0. This sentence ought to be qualified since, as we have remarked in Sect. 1.4, a Brownian motion which starts from level 0 instantaneously returns there, and does so infinitely often in any interval $(0, \varepsilon)$ with $\varepsilon > 0$. Still, this quantity is well defined for QBDs and Fluid flows, it has proved to be very useful in these contexts, and we wish to define for MMBMs a quantity that is similar, if not exactly the same. The parenthesis below is meant to give some intuitive understanding for the rationale behind the definition given in (4.23).

Parenthesis 4.7. Discrete State Space—QBDs ________________

Quasi-birth-and-death processes (QBDs) are two-dimensional processes similar to MMBMs, the main difference being that the level component takes integer values, and increases or decreases by 1 only. QBDs may be seen as Markov additive processes with increments limited to $\{-1, 0, 1\}$, and they have been much studied since the end of the 1960s as Markov chains on the state space $\mathbb{Z} \times \mathcal{E}$. For an introduction, we refer to Latouche and Ramaswami [106], an earlier reference is Neuts [116].

The transition matrix is block tridiagonal:

$$Q = \begin{bmatrix} \ddots & \ddots & & & \\ \ddots & A_0 & A_1 & & \\ & A_{-1} & A_0 & A_1 & \\ & & A_{-1} & A_0 & \ddots \\ & & & \ddots & \ddots \end{bmatrix},$$

where A_{-1}, A_0 and A_1 are $m \times m$ matrices, and their entries are transition rates.

For QBDs, the local time $\ell_j(x, t)$ is defined as

$$\ell_j(x, t) = \int_0^t \mathbb{1}\{X(s) = x, \varphi(s) = j\} \, \mathrm{d}s, \tag{4.17}$$

for integer values of x; in other words, $\ell_j(x, t)$ is the sum of the intervals of time spent by $\{X, \varphi\}$ in state (x, j) up to time t.

Two matrices play a major role similar to e^U and $e^{\widehat{U}}$ in the analysis of first passage times across levels. These are denoted as G and $\widehat{G}$ defined as

$$(G)_{ij} = \mathrm{P}_{0,i}[\tau_{-1} < \infty, \varphi(\tau_{-1}) = j] \qquad \text{and} \qquad (\widehat{G})_{ij} = \mathrm{P}_{0,i}[\tau_1 < \infty, \varphi(\tau_1) = j],$$

for i, j in $\mathcal{E}$. In continuous time, they are respectively the minimal nonnegative solutions of the matrix-quadratic equations

$$A_{-1} + A_0 X + A_1 X^2 = 0 \tag{4.18}$$

for G, and

$$A_1 + A_0 X + A_{-1} X^2 = 0 \tag{4.19}$$

for $\widehat{G}$, equivalent in the context of QBDs to the equations (2.23, 2.24) for U and $\widehat{U}$.

There are two other matrices of note, usually denoted as R and $\widehat{R}$, which are the minimal nonnegative solutions of

$$A_1 + X A_0 + X^2 A_{-1} = 0 \tag{4.20}$$

for R and

$$A_{-1} + X A_0 + X^2 A_1 = 0 \tag{4.21}$$

for $\widehat{R}$. Observe that G and $\widehat{R}$ are right- and left-matrix solutions of the matrix-quadratic equation $A_{-1} + A_0 z + A_1 z^2 = 0$, while $\widehat{G}$ and R are right- and left-matrix solutions of $A_{-1} z^2 + A_0 z + A_1 = 0$.

Define

$$\tau' = \inf\{t > 0 : (X(t), \varphi(t)) \neq (X(0), \varphi(0))\}$$

to be the first jump time of the QBD and

$$\tau_0^* = \inf\{t \geq \tau' : X(t) = X(0)\}$$

to be the first jump time to the initial level. The last remark in Neuts [116, Section 1.7] gives the physical interpretation

$$(R^k)_{ij} = \frac{\mathrm{E}_{0,i}[\ell_j(k, \tau_0^*)]}{\mathrm{E}_{0,i}[\ell_i(0, \tau_0^*)]}, \tag{4.22}$$

for $k \geq 0$, and $i, j \in \mathcal{E}$. That is, $(R^k)_{ij}$ is the expected sojourn time in (k, j) per unit of time spent in $(0, i)$ during the interval $(0, \tau_0^*)$. In a similar manner,

$$(\widehat{R}^k)_{ij} = \frac{\mathrm{E}_{0,i}[\ell_j(-k, \tau_0^*)]}{\mathrm{E}_{0,i}[\ell_i(0, \tau_0^*)]},$$

for $k \geq 0$.

△

For MMBMs, we define the matrix function $R(x)$ with entries

$$R_{ij}(x) = \frac{(L_0[x, t_1^*])_{ij}}{(L_0[0, t_1^*])_{ii}}, \tag{4.23}$$

for $x > 0$, and i, j in $\mathcal{E}$, where t_1^* defined in (4.7) is the first return to level 0 after the first jump time t_1' of the phase Markov chain.

Both t_1' and t_1^* are strictly positive. Also, there is no visit to level 0 in the time interval (t_1', t_1^*), and $(L_0[0, t_1^*])_{ii} = (L_0[0, t_1'])_{ii}$ which is strictly positive by (4.11). Therefore, $R_{ij}(x)$ is well defined. A difference between (4.23) and (4.22) is that the QBD does not leave the initial level during the interval $(0, \tau')$, while the MMBM, being driven by Brownian motion, oscillates around the initial level during $(0, t_1')$.

We show in this section that $R(x)$ has properties similar to the matrix R^k of QBDs. First, we establish in Theorem 4.8 a connection between $R(x)$, the local time at level 0 and the local time at level x beyond time t_1^*. Next, we show in Theorem 4.9 that there exists a matrix R such that $R(x) = e^{Rx}$. Finally, we prove in Theorem 4.14 that R is a solution of a matrix-quadratic equation similar to $\widehat{U}$. Needless to say, an equivalent set of properties hold for the matrix $\widehat{R}(x)$, with

$$\widehat{R}_{ij}(x) = \frac{(L_0[-x, t_1^*])_{ij}}{(L_0[0, t_1^*])_{ii}}, \qquad \text{for } x > 0. \tag{4.24}$$

Consider $x > 0$, arbitrary but fixed. Define two interleaved sequences $\{\eta_k : k \geq 1\}$ and $\{\zeta_k : k \geq 1\}$ of passage times to level 0 and level x, respectively:

$$\eta_1 = \inf\{t > \tau_x : X(t) = 0\} \tag{4.25}$$

is the first epoch when the process moves to level 0 *after* a visit to level x, and in general

$$\zeta_k = \inf\{t > \eta_k : X(t) = x\} \tag{4.26}$$

$$\eta_{k+1} = \inf\{t > \zeta_k : X(t) = 0\} \tag{4.27}$$

for $k \geq 1$. We make the usual assumption that $\zeta_k = \infty$ or $\eta_k = \infty$ if there is no time t at which their definition is satisfied.[1]

It would be more precise to denote these epochs as the functions $\eta_k(x)$ and $\zeta_k(x)$, but our slight abuse of notation simplifies our presentation. We briefly discuss this after Remark 4.18.

Theorem 4.8. *For $x > 0$, arbitrary but fixed, we have*

$$L_0[x, \eta_k] = e^{\widehat{U}x} L_x[x, \eta_k] = L_0[0, \eta_k]\, R(x), \tag{4.28}$$

for $k \geq 1$, where η_k defined in (4.27) is the kth return to 0 following a visit to x.

Furthermore, if $\boldsymbol{\pi}^{\mathrm{T}}\boldsymbol{\mu} \neq 0$, then

$$L_0[x, \infty] = e^{\widehat{U}x} L_0[0, \infty] = L_0[0, \infty] R(x). \tag{4.29}$$

Proof. If $X(0) = 0$, the local time at level x is equal to 0 until the first visit there. Therefore, it is obvious that $L_0[x, \eta_k] = e^{\widehat{U}x} L_x[x, \eta_k]$, and we may focus our attention on proving that

$$L_0[x, \eta_k] = L_0[0, \eta_k] R(x).$$

We begin by proving this for $k = 1$. We use the sequence $\{t_n^*\}$ introduced in (4.7), of all epochs where the process returns to level 0 after a change of phase, and we define the Markov chain $\{S_n : n \geq 0\}$ on $\mathcal{E} \cup \{\dagger\}$, where $\dagger$ is an absorbing state, with $S_n = \varphi(t_n^*)$ if t_n^* is finite and less than η_1, and $S_n = \dagger$ otherwise. The transition probability matrix among states in $\mathcal{E}$ is denoted as $\widetilde{K}$.

By the definition (4.25) of η_1, we know that $t_n^* < \eta_1$ iff $t_n^* < \tau_x$, and we note for future reference that $\widetilde{K}$ may be expressed as follows:

$$\widetilde{K}_{i,j} = \mathrm{P}_i[t_1^* < \tau_x, \varphi(t_1^*) = j], \qquad \text{for } i, j \text{ in } \mathcal{E}. \tag{4.30}$$

If $\boldsymbol{\pi}^{\mathrm{T}}\boldsymbol{\mu} \neq 0$, then $\{X\}$ is transient, and the total number of returns to 0 is finite, if $\boldsymbol{\pi}^{\mathrm{T}}\boldsymbol{\mu} = 0$, then $\{X\}$ is null recurrent and η_1 is finite almost surely; in both cases,

---

[1] It is customary to define such sequences with $k = 0$ as starting points but that is not convenient here.

the index $N^* = \sup\{n : S_n \in \mathcal{E}\}$ is finite with probability 1, and $\widetilde{K}$ is a substochastic matrix. We condition on the first return to level 0 and write that

$$\mathcal{L}_0[x, \eta_1]_{ij} = \mathcal{L}_0[x, t_1^*]_{ij} + \sum_{u \in \mathcal{E}} \widetilde{K}_{iu} \mathcal{L}_0[x, \eta_1]_{uj} \tag{4.31}$$

$$= \sum_{u \in \mathcal{E}} (I - \widetilde{K})^{-1}_{iu} \mathcal{L}_0[x, t_1^*]_{uj}. \tag{4.32}$$

By (4.23), this becomes

$$\mathcal{L}_0[x, \eta_1]_{ij} = \sum_{u \in \mathcal{E}} (I - \widetilde{K})^{-1}_{iu} \mathcal{L}_0[0, t_1^*]_{uu} R_{uj}(x).$$

Furthermore, if we set $x = 0$ in (4.32), we find that

$$\mathcal{L}_0[0, \eta_1]_{ij} = (I - \widetilde{K})^{-1}_{ij} \mathcal{L}_0[0, t_1^*]_{jj}$$

since $\mathcal{L}_0[0, t_1^*]_{uj} = 0$ for $u \neq j$. Therefore,

$$\mathcal{L}_0[x, \eta_1]_{ij} = \sum_{u \in \mathcal{E}} \mathcal{L}_0[0, \eta_1]_{iu} R_{uj}(x), \tag{4.33}$$

which concludes the proof of (4.28) for $k = 1$. For $k \geq 1$, we condition on the phase at time η_1:

$$\begin{aligned}
\mathcal{L}_0[x, \eta_k] &= \mathcal{L}_0[x, \eta_1] + e^{\widehat{U}x} e^{Ux} \mathcal{L}_0[x, \eta_{k-1}] \\
&= \sum_{0 \leq n \leq k-1} (e^{\widehat{U}x} e^{Ux})^n \; \mathcal{L}_0[x, \eta_1] \\
&= \sum_{0 \leq n \leq k-1} (e^{\widehat{U}x} e^{Ux})^n \; \mathcal{L}_0[0, \eta_1] R(x) \\
&= \mathcal{L}_0[0, \eta_k] R(x).
\end{aligned} \tag{4.34}$$

Finally, to extend (4.28) over the whole history of the process, we take the limit as k tends to ∞. If $\boldsymbol{\pi}^{\mathrm{T}} \boldsymbol{\mu} \neq 0$, the matrix $e^{\widehat{U}x} e^{Ux}$ is substochastic, and the series $\sum_{n \geq 0} (e^{\widehat{U}x} e^{Ux})^n$ converges. This completes our proof. □

Once more, it appears that things are more involved when $\boldsymbol{\pi}^{\mathrm{T}} \boldsymbol{\mu} = 0$, and we shall examine in Sect. 4.3 the form that Eq. (4.29) takes in that case.

Theorem 4.9. *There exists a matrix R such that*

$$R(x) = e^{Rx}. \tag{4.35}$$

Proof. To begin with, assume that $\{X\}$ is transient, that is, $\boldsymbol{\pi}^{\mathrm{T}} \boldsymbol{\mu} \neq 0$. By (4.29), we have

$$\mathcal{L}_0[0, \infty] R(x + y) = e^{\widehat{U}(x+y)} \mathcal{L}_0[0, \infty] = e^{\widehat{U}x} e^{\widehat{U}y} \mathcal{L}_0[0, \infty]. \tag{4.36}$$

We saw with Theorem 4.4 that $L_0[0, \infty]$ is nonsingular, therefore

$$e^{\widehat{U}x} = L_0[0, \infty]R(x)L_0[0, \infty]^{-1},$$

from which (4.36) becomes

$$L_0[0, \infty]R(x + y) = L_0[0, \infty]R(x)R(y),$$

and so

$$R(x + y) = R(x)R(y). \tag{4.37}$$

As $R(0) = I$, this proves (4.35).

If $\boldsymbol{\pi}^{\mathrm{T}}\boldsymbol{\mu} = 0$, then $L_0[0, \infty]$ is not finite, and we need another argument. We follow the procedure suggested in Sect. 1.8, and we introduce an artificial killing mechanism with rate α. Instead of (4.23), we define

$$(R^{(\alpha)}(x))_{ij} = \frac{(L_0[x, t_1^* \wedge e_\alpha])_{ij}}{(L_0[0, t_1^* \wedge e_\alpha])_{ii}},$$

where e_α is a random variable, exponentially distributed, with parameter α, independent of the MMBM. We repeat the argument that leads to (4.29) and to (4.37), to conclude that

$$R^{(\alpha)}(x + y) = R^{(\alpha)}(x)R^{(\alpha)}(y),$$

and the semi-group property (4.37) follows in the limit as α tends to 0, by continuity of $R^{(\alpha)}(x)$. □

Remark 4.10. In short, for $x > 0$, e^{Rx} gives the expected local time at level x per unit of local time at level 0. This is the interpretation we were looking for at the end of parenthesis 4.7, it will be useful in Chap. 5. In the same manner, $e^{\widehat{R}x}$ gives the expected local time at level $-x$ per unit of local time at level 0.

Remark 4.11. If we get back to the definition (4.23) of $R(x)$, we may rewrite the semi-group property (4.37) as

$$L_0[x + y, t_1^*]_{ij} = \sum_{u \in \mathcal{E}} L_0[x, t_1^*]_{iu} R_{uj}(y), \tag{4.38}$$

and this has a very plausible physical interpretation: Starting from level 0, it is obviously necessary for $\{X\}$ to move through level x before it reaches level $x + y$, and the meaning we give to (4.38) is that the expected local time in $(x + y, j)$ is the sum of the expected time spent in the various phases at level x, weighted by the expected time spent y units higher per unit of local time at level x.

It is tempting to use this interpretation as a starting point for the proof of Theorem 4.9 and to repeat the approach followed in [106, Theorem 6.2.7]. Unfortunately, the formal argument in [106] is designed for QBDs, and it relies on the set of negative levels being taboo. Here, on the contrary, the MMBM is allowed to visit both positive and negative levels, without restriction, during the initial interval of time $(0, t_1')$.

Remark 4.12. In view of (4.35), we may rewrite (4.29) as

$$e^{\widehat{U}x} L_0[0,\infty] = L_0[0,\infty] e^{Rx}, \tag{4.39}$$

if $\boldsymbol{\pi}^{\mathrm{T}}\boldsymbol{\mu} \neq 0$. Taking the derivative with respect to x and evaluating it at $x = 0$ leads to

$$\widehat{U} L_0[0,\infty] = L_0[0,\infty] R. \tag{4.40}$$

Obviously, similar equations hold, to connect U and $\widehat{R}$:

$$e^{Ux} L_0[0,\infty] = L_0[0,\infty] e^{\widehat{R}x} \tag{4.41}$$

and

$$U L_0[0,\infty] = L_0[0,\infty] \widehat{R}. \tag{4.42}$$

We shall see in the next section what these become when $\boldsymbol{\pi}^{\mathrm{T}}\boldsymbol{\mu} = 0$.

Example 4.13. We are interested in the coefficients $L_x[x, \eta_n]$ of $e^{\widehat{U}x}$ and $L_0[0, \eta_n]$ of e^{Rx} in (4.28), in case $\boldsymbol{\pi}^{\mathrm{T}}\boldsymbol{\mu} \neq 0$. As we have seen, they both converge to the same matrix $L_0[0,\infty]$ as $n \to \infty$, that is, they converge to the expected local time at level 0 over the whole history of the process.

We know from the argument that leads to (4.34) that

$$L_0[0, \eta_n] = \sum_{0 \le \nu \le n-1} (e^{\widehat{U}x} e^{Ux})^{\nu} L_0[0, \eta_1].$$

Also, $L_0[0, \eta_1] = L_0[0, \tau_x]$ since η_1 is the first return to 0 after τ_x. Finally, with the expression for $L_0[0, \tau_x]$ in Corollary 4.5, we obtain

$$L_0[0, \eta_n] = (I - (e^{\widehat{U}x} e^{Ux})^n) L_0[0, \infty],$$

and, by a similar argument,

$$L_x[x, \eta_n] = (I - (e^{Ux} e^{\widehat{U}x})^n) L_0[0, \infty].$$

The difference between the coefficients of $e^{\widehat{U}x}$ and e^{Rx} in (4.28) is

$$L_x[x, \eta_n] - L_0[0, \eta_n] = ((e^{\widehat{U}x} e^{Ux})^n - (e^{Ux} e^{\widehat{U}x})^n) L_0[0, \infty],$$

and its convergence to 0 is asymptotically geometric, with rate equal to the spectral radius of the product $e^{\widehat{U}x} e^{Ux}$. In this example, we proceed to a numerical evaluation of the difference for small values of n.

We use the same noisy M/E$_k$/1 queue as in Example 3.5 and we plot in Fig. 4.1 the norm $\|L_x[x, \eta_n] - L_0[0, \eta_n]\|_\infty$ as a function of n; note the difference of scale on the horizontal axes. The parameters for Fig. 4.1a on the left are the following: The Erlang service time distribution has $k = 9$ stages, so that the number m of phases is

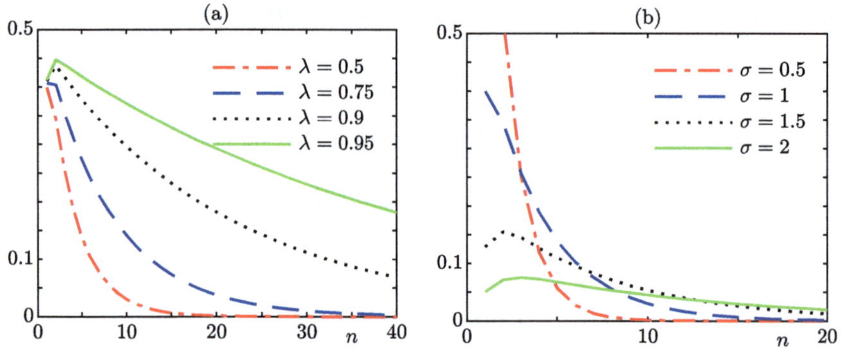

Fig. 4.1 Plot of $\|L_x[x, \eta_n] - L_0[0, \eta_n]\|_\infty$ as a function of n. In (**a**) on the left, the examples differ by the value of λ, in (**b**), they differ by the value of $\boldsymbol{\sigma}$

equal to 10, $\boldsymbol{\sigma} = \left[1, \ldots, 1\right]$, $x = 0.5$. The arrival rate λ of new customers takes the four values in the table below; we give in the second row the corresponding values of the stationary drift and in the third row the asymptotic decay rate $\widehat{\rho} = \mathrm{sp}(e^{\widehat{U}x} e^{Ux})$.

λ	0.5	0.75	0.9	0.95
$\boldsymbol{\pi}^{\mathsf{T}}\boldsymbol{\mu}$	−0.3333	−0.1429	−0.0526	−0.0256
$\widehat{\rho}$	0.7361	0.8764	0.9523	0.9765

The shape of the four curves is largely dictated by the decay rate $\widehat{\rho}$: As λ increases and gets closer to 1, the stationary drift gets closer to 0 and the matrix $e^{\widehat{U}}$ gets closer to being stochastic. On the other hand, the matrix e^{U} is stochastic in all cases. Thus, the product $e^{\widehat{U}x} e^{Ux}$ gets closer to being stochastic and this is a simple explanation for $\widehat{\rho}$ to approach 1. The four curves are nearly the same for $n = 1$, and it is not surprising that the difference converges more slowly to 0 for large values of λ.

In Fig. 4.1b on the right, $\boldsymbol{\sigma} = c\left[1, \ldots, 1\right]$, c takes the four values in the table below, and the arrival rate λ is equal to 0.5. The stationary drift is $\boldsymbol{\pi}^{\mathsf{T}}\boldsymbol{\mu} = -0.333$ for all cases.

c	0.5	1	1.5	2
$\widehat{\rho}$	0.4749	0.7361	0.8655	0.9209

We see that the influence of $\boldsymbol{\sigma}$ is two-fold: The decay rate $\widehat{\rho}$ is larger for larger values of c (and the asymptotic decay is slower), and the difference is larger for $n = 1$.

We show with the next theorem that R is a left solution of the matrix equation $\mathcal{F}(\cdot) = 0$. In the proof, we need an equation that will be useful later, and for that reason we display it now:

$$\left.\frac{\partial}{\partial x} L_0[0, \tau_x]\right|_{x\downarrow 0} = \left.\frac{\partial}{\partial x} L_0[0, \tau_{-x}]\right|_{x\downarrow 0} = 2\Delta_{\sigma}^{-2}, \tag{4.43}$$

where we use the shorthand notation $\left.\frac{\partial}{\partial x} f(x)\right|_{x\downarrow 0}$ for $\lim_{x \overset{\geq}{\to} 0} \frac{\partial}{\partial x} f(x)$.

When $\boldsymbol{\pi}^{\mathrm{T}}\boldsymbol{\mu} \neq 0$, (4.43) is an immediate consequence of (4.14): We have by Corollary 4.5 that

$$\frac{\partial}{\partial x} L_0[0, \tau_x] = -(\widehat{U} e^{\widehat{U}x} e^{Ux} + e^{\widehat{U}x} e^{Ux} U) L_0[0, \infty]$$

and

$$\frac{\partial}{\partial x} L_0[0, \tau_{-x}] = -(U e^{Ux} e^{\widehat{U}x} + e^{Ux} e^{\widehat{U}x} \widehat{U}) L_0[0, \infty]$$

for $x > 0$. We obtain (4.43) in the limit as x tends to 0 from above.

When $\boldsymbol{\pi}^{\mathrm{T}}\boldsymbol{\mu} = 0$, we use the killing mechanism and replace $L_0[0, \infty]$ with $L_0[0, e_\alpha]$, where e_α is an exponentially distributed random variable with parameter α, independent of the process. Equation (4.14) is replaced with

$$L_0[0, e_\alpha] = -2(U^{(\alpha)} + \widehat{U}^{(\alpha)})^{-1} \Delta_\sigma^{-2},$$

where $U^{(\alpha)}$ and $\widehat{U}^{(\alpha)}$ are respectively the solutions of (2.23) and (2.24) with $Q - \alpha I$ to replace Q, as was mentioned in Sect. 1.8. Corollary 4.5 is replaced with

$$L_0[0, e_\alpha \wedge \tau_x] = (I - e^{\widehat{U}^{(\alpha)}x} e^{U^{(\alpha)}x}) L_0[0, e_\alpha]$$
$$L_0[0, e_\alpha \wedge \tau_{-x}] = (I - e^{U^{(\alpha)}x} e^{\widehat{U}^{(\alpha)}x}) L_0[0, e_\alpha],$$

and we obtain

$$\left.\frac{\partial}{\partial x} L_0[0, \tau_x \wedge e_\alpha]\right|_{x \downarrow 0} = \left.\frac{\partial}{\partial x} L_0[0, \tau_{-x} \wedge e_\alpha]\right|_{x \downarrow 0} = 2\Delta_\sigma^{-2},$$

independently of α. Taking the limit as $\alpha \to 0$ gives us (4.43).

Theorem 4.14. *The matrix R defined in Theorem 4.9 is given by*

$$R = \Delta_\sigma^2 \, U \, \Delta_\sigma^{-2} + 2\Delta_\mu \, \Delta_\sigma^{-2}. \tag{4.44}$$

It is a solution of the matrix-quadratic equation

$$\frac{1}{2} X^2 \Delta_\sigma^2 - X \Delta_\mu + Q = 0. \tag{4.45}$$

Proof. Once again, we condition on the state of the system at the epoch t_1' when there is a first change of phase. After this setup, the remainder of the proof consists of simple technical manipulations.

With t_1' and t_1^* defined in (4.6, 4.7), it is clear that $L_0[0, t_1^*]_{ii} = L_0[0, t_1']_{ii}$, and so the denominator of (4.23) is equal to d_i by Lemma 4.2. We decompose the numerator as the sum

$$L_0[x, t_1^*]_{ij} = L_0[x, t_1']_{ii} \, \delta_{ij} + L_0[x, (t_1', t_1^*)]_{ij}, \qquad x > 0,$$

where $L_0[x, (t_1', t_1^*)]_{ij}$ is the expected local time in (x, j) during the interval of time (t_1', t_1^*), given that at time 0, the process starts in $(0, i)$.

By Harrison [80, Page 45], $L_0[x, t_1']_{ii} = e^{-a_i x} d_i$, with a_i given in (2.18). In matrix notation, therefore,

$$R(x) = \Delta_d^{-1}\{\Delta_d \, e^{-\Delta_a x} + L_0[x, (t_1', t_1^*)]\}. \tag{4.46}$$

To determine the state of the MMBM at time t_1', we use the Wiener-Hopf factorisation Lemma 2.6 and write that

$$R(x) = \Delta_d^{-1}\{\Delta_d \, e^{-\Delta_a x} + \int_0^\infty \int_0^v V_{ab}(v, y)\widetilde{P} \, L_{v-y}[x, \tau_0] \, dv \, dy\}, \tag{4.47}$$

in the integrals, v is the maximum reached in the time interval $(0, t_1')$, and the integration bounds are such that $X(t_1') = v - y > 0$.

As $R(x) = e^{Rx}$, we shall obtain an explicit expression for R by taking the derivative with respect to x of the right-hand side of (4.47), evaluated at $x = 0$, and to that end, we only need to determine the derivative of $L_{v-y}[x, \tau_0]$.

By the strong Markov property, we have that

$$(L_{v-y}[x, \tau_0])_{ij} = \sum_{u \in \mathcal{E}} \mathrm{P}_{v-y,i}[\tau_x < \tau_0, \varphi(\tau_x) = u] \, (L_x[x, \tau_0])_{uj}, \tag{4.48}$$

for i, j in $\mathcal{E}$. The first factor in the sum is equal to

$$(e^{U(v-y-x)})_{iu} \qquad \text{if } v - y \geq x, \quad \text{or to} \qquad (\widehat{\Psi}^{(x)}(v-y))_{iu} \qquad \text{if } v - y < x.$$

We deal with limits as $x \to 0$ and so x will eventually be smaller than $v - y$ for any given $v - y$. Thus, we only need to remember that

$$L_{v-y}[x, \tau_0] = e^{U(v-y-x)} L_x[x, \tau_0] \qquad \text{for } x < v - y.$$

Obviously, $L_0[0, \tau_0] = 0$ since $\tau_0 = 0$ when $X(0) = 0$, and $L_x[x, \tau_0] = L_0[0, \tau_{-x}]$ by the spatial homogeneity of MMBMs. Therefore,

$$\begin{aligned} \frac{\partial}{\partial x} L_{v-y}[x, \tau_0]\Big|_{x \downarrow 0} &= e^{U(v-y)} \frac{\partial}{\partial x} L_0[0, \tau_{-x}]\Big|_{x \downarrow 0} \\ &= 2e^{U(v-y)} \Delta_\sigma^{-2} \qquad \text{by (4.43).} \end{aligned}$$

Eventually, we find that

$$\begin{aligned} R &= \Delta_d^{-1}\{ - \Delta_d \Delta_a + 2 \int_0^\infty \int_0^v V_{ab}(v, y)\widetilde{P} \, e^{U(v-y)} \Delta_\sigma^{-2} \, dv \, dy\} \\ &= -\Delta_a + 2\Delta_d^{-1} \int_0^\infty \int_0^\infty V_{ab}(z + y, y)\widetilde{P} \, e^{Uz} \Delta_\sigma^{-2} \, dy \, dz \\ &= -\Delta_a + 2\Delta_d^{-1}(\Delta_a + \Delta_b)^{-1} \Delta_a \Delta_b \{ \int_0^\infty e^{-\Delta_a z} \widetilde{P} \, e^{Uz} \, dz\} \Delta_\sigma^{-2} \end{aligned}$$

by the definition (2.17) of V_{ab} and

$$R = -\Delta_a + 2\Delta_d^{-1}(\Delta_a + \Delta_b)^{-1}(U + \Delta_b)\Delta_\sigma^{-2}$$

by the fixed point equation (2.19) for U. The proof of (4.44) follows from (2.13, 2.14, 2.15, 2.18, 4.12) after simple manipulations.

The proof of the second statement is by direct verification: We replace X in (4.45) with its expression (4.44) and find that the matrix-quadratic expression is equal to 0. □

Corollary 4.15. *The matrix $\widehat{R}(x)$ defined in (4.24) is equal to $e^{\widehat{R}x}$ with*

$$\widehat{R} = \Delta_\sigma^2\, \widehat{U}\, \Delta_\sigma^{-2} - 2\Delta_\mu\, \Delta_\sigma^{-2}. \tag{4.49}$$

It is a solution of the matrix-quadratic equation

$$\frac{1}{2}\, X^2\, \Delta_\sigma^2 + X\, \Delta_\mu + Q = 0. \tag{4.50}$$

Proof. By [80, Page 45], $L_0[-x, t_1']_{ii} = e^{-b_i x} d_i$, for $x \geq 0$, with b_i given in (2.18). Instead of (4.47), we write that

$$\widehat{R}(x) = \Delta_d^{-1}\left\{\Delta_d\, e^{-\Delta_b x} + \int_0^\infty \int_0^v V_{ab}(y, v)\widetilde{P}\, L_{y-v}[-x, \tau_0]\, \mathrm{d}v\, \mathrm{d}y\right\} : \tag{4.51}$$

We use here the interpretation of $V_{ab}(y, v)$ such that v is the absolute value of the minimum reached in the interval $(0, t_1')$ and that $X(t_1') = y - v$, and the integration bounds are such that $y - v < 0$. From this point, we repeat the argument of Theorem 4.14. □

The matrices R and $\widehat{R}$ were already mentioned in Eq. (2.38), and Theorem 4.14 together with its Corollary may be seen as the justification of the claim made then. As a matter of fact, a conclusion to be drawn from this section is that there is a whole menagerie of equations to connect the four matrices U, $\widehat{U}$, R and $\widehat{R}$. Some, like (4.40), require that the MMBM should be transient and others, like (4.44), hold also when the process is null recurrent.

Another example of equation that holds for any value of the stationary drift is

$$e^{\widehat{U}x} L_0[0, \tau_{-x}] = L_0[0, \tau_x]e^{Rx}. \tag{4.52}$$

Its justification goes as follows: If $X(0) = x$, then $\tau_x = 0$, and the first passage time η_1 to level 0 after a visit to level x is identical to τ_0, thus,

$$L_0[0, \tau_{-x}] = L_x[x, \tau_0] = L_x[x, \eta_1], \tag{4.53}$$

the first equality was justified in the proof of Theorem 4.14. Moreover, there is no sojourn in level 0 during the interval (τ_x, η_1), so that

$$L_0[0,\tau_x] = L_0[0,\eta_1]. \quad (4.54)$$

It is easy to obtain (4.52) from (4.28) and these two equations. As the coefficients $L_0[0,\tau_{-x}]$ and $L_0[0,\tau_x]$ of $e^{\widehat{U}x}$ and e^{Rx} are different, (4.52) tells us a different story from (4.29).

A last example is

$$\Delta_\sigma^2(U+\widehat{U})\,\widehat{U} = R\,\Delta_\sigma^2(U+\widehat{U}). \quad (4.55)$$

This equation follows from (4.14, 4.40) when $\boldsymbol{\pi}^\top\boldsymbol{\mu} \neq 0$, but it holds without restriction, as one may verify by replacing R with its expression (4.44). More generally, (4.55) implies that

$$\Delta_\sigma^2(U+\widehat{U})\,\widehat{U}^k = R^k\,\Delta_\sigma^2(U+\widehat{U})$$

for all $k \geq 0$ and that $\Delta_\sigma^2(U+\widehat{U})e^{\widehat{U}x} = e^{Rx}\Delta_\sigma^2(U+\widehat{U})$ for all x.

These equations do not have all the same importance, and we shall make no attempt to uncover them all in this book. We do, however, pay a particular attention to the relations (4.39) between $\widehat{U}$ and R, and (4.41) between U and $\widehat{R}$, and we show in the next section how they are transformed when $\boldsymbol{\pi}^\top\boldsymbol{\mu} = 0$.

4.3 $e^{\widehat{U}x}$ and e^{Rx} — Zero Drift

We intend in this section to complete the statement of Theorem 4.8 in case $\boldsymbol{\pi}^\top\boldsymbol{\mu} = 0$. Assume temporarily that $\boldsymbol{\pi}^\top\boldsymbol{\mu} \neq 0$, and recall (4.29) and its equivalent (4.39):

$$e^{\widehat{U}x}L_0[0,\infty] = L_0[0,\infty]e^{Rx}. \quad (4.56)$$

This may be rewritten, using Corollary 4.5, as

$$e^{\widehat{U}x}(I - e^{Ux}e^{\widehat{U}x})^{-1}L_0[0,\tau_{-x}] = (I - e^{\widehat{U}x}e^{Ux})^{-1}L_0[0,\tau_x]e^{Rx}. \quad (4.57)$$

When $\boldsymbol{\pi}^\top\boldsymbol{\mu} = 0$, the problem with (4.56) is that the coefficient $L_0[0,\infty]$ of $e^{\widehat{U}x}$ and e^{Rx} is a matrix with infinite entries. The formulation used in (4.57) is not much better, since $e^{Ux}e^{\widehat{U}x}$ and $e^{\widehat{U}x}e^{Ux}$ are stochastic matrices both, and $I - e^{Ux}e^{\widehat{U}x}$ and $I - e^{\widehat{U}x}e^{Ux}$ are singular. However, we may guess from the last equation that we might replace the two inverse matrices with group inverses. Such is the case, indeed, as we show in Theorem 4.17. In addition, we show how the coefficients of the matrices $e^{\widehat{U}x}$ and e^{Rx} may be interpreted in terms of the expected local time at level 0 over the *whole history* of the MMBM, in a manner not as transparent, of course, as in Eq. (4.56).

The proof of Theorem 4.17 depends on a technical lemma of independent interest, it offers a generalisation of the equality $M_1(I - M_2M_1)^{-1} = (I - M_1M_2)^{-1}M_1$, which is obvious when both $I - M_1M_2$ and $I - M_2M_1$ are nonsingular.

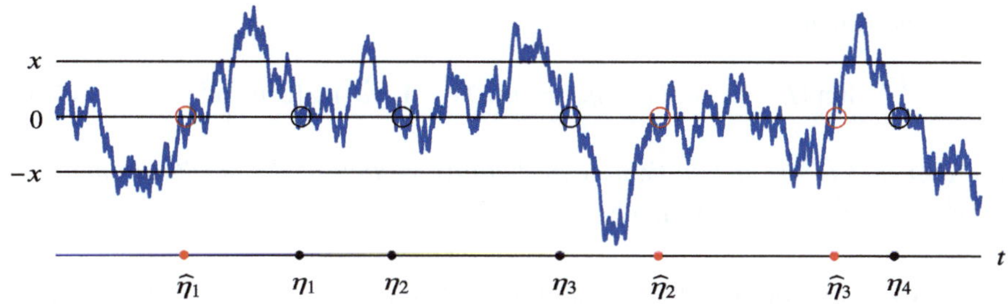

Fig. 4.2 Interleaving of the sequences $\{\eta_n\}$ and $\{\widehat{\eta}_n\}$ defined in (4.27, 4.58)

Lemma 4.16. *Assume that M_1 and M_2 are two stochastic matrices. Assume that the products M_1M_2 and M_2M_1 are both irreducible. We have*

$$M_1(I - M_2M_1)^{\#} = (I - M_1M_2)^{\#} M_1.$$

Proof. Both M_1M_2 and M_2M_1 are stochastic, and we use the relation (1.39) between the group inverse and the fundamental matrix for discrete-time Markov chains.

Define $\boldsymbol{v}_1$ and $\boldsymbol{v}_2$ to be the stationary probability vectors of M_1M_2 and M_2M_1, respectively. We have

$$\mathbf{1} \cdot \boldsymbol{v}_1^{\mathrm{T}} M_1 = \mathbf{1} \cdot \boldsymbol{v}_2^{\mathrm{T}} = M_1 \mathbf{1} \cdot \boldsymbol{v}_2^{\mathrm{T}}$$

and successively

$$\begin{aligned}
(I - M_1M_2 + \mathbf{1} \cdot \boldsymbol{v}_1^{\mathrm{T}})M_1 &= M_1(I - M_2M_1 + \mathbf{1} \cdot \boldsymbol{v}_2^{\mathrm{T}}) \\
M_1(I - M_2M_1 + \mathbf{1} \cdot \boldsymbol{v}_2^{\mathrm{T}})^{-1} &= (I - M_1M_2 + \mathbf{1} \cdot \boldsymbol{v}_1^{\mathrm{T}})^{-1} M_1 \\
M_1 \left\{(I - M_2M_1)^{\#} + \mathbf{1} \cdot \boldsymbol{v}_2^{\mathrm{T}}\right\} &= \left\{(I - M_1M_2)^{\#} + \mathbf{1} \cdot \boldsymbol{v}_1^{\mathrm{T}}\right\} M_1 \qquad \text{by (1.39)},
\end{aligned}$$

and the lemma is proved. □

In addition to the sequence $\{\eta_k\}$ defined in (4.27), of passage times to level 0 separated by visits to level x, we define the sequence $\{\widehat{\eta}_k\}$ of visits to level 0 separated by visits to level $-x$, with $x > 0$:

$$\begin{aligned}
\widehat{\eta}_1 &= \inf\{t > \tau_{-x} : X(t) = 0\}, \\
\zeta_k' &= \inf\{t > \widehat{\eta}_k : X(t) = -x\}, \qquad x > 0, \\
\widehat{\eta}_{k+1} &= \inf\{t > \zeta_k' : X(t) = 0\}.
\end{aligned} \tag{4.58}$$

The two sequences are illustrated in Fig. 4.2.

Theorem 4.17. *If the MMBM is null recurrent, then*

$$e^{\widehat{U}x}(I - e^{Ux}e^{\widehat{U}x})^{\#} L_0[0, \tau_{-x}] = (I - e^{\widehat{U}x}e^{Ux})^{\#} L_0[0, \tau_x]\, e^{Rx}. \tag{4.59}$$

This is equivalent to

$$e^{\widehat{U}x} \lim_{n\to\infty} (I - \mathbf{1} \cdot \omega^{\mathrm{T}}) L_0[0, \widehat{\eta}_n] = \lim_{n\to\infty} (I - \mathbf{1} \cdot \widehat{\omega}^{\mathrm{T}}) L_0[0, \eta_n]\, e^{Rx}, \tag{4.60}$$

where ω and $\widehat{\omega}$ are the stationary probability vectors of the stochastic matrices $e^{Ux} e^{\widehat{U}x}$ and $e^{\widehat{U}x} e^{Ux}$, respectively.

Proof. We start from (4.52), pre-multiply both sides by $(I - e^{\widehat{U}x} e^{Ux})^{\#}$ and apply Lemma 4.16 to the expression on the left, thereby proving (4.59).

Next, we use the Markov-regenerative process embedded at the epochs η_n, we condition on the phase at time η_1, and we repeat an argument used in the proof of (4.34):

$$\begin{aligned} L_0[0, \eta_n] &= L_0[0, \eta_1] + e^{\widehat{U}x} e^{Ux} L_0[0, \eta_{k-1}] \\ &= \sum_{0 \le \nu \le n-1} (e^{\widehat{U}x} e^{Ux})^{\nu} L_0[0, \eta_1]. \end{aligned} \tag{4.61}$$

In a similar manner, the process embedded at the epochs $\widehat{\eta}_n$ when the MMBM returns to level 0 after a visit to level $-x$ gives us that

$$L_0[0, \widehat{\eta}_n] = \sum_{0 \le \nu \le n-1} (e^{Ux} e^{\widehat{U}x})^{\nu} L_0[0, \widehat{\eta}_1]. \tag{4.62}$$

If M is an irreducible stochastic matrix with stationary probability vector $\boldsymbol{m}$, then by (1.38)

$$\begin{aligned} (I - M)^{\#} &= \lim_{n\to\infty} \sum_{0 \le \nu \le n-1} (M^{\nu} - \mathbf{1} \cdot \boldsymbol{m}^{\mathrm{T}}) \\ &= \lim_{n\to\infty} (I - \mathbf{1} \cdot \boldsymbol{m}^{\mathrm{T}}) \sum_{0 \le \nu \le n-1} M^{\nu}, \end{aligned} \tag{4.63}$$

and so we rewrite (4.59) as

$$\begin{aligned} &e^{\widehat{U}x} \lim_{n\to\infty} (I - \mathbf{1} \cdot \omega^{\mathrm{T}}) \sum_{0 \le \nu \le n-1} (e^{Ux} e^{\widehat{U}x})^{\nu} L_0[0, \tau_{-x}] \\ &= \lim_{n\to\infty} (I - \mathbf{1} \cdot \widehat{\omega}^{\mathrm{T}}) \sum_{0 \le \nu \le n-1} (e^{\widehat{U}x} e^{Ux})^{\nu} L_0[0, \tau_x]\, e^{Rx}. \end{aligned}$$

Finally, we observe that $L_0[0, \tau_{-x}] = L_0[0, \widehat{\eta}_1]$ since there is no visit to level 0 during the interval of time $(\tau_{-x}, \widehat{\eta}_1)$. This, together with (4.54, 4.61, 4.62), proves (4.60). □

Remark 4.18. There is some similarity, but much difference, between Eq. (4.39) and Eq. (4.60). To simplify the discussion below, we denote as $C_{\widehat{U}}$ and C_R the coefficients of $e^{\widehat{U}x}$ and of e^{Rx}, respectively.

In both equations, $C_{\widehat{U}}$ and C_R are related to the expected local time at level 0 over the whole history of the MMBM. In (4.39), the limit of $L_0[0,t]$ is taken as t goes to ∞ in any which way. In (4.60), however, t goes to ∞ along two specific sequences: For $C_{\widehat{U}}$, it is the sequence $\{\widehat{\eta}_n\}$ of returns to 0 between visits to $-x$, and for C_R, it is the sequence $\{\eta_n\}$ of returns to 0 between visits to x.

There is, of course, the possibility that $C_{\widehat{U}}$ and C_R might be equal if $\boldsymbol{\pi}^{\mathrm{T}}\boldsymbol{\mu} = 0$, despite their having different representations, but they are different, indeed. As a simple example, take $\mathcal{E} = \{1, 2\}$,

$$\boldsymbol{\mu}^{\mathrm{T}} = \begin{bmatrix}1 & -2\end{bmatrix}, \qquad \boldsymbol{\sigma}^{\mathrm{T}} = \begin{bmatrix}1 & 1\end{bmatrix}, \qquad \text{and} \qquad Q = \begin{bmatrix}-1 & 1 \\ 2 & -2\end{bmatrix},$$

for which it is obvious that $\boldsymbol{\pi}^{\mathrm{T}}\boldsymbol{\mu} = 0$. For $x = 1$,

$$C_{\widehat{U}} = \begin{bmatrix}0.038 & -0.006 \\ -0.548 & 0.088\end{bmatrix} \qquad \text{and} \qquad C_R = \begin{bmatrix}0.038 & -0.274 \\ -0.012 & 0.088\end{bmatrix}.$$

Another difference is that the coefficients in (4.60) are limits of the *difference* between the expected local time at 0 if $\varphi(0)$ is a specific phase, and the expected local time if $\varphi(0)$ is chosen with the stationary distribution of the phase at the epochs of return to level 0, from $-x$ or from x.

We need to remember here that the η_ks and the $\widehat{\eta}_k$s are functions of x. For this reason, we would not get an expression connecting directly the matrices $\widehat{U}$ and R simply by taking the derivative on both sides of (4.60), like we did to obtain (4.40) in the transient case.

To conclude this section, we give, without much proof, the equations to connect e^{Ux} and $e^{\widehat{R}x}$ when $\boldsymbol{\pi}^{\mathrm{T}}\boldsymbol{\mu} = 0$. Briefly stated: we start from

$$e^{Ux} L_0[0, \tau_x] = L_0[0, \tau_{-x}] e^{\widehat{R}x}$$

instead of (4.52), and we adapt the steps in the proof of Theorem 4.17.

Corollary 4.19. *If the MMBM is null recurrent, then*

$$e^{Ux} (I - e^{\widehat{U}x} e^{Ux})^{\#} L_0[0, \tau_x] = (I - e^{Ux} e^{\widehat{U}x})^{\#} L_0[0, \tau_{-x}]\, e^{\widehat{R}x}.$$

This is equivalent to

$$e^{\widehat{U}x} \lim_{n\to\infty} (I - \mathbf{1}\cdot\widehat{\boldsymbol{\omega}}^{\mathrm{T}}) L_0[0, \eta_n] = \lim_{n\to\infty} (I - \mathbf{1}\cdot\boldsymbol{\omega}^{\mathrm{T}}) L_0[0, \widehat{\eta}_n]\, e^{\widehat{R}x},$$

where $\boldsymbol{\omega}$ and $\widehat{\boldsymbol{\omega}}$ are the stationary probability vectors of the stochastic matrices $e^{Ux}e^{\widehat{U}x}$ and $e^{\widehat{U}x}e^{Ux}$, respectively. □

4.4 Wiener-Hopf Factorisation

We proved with Theorem 2.11 and Corollary 2.12 that the matrix D defined in (2.30) has m eigenvalues in $\mathbb{C}_{\le}$ which are the eigenvalues of U, and m eigenvalues in $\mathbb{C}_{\ge}$ which are the eigenvalues of $-\widehat{U}$, and that both U and $\widehat{U}$ have one eigenvalue equal to 0 when $\boldsymbol{\pi}^{\mathsf{T}}\boldsymbol{\mu} = 0$. We bring R and $\widehat{R}$ in the picture and obtain with the next theorem the Wiener-Hopf factorisation of the matrix polynomial $\mathcal{F}(z)$.

Theorem 4.20. *The matrix Laplace exponent (1.29) may be written as*

$$\mathcal{F}(z) = (zI + R)(\frac{1}{2}\Delta_\sigma^2)(zI - U), \tag{4.64}$$

and also as

$$\mathcal{F}(z) = (zI - \widehat{R})(\frac{1}{2}\Delta_\sigma^2)(zI + \widehat{U}). \tag{4.65}$$

Proof. We start from the expression (1.29) for $\mathcal{F}(z)$ and use the fact that U is a right-matrix solution of $\mathcal{F}(U) = 0$. We write that

$$\begin{aligned}\mathcal{F}(z) &= \frac{1}{2}\Delta_\sigma^2(z^2 I + 2\Delta_\sigma^{-2}\Delta_\mu z + 2\Delta_\sigma^{-2}Q)\\ &= \frac{1}{2}\Delta_\sigma^2(zI + U + 2\Delta_\sigma^{-2}\Delta_\mu)(zI - U),\end{aligned}$$

and (4.64) readily follows from (4.44). A similar argument leads to (4.65). □

This theorem gives a nice indication that U and $\widehat{R}$, on the one hand, $\widehat{U}$ and R, on the other hand, share the same sets of eigenvalues:

$$\rho(U) = \rho(\widehat{R}) = \{\lambda_1, \ldots, \lambda_m\} \tag{4.66}$$

$$\rho(\widehat{U}) = \rho(R) = \{-\lambda_{m+1}, \ldots, -\lambda_{2m}\}, \tag{4.67}$$

where $\lambda_1, \ldots, \lambda_{2m}$ are the roots of $a(z) = \det(2\Delta_\sigma^{-2}\mathcal{F}(z))$, see Theorem 2.11. Actually, we already knew from (4.40, 4.42) that U is similar to $\widehat{R}$ and $\widehat{U}$ is similar to R when $\boldsymbol{\pi}^{\mathsf{T}}\boldsymbol{\mu} \ne 0$. With Theorem 4.20, we know that it holds true always.

Observe that the two matrices R and $\widehat{R}$ may be determined independently of U and $\widehat{U}$, as we may directly solve the quadratic equation (4.45) or (4.50). They are characterised with the next theorem, a straightforward adaptation of Lemma 2.4 and Theorems 2.11 and 2.13.

Theorem 4.21. *The matrices R and $\widehat{R}$ are identified as follows among the solutions of (4.50).*

(a) If $\boldsymbol{\pi}^{\mathsf{T}}\boldsymbol{\mu} > 0$, then $-R$ is the unique solution with one eigenvalue equal to 0 and $m-1$ eigenvalues in $\mathbb{C}_{>}$, the matrix $\widehat{R}$ is the unique solution with all eigenvalues in $\mathbb{C}_{<}$.

(b) *If $\boldsymbol{\pi}^{\mathrm{T}}\boldsymbol{\mu} < 0$, then $-R$ is the unique solution with all eigenvalues in $\mathbb{C}_{>}$, the matrix $\widehat{R}$ is the unique solution with one eigenvalue equal to 0 and $m-1$ eigenvalues in $\mathbb{C}_{<}$.*

(c) *If $\boldsymbol{\pi}^{\mathrm{T}}\boldsymbol{\mu} = 0$, then $-R$ is the unique solution associated with the antistable left canonical invariant subspace of $a(z)$, while $\widehat{R}$ is the unique solution associated with the stable left canonical invariant subspace, both have one eigenvalue equal to 0.* □

4.5 Local Time Before Crossings

We have given in Corollary 4.5 two expressions for the expected local time at level 0 before the first ladder epoch τ_x, one for positive and one for negative values of x. This requires that the stationary drift should be different from 0. We do not need that assumption in the next theorem.

Theorem 4.22. *The matrix of expected local time at level 0, before the first passage to a different level, is given by*

$$L_0[0, \tau_x] = \int_0^x e^{\widehat{U}v}(2\Delta_\sigma^{-2})e^{\widehat{R}v}\,\mathrm{d}v, \tag{4.68}$$

and

$$L_0[0, \tau_{-x}] = \int_0^x e^{Uv}(2\Delta_\sigma^{-2})e^{Rv}\,\mathrm{d}v, \tag{4.69}$$

for $x > 0$.

Proof. The local time at level 0 during the interval of time $(0, \tau_x)$ is the sum of two contributions: the local time at level 0 during $(0, \tau_y)$ for some $y < x$ and, renaming level y as the new level 0, the local time at the new level $-y$ during the remaining interval of time (τ_y, τ_x).

With this in mind, we partition the interval $(0, x)$ into some number N of subintervals of size $h = x/N$ and write that

$$L_0[0, \tau_x] = L_0[0, \tau_h] + e^{\widehat{U}h}L_0[-h, \tau_{x-h}]. \tag{4.70}$$

Next, we adapt part of the proof of Theorem 4.8, to conclude that

$$L_0[-y, \tau_x] = L_0[0, \tau_x]e^{\widehat{R}y}, \tag{4.71}$$

for x, $y > 0$: Like in that proof, we use the time sequence $\{t_n^*\}$ defined in (4.7), of visits to level 0 separated by a change of phase, also the transition matrix $\widetilde{K}$ defined in (4.30), of return probabilities to level 0 before level x, and the sequence $\{\eta_k\}$ defined in (4.27), of returns to level 0 separated by visits to level x. As there is no visit to the negative levels during the time interval (τ_x, η_1), therefore $L_0[-y, \tau_x]$ is equal to $L_0[-y, \eta_1]$, and we may repeat the sequence of steps between (4.31) and (4.33), eventually to conclude (4.71). We replace (4.70) with

$$\begin{aligned} L_0[0, \tau_x] &= L_0[0, \tau_h] + e^{\widehat{U}h} L_0[0, \tau_{x-h}] e^{\widehat{R}h} \\ &= \sum_{0 \le \nu \le N-1} e^{\widehat{U}\nu h} L_0[0, \tau_h] e^{\widehat{R}\nu h}. \end{aligned} \tag{4.72}$$

By (4.43), we know that

$$L_0[0, \tau_h] = 2\Delta_\sigma^{-2} h + o(h) \tag{4.73}$$

for $h > 0$ sufficiently small. We take the limit in (4.72) as $N \longrightarrow \infty$ and so prove (4.68). The proof of (4.69) is similar. □

It is easy to justify (4.68) with the interpretation given in Remark 4.10: The integral means that we take into consideration every level v between 0 and x, the factor $e^{\widehat{U}v}$ gives the distribution of the phase upon reaching level v for the first time, $2\Delta_\sigma^{-2}\, \mathrm{d}v$ is the expected local time at level v during the interval $(\tau_v, \tau_v + \mathrm{d}v)$, and $e^{\widehat{R}v}$ is the expected local time at level 0 per unit of local time at level v.

There is also a purely algebraic justification of Theorem 4.22 that we may turn to when $\boldsymbol{\pi}^{\mathrm{T}}\boldsymbol{\mu} \neq 0$: For (4.68) to hold, one needs to show that

$$\frac{\partial}{\partial x} L_0[0, \tau_x] = e^{\widehat{U}x} (2\Delta_\sigma^{-2}) e^{\widehat{R}x}. \tag{4.74}$$

We start from (4.15) and successively obtain that

$$\begin{aligned} \frac{\partial}{\partial x} L_0[0, \tau_x] &= -e^{\widehat{U}x} (\widehat{U} + U) e^{Ux} L_0[0, \infty], \\ &= -e^{\widehat{U}x} (\widehat{U} + U) L_0[0, \infty] e^{\widehat{R}x} \end{aligned}$$

by (4.41), and this becomes (4.74) by Theorem 4.4.

Different characterisations of $L_0[0, \tau_x]$ follow from (4.68) (obviously, we might draw from (4.69) similar characterisations of $L_0[0, \tau_{-x}]$). We may, for instance, premultiply both sides of (4.68) by $\widehat{U}$ and integrate by part, obtaining that $L_0[0, \tau_x]$ is a solution of the Sylvester equation

$$\widehat{U} X + X \widehat{R} = e^{\widehat{U}x} (2\Delta_\sigma^{-2}) e^{\widehat{R}x} - 2\Delta_\sigma^{-2}.$$

If $\boldsymbol{\pi}^{\mathrm{T}}\boldsymbol{\mu} \neq 0$, Theorems 2.11 and 4.21 stipulate that the eigenvalues of $\widehat{U}$ and those of $-\widehat{R}$ are in different subsets of $\mathbb{C}$, therefore, this linear equation is nonsingular by Lemma 2.18, and it may be considered to be a characteristic equation for $L_0[0, \tau_x]$. If $\boldsymbol{\pi}^{\mathrm{T}}\boldsymbol{\mu} = 0$, however, $\widehat{U}$ and $\widehat{R}$ share the eigenvalue 0, and the system is singular.

We may also write (4.68) as

$$L_0[0, \tau_x] = \int_0^x e^{\widehat{U}v}(2\Delta_\sigma^{-2})e^{\widehat{R}(v-x)}\, e^{\widehat{R}x}\, \mathrm{d}v$$

and, following Higham [84, Eq. 10.40], write that equation as

$$L_0[0, \tau_x] = \begin{bmatrix} I & 0 \end{bmatrix} e^{Vx} \begin{bmatrix} 0 \\ I \end{bmatrix} e^{\widehat{R}x}, \tag{4.75}$$

for $x \geq 0$, where

$$V = \begin{bmatrix} \widehat{U} & 2\Delta_\sigma^{-2} \\ 0 & -\widehat{R} \end{bmatrix}.$$

A nice feature of (4.75) is that it holds for all values of $\boldsymbol{\pi}^\mathrm{T}\boldsymbol{\mu}$. Observe, however, that half the eigenvalues of the matrix V are in $\mathbb{C}_\geq$, and, since the exponential function $e^{\lambda x}$ grows very fast with x for positive λs, the matrix function e^{Vx} will automatically be growing fast, and we should expect that (4.75) is not convenient for computational purposes when x takes large values.

Once we have got the expected local time at 0 under taboo of some level x, it is easy to determine the expected local time $L_0[y, \tau_x]$ at any level y different from 0. We assume in Lemma 4.23 that $y < x$, the case $x < y$ easily follows.

Lemma 4.23.
Case 1. If $0 < y < x$, then the expected local time at level y before the first passage to level x is

$$L_0[y, \tau_x] = e^{\widehat{U}y} L_0[0, \tau_{x-y}].$$

Case 2. If $y < 0 < x$, then

$$L_0[y, \tau_x] = \Psi^{(|y|,x)} L_0[0, \tau_{x+|y|}]. \tag{4.76}$$

Case 3. If $y < x < 0$, then $L_0[y, \tau_x] = 0$.

Proof. The third case is the most obvious: Starting from level 0, the process cannot visit level y without crossing level x first.

In Case 1, the process cannot but cross level y first. The distribution of the phase at time τ_y is given by $e^{\widehat{U}y}$, we re-label the levels so that y is the new level 0, and the expected local time in level y during the interval (τ_y, τ_x) is equal to $L_0[0, \tau_{x-y}]$ by the Markov property.

In Case 2, we use the same argument, but now the phase distribution at time τ_y is $\Psi^{(|y|,x)}$ since it might be possible for the MMBM to cross level x before level y. □

Remark 4.24. Incidentally, with $L_0[0, \tau_x]$ being nonsingular by Corollary 4.5, Eq. (4.76) gives us another expression for the exit probability matrix through the upper boundary of a finite interval:

$$\Psi^{(a,b)} = L_0[-a, \tau_b](L_0[0, \tau_{b+a}])^{-1}, \tag{4.77}$$

for a, $b > 0$. Contrary to (3.6), this equation holds for all $\boldsymbol{\pi}^{\mathrm{T}}\boldsymbol{\mu}$. Obviously, we have by a similar argument

$$\widehat{\Psi}^{(a,b)} = L_0[b, \tau_{-a}](L_0[0, \tau_{-(b+a)}])^{-1}. \tag{4.78}$$

For a last illustration of the usefulness of expected local times, we refer to Sect. 3.3 where we analysed the expected time spent in an interval $(-a, b)$ before exiting through either of the two boundaries. In Theorem 3.11, we had to take into account the sign of $\boldsymbol{\pi}^{\mathrm{T}}\boldsymbol{\mu}$; here, we shall use the fact that $L_0[y, \tau_x]$ is finite always.

Lemma 4.25. *Assume that $X(0) = 0$ and that $\boldsymbol{\alpha}$ is the distribution of $\varphi(0)$. Assume that $a \geq 0$, $b \geq 0$, $a + b > 0$.*

The expected sojourn time in the interval $(-a, b)$ before the first exit is

$$\begin{aligned} \mathrm{E}[\tau_{-a} \wedge \tau_b] = \boldsymbol{\alpha}^{\mathrm{T}}\Big\{ &\int_0^a (\Psi^{(y,b)} - \Psi^{(a,b)} e^{\widehat{U}(a-y)}) L_0[0, \tau_{b+y}]\,\mathrm{d}y \\ &+ \int_0^b (e^{\widehat{U}y} - \Psi^{(a,b)} e^{\widehat{U}(a+y)}) L_0[0, \tau_{b-y}]\,\mathrm{d}y \Big\}\mathbf{1}. \end{aligned} \tag{4.79}$$

Proof. Our argument here is somewhat similar to the one we used to prove Theorem 3.11, the main difference is that we specify the local time at *level* x and integrate over x in $(-a, b)$:

$$\mathrm{E}[\tau_{-a} \wedge \tau_b] = \boldsymbol{\alpha}^{\mathrm{T}} \int_{-a}^{b} L_0[x, \tau_{-a} \wedge \tau_b]\mathbf{1}\,\mathrm{d}x. \tag{4.80}$$

The local time in x until $\tau_{-a} \wedge \tau_b$ is the difference of the local time until τ_b and the time during (τ_{-a}, τ_b), in case the process hits the lower boundary first. Thus, we may write the integral in (4.80) as

$$\begin{aligned} \mathcal{L}^{(a,b)} &= \int_{-a}^{b} L_0[x, \tau_{-a} \wedge \tau_b]\,\mathrm{d}x \\ &= \int_{-a}^{b} (L_0[x, \tau_b] - \Psi^{(a,b)} L_0[x + a, \tau_{b+a}])\,\mathrm{d}x. \end{aligned}$$

By Lemma 4.23, this becomes

$$\begin{aligned} \mathcal{L}^{(a,b)} = &\int_{-a}^{0} \Psi^{(|x|,b)} L_0[0, \tau_{b+|x|}]\,\mathrm{d}x + \int_0^b e^{\widehat{U}x} L_0[0, \tau_{b-x}]\,\mathrm{d}x \\ &- \Psi^{(a,b)} \int_0^{b+a} e^{\widehat{U}z} L_0[0, \tau_{b+a-z}]\,\mathrm{d}z. \end{aligned}$$

The right-hand side may be reorganised in different ways, the one chosen here is to separate the interval $(-a, 0)$ from $(0, b)$:

$$\mathcal{L}^{(a,b)} = \int_0^a \Psi^{(y,b)} L_0[0, \tau_{b+y}]\,\mathrm{d}y - \Psi^{(a,b)} \int_0^a e^{\widehat{U}z} L_0[0, \tau_{b+a-z}]\,\mathrm{d}z$$
$$+ \int_0^b e^{\widehat{U}x} L_0[0, \tau_{b-x}]\,\mathrm{d}x - \Psi^{(a,b)} \int_a^{b+a} e^{\widehat{U}z} L_0[0, \tau_{b+a-z}]\,\mathrm{d}z,$$

which completes the proof. □

4.6 Time Reversal

Time reversal is a tool often used in the analysis of Markovian processes (see for instance Asmussen [14, Section II.5], Norris [122, Section 1.9], Resnick [132, Section 5.9]) and has led to duality properties identified in Asmussen and Ramaswami [18].

Consider an MMBM $\{X, \varphi : t \in \mathbb{R}\}$ and assume that the phase process is stationary, so that $\varphi(t)$ has the stationary distribution $\boldsymbol{\pi}$ for any given t. For an arbitrary $T > 0$, we define $\{\widetilde{\varphi}(t) : t \in \mathbb{R}\}$ by

$$\widetilde{\varphi}(s) = \varphi((T - s)-), \tag{4.81}$$

using the left limit at jump times, so that $\{\widetilde{\varphi}\}$ is right continuous. This is a stationary, time-homogeneous Markov chain with generator $\widetilde{Q} = \Delta_\pi^{-1} Q^{\mathrm{T}} \Delta_\pi$.

We also define

$$\widetilde{X}(t) = X(T) - X(T - t).$$

Intervals where $\varphi(t)$ is constant, and equal to i, say, become intervals where $\widetilde{\varphi}(t)$ is constant and equal to i as well, and where $\{\widetilde{X}\}$ behaves like a Brownian motion with parameters μ_i and σ_i.

Thus, $(\widetilde{X}, \widetilde{\varphi})$ is an MMBM, its characteristic Laplace exponent is

$$\widetilde{\mathcal{F}}(s) = \Delta_\pi^{-1} \mathcal{F}(s)^{\mathrm{T}} \Delta_\pi = \frac{1}{2} \Delta_\sigma^2\, s^2 + \Delta_\mu\, s + \Delta_\pi^{-1} Q^{\mathrm{T}} \Delta_\pi, \tag{4.82}$$

and the stationary drift of the new process remains equal to $\boldsymbol{\pi}^{\mathrm{T}} \boldsymbol{\mu}$.

The process $\{\widetilde{X}, \widetilde{\varphi}\}$ is the *time-reversed* version of $\{X, \varphi\}$. The characteristic features of $\{X, \varphi\}$ and $\{\widetilde{X}, \widetilde{\varphi}\}$ are very simply related through the so-called *duality relations*, and we summarise some of these in the next lemma.

Lemma 4.26. *Let U_{tr}, $\widehat{U}_{\mathrm{tr}}$, R_{tr}, $\widehat{R}_{\mathrm{tr}}$ and $L_{\mathrm{tr},0}[0, \infty]$ be, for the time-reversed process with Laplace exponent $\widetilde{F}(s)$ in (4.82), the analogues of the matrices U, $\widehat{U}$, R, $\widehat{R}$ and $L_0[0, \infty]$. The two sets of matrices are related as follows:*

$$U_{\mathrm{tr}} = \Delta_\pi^{-1} \widehat{R}^{\mathrm{T}} \Delta_\pi, \qquad R_{\mathrm{tr}} = \Delta_\pi^{-1} \widehat{U}^{\mathrm{T}} \Delta_\pi, \tag{4.83}$$

$$\widehat{U}_{\mathrm{tr}} = \Delta_\pi^{-1} R^{\mathrm{T}} \Delta_\pi, \qquad \widehat{R}_{\mathrm{tr}} = \Delta_\pi^{-1} U^{\mathrm{T}} \Delta_\pi \tag{4.84}$$

and, if the process is transient,

$$\widetilde{L}_{\mathrm{tr},0}[0,\infty] = \Delta_\pi^{-1}(L_0[0,\infty])^{\mathrm{T}}\Delta_\pi.$$

Proof. The proof of (4.83, 4.84) is by direct verification that U_{tr}, $\widehat{U}_{\mathrm{tr}}$, R_{tr} and $\widehat{R}_{\mathrm{tr}}$ are solutions of their respective matrix-quadratic equations. For instance, we start from (4.45), replace the unknown X with the solution R, take the transpose of both sides, pre-multiply by Δ_π^{-1} and post-multiply by Δ_π:

$$\frac{1}{2}\Delta_\sigma^2\Delta_\pi{}^{-1}(R^{\mathrm{T}})^2\Delta_\pi - \Delta_\mu\Delta_\pi^{-1}R^{\mathrm{T}}\Delta_\pi + \Delta_\pi^{-1}Q^{\mathrm{T}}\Delta_\pi = 0,$$

so that the matrix $Y = \Delta_\pi^{-1}R^{\mathrm{T}}\Delta_\pi$ is a right solution of $\widetilde{F}(\cdot) = 0$. It is similar to R^{T} and shares its eigenvalues. Thus, if we put together the characterising Theorems 2.13 and 4.21, we conclude that Y is the matrix $\widehat{U}_{\mathrm{tr}}$.

To prove the last statement, we apply Theorem 4.4 to the MMBM $\{\widetilde{X}, \widetilde{\varphi}\}$ and replace U_{tr} and $\widehat{U}_{\mathrm{tr}}$ in the expression of $L_0[0,\infty]$ with their expressions in (4.83, 4.84):

$$\begin{aligned} L_{\mathrm{tr},0}[0,\infty] &= -2(U_{\mathrm{tr}} + \widehat{U}_{\mathrm{tr}})^{-1}\Delta_\sigma^{-2} \\ &= -2\{\Delta_\pi^{-1}(\widehat{R}^{\mathrm{T}} + R^{\mathrm{T}})^{-1}\Delta_\pi\}\Delta_\sigma^{-2} \\ &= -2\Delta_\pi^{-1}\Delta_\sigma^{-2}(\widehat{U}^{\mathrm{T}} + U^{\mathrm{T}})^{-1}\Delta_\pi \end{aligned}$$

by the expressions (4.44, 4.49) of R and $\widehat{R}$ as functions of U and $\widehat{U}$,

$$= \Delta_\pi^{-1}(L_0[0,\infty])^{\mathrm{T}}\Delta_\pi,$$

as claimed. □

Remark 4.27. Lemma 4.26 is slightly different from the duality property described in Ramaswami [129] and Ahn and Ramaswami [7]. There, the dual process $\{X_{\mathrm{d}}, \varphi_{\mathrm{d}} : t \in \mathbb{R}\}$ is defined to be the MMBM with parameters $Q_{\mathrm{d}} = \Delta_\pi^{-1}Q^{\mathrm{T}}\Delta_\pi$, $\boldsymbol{\mu}_{\mathrm{d}} = -\boldsymbol{\mu}$ and $\boldsymbol{\sigma}_{\mathrm{d}} = \boldsymbol{\sigma}$. In other words, it is obtained by reversing the order of visit of the phases *and changing the sign* of the drift in each phase. In that case, with U_{d}, R_{d}, ... denoting the matrices analogous to U, R, ..., we find that

$$\begin{aligned} U_{\mathrm{d}} &= \Delta_\pi^{-1}R^{\mathrm{T}}\Delta_\pi, & R_{\mathrm{d}} &= \Delta_\pi^{-1}U^{\mathrm{T}}\Delta_\pi, \\ \widehat{U}_{\mathrm{d}} &= \Delta_\pi^{-1}\widehat{R}^{\mathrm{T}}\Delta_\pi, & \widehat{R}_{\mathrm{d}} &= \Delta_\pi^{-1}\widehat{U}^{\mathrm{T}}\Delta_\pi, \end{aligned}$$

instead of (4.83, 4.84): The generator U of the ladder process to lower levels is exchanged with the intensity matrix R for the time spent at higher levels, and $\widehat{U}$ is exchanged with $\widehat{R}$.

4.7 Notes

There is a fair amount of original material in this chapter: new proofs for some results which are part of the well-established literature, as well as new results. I believe that various explicit expressions, such as those for R and $\widehat{R}$, etc. are new, at least under the form that I give them. The expression (4.14) for the total expected local time $L_0[0, \infty]$ at level 0 is found in Breuer [46, Remark 2].

In Sect. 4.2, the sequence of steps through Theorems 4.8 and 4.9, to the quadratic equations (4.45, 4.50) for R and $\widehat{R}$, is new and asks for some comment. The quadratic equations have been obtained long ago in Asmussen [13]. Asmussen's argument is based on the time-reversal transformation described in Sect. 4.6, and this has been the standard approach followed since then. For a long time I have wondered about arguments based on properties of the MMBM itself, like we have had for QBDs and Fluid queues, that would not require to time-reverse the process. A few years ago, Jevgenijs Ivanovs told me of his interpretation (4.23) for the matrix R, and this was my *eureka* moment. My last step, the proof of Theorem 4.14, was inspired by a conversation with Matthieu Simon. Incidentally, Ivanovs has pointed out that one might use, in the definition (4.23) of $R(x)$, any other stopping time $\widetilde{t} \leq t_1^*$ such that $X_{\widetilde{t}} = 0$ when $\widetilde{t} < \infty$, and $(L_0[0, \widetilde{t}])_{ii} > 0$ for all i. Any such stopping time leads to the same matrix $R(x)$.

It has been customary, traditionally, to avoid dealing with null recurrent MMBMs, many a theorem beginning with the assumption that $\boldsymbol{\pi}^{\mathrm{T}}\boldsymbol{\mu} \neq 0$ or that the generator Q is defective. Sometimes, an artificial killing mechanism has been introduced, with the suggestion to let the killing rate go to 0 and to trust in continuity properties. Notable exceptions are D'Auria et al. [55] and Hajek [79]. In Theorem 4.17, the assumption that $\boldsymbol{\pi}^{\mathrm{T}}\boldsymbol{\mu} = 0$ is met head-on and the analysis deals with the fact that both e^{Ux} and $e^{\widehat{U}x}$ are stochastic matrices. As far as I know, Latouche and Taylor [108, Section 4] is the first mention of that result.

Getting back to Parenthesis 4.7: assume for the time being that the QBD is regulated at 0 and is positive recurrent, we know that the process has a stationary probability distribution of the form $\boldsymbol{v}^{\mathrm{T}} \begin{bmatrix} I & R & R^2 & R^3 & \ldots \end{bmatrix}$ for some vector $\boldsymbol{v}$ of size m. Actually, this role of the matrix R was the initial focus of Matrix-analytic methods in the published literature (Evans [65], Neuts [115], see also Wallace [142]).

The expression *Wiener-Hopf factorisation* is used in different contexts in the literature. Its use in Lemma 2.6 refers to the equation $\lambda/(\lambda + \psi(s)) = \psi_\lambda^+(s)\psi_\lambda^-(s)$, where λ is the parameter of an exponential random variable T, $\psi(s)$ is the characteristic exponent (1.21) of a Lévy process $\{X\}$, $\psi_\lambda^+(s)$ is the characteristic function of the running maximum $M_X(T)$, and $\psi_\lambda^-(s)$ is the characteristic function of $X(T) - M_X(T)$. This equation holds for any Lévy process (Bertoin [32, Page 165]), it is proved in Ivanovs et al. [91, Lemma 7.4] for Markov-modulated M/G/1-type Markov chains. Secondly, Eqs. (2.39) and (4.64) are two examples of Wiener-Hopf factorisation of matrix polynomials: The object there is to split the polynomial roots into two subsets of $\mathbb{C}$, see Adukov et al. [2] for a general formulation. In the present case, the two subsets are $\mathbb{C}_\leq$ and $\mathbb{C}_\geq$, while for discrete-time QBDs the two subsets are the unit disk and its outside (Bini et al. [36, Chapter 3]). Finally, Barlow et al. [22] and Rogers [133] are motivated by the physical significance of the matrices U, $\widehat{U}$, Ψ_{+-}, and Ψ_{-+} to define (2.56) as the Wiener-Hopf factorisation of the matrix $C^{-1}T$. For other uses of the expression, I would direct the reader to the bibliography in Klusik and Palmowski [96].

A last comment, *à propos* the several characterisations of $L_0[0, \tau_x]$: I like the one in Corollary 4.5 best. It is both elegant and easy to use in numerical evaluations, it is just a pity that it fails when the stationary drift $\boldsymbol{\pi}^{\mathsf{T}}\boldsymbol{\mu}$ is equal to zero.

Chapter 5
Regulated Processes

⋄ *Stationary distribution* ⋄ *Markov-regenerative process* ⋄ *One-sided Skorokhod regulation* ⋄ *Occupation time* ⋄ *Caudal characteristic* ⋄ *Two-sided Skorokhod regulation* ⋄ *Change of regime* ⋄ *Regeneration intervals* ⋄ *Sticky boundary* ⋄ *Change of clock* ⋄

Our attention in this chapter turns to MMBMs with one (or two) regulated boundary. This gives us the opportunity to show the importance of the matrix R for the analysis of the stationary distribution when $\boldsymbol{\pi}^{\mathrm{T}}\boldsymbol{\mu} < 0$. In Sect. 5.1, there is one regulated boundary so that the level is forced to remain positive, and in Sect. 5.2, the level is forced to remain in a finite interval $[0, b]$. In Sect. 5.3, level 0 is a reactive boundary such that the controlling phase process changes instantaneously when the level returns to 0. Finally, we give in Sect. 5.4 a model for which level 0 is a sticky boundary.

In application studies, MMBMs serve as models for queueing, production or inventory systems, as well as buffers in high-speed telecommunication networks. Other applications are found in the area of insurance mathematics, to solve dividend problems and determine insurance risks, and in mathematics of finance where they serve as models for option pricing. In most cases, the level may not take just any value in $\mathbb{R}$. Often, it represents a physical quantity which may not become negative, sometimes it is bounded from above as well.

We call such systems *regulated* MMBMs and use the notation $\{Z(t), \varphi(t) : t \geq 0\}$ to distinguish them from their unregulated cousin $\{X(t), \varphi(t) : t \geq 0\}$ defined in Sect. 1.7. We have met regulated MMBMs already in Sect. 3.4 where Z was defined as

$$Z(t) = X(t) - m_X(t) \tag{5.1}$$

for $t \geq 0$, and where we recognised $-m_X(\cdot)$ to be the Skorokhod regulator. This definition of $\{Z, \varphi\}$ is used in Sect. 5.1, and other definitions will be used in later sections.

© The Author(s), under exclusive license to Springer Nature Switzerland AG 2026

G. Latouche, *Markov-Modulated Brownian Motion*, Springer Series in Operations Research and Financial Engineering, https://doi.org/10.1007/978-3-031-98576-8_5

We assume throughout this chapter that level 0 is a lower boundary and we have in mind queueing-like models. One expects such systems to have built-in stability features, so that the stationary distribution of Z is a feature of primary interest. For that reason, we shall (nearly systematically) assume that the stationary drift $\boldsymbol{\pi}^{\mathsf{T}}\boldsymbol{\mu}$ is negative, ensuring that the process repeatedly returns to level 0. The one exception to this assumption is in Sect. 5.2, where the process is regulated on two sides and forced to remain in some finite interval $[0, b]$.

We denote by $\boldsymbol{G}(x)$ the stationary distribution function of Z, when it exists:

$$G_i(x) = \lim_{t\to\infty} \mathrm{P}[Z(t) \le x, \varphi(t) = i].$$

To determine $\boldsymbol{G}$, we follow a standard Markov-regenerative approach (see Çinlar [49, Chapter 10]). We start with a sequence $\{t_n : n \ge 0\}$ of epochs such that

- the sequence $\{Z(t_n), \varphi(t_n) : n \ge 0\}$ is a Markov chain on some finite state space $\widehat{\mathcal{E}}$,
- the process $\{Z(t), \varphi(t) : t \ge t_n\}$ after time t_n is conditionally independent of $\{Z(t), \varphi(t) : 0 \le t \le t_n\}$, given $(Z(t_n), \varphi(t_n))$, and
- the process $\{Z(t), \varphi(t) : t \ge t_n\}$, given $(Z(t_n), \varphi(t_n)) = (x, i)$, has the same law as $\{Z(t), \varphi(t) : t \ge 0\}$ given $(Z(0), \varphi(0)) = (x, i)$, for $n \ge 0$, (x, i) in $\widehat{\mathcal{E}}$.

Roughly speaking, we identify regeneration times such that the future of $\{Z, \varphi\}$ after a regeneration occurs may depend on the state at that time but is otherwise independent of the history before then. The nice feature of Markov-regenerative processes is that one may determine their long-term behaviour from the analysis of their short-term properties.

5.1 One-Sided Skorokhod Regulation

In this section, the regulated process is defined by (5.1), and we take as regeneration times the sequence $\{\theta_n^* : n \ge 0\}$ of successive epochs when the regulated process returns to level 0 after a change of phase, similar to the sequence (4.7) used in the analysis of $L_0[x, \infty]$:

$$\theta_0^* = \inf\{t \ge 0 : Z(t) = 0\}, \tag{5.2}$$

$$\theta_n' = \inf\{t > \theta_{n-1}^* : \varphi(t) \ne \varphi(\theta_{n-1}^*)\}, \tag{5.3}$$

$$\theta_n^* = \inf\{t > \theta_n' : Z(t) = 0\}, \tag{5.4}$$

for $n \ge 1$. The θ_n^*s are finite a.s. for all n if $\boldsymbol{\pi}^{\mathsf{T}}\boldsymbol{\mu} \le 0$, and they form a Markov-regenerative set of epochs, in direct consequence of the Markov property in Theorem 1.12.

We proceed in three steps. To begin with, we determine the state transition probability matrix at regeneration times. Next, we determine the expected sojourn time of $\{Z\}$ in any interval $[0, x]$ during a regeneration interval $[\theta_n^*, \theta_{n+1}^*)$; finally, we bring everything together to give a simple expression for the distribution function $\boldsymbol{G}$.

The set of states at regeneration epochs is $\widehat{\mathcal{E}} = \{0\} \times \mathcal{E}$, and, to simplify the expressions to follow, we remove $Z(\theta_n^*)$ from the state description since it is a constant. We denote $\varphi(\theta_n^*)$ as φ_n^*, the process $\{\varphi_n^* : n \geq 0\}$ is a Markov chain and we determine its transition probability matrix in the next lemma. In its proof, we track the evolution of the phase process over changes of the *level* instead of *time*, as one might be tempted to do.

Lemma 5.1. *The transition probability matrix of the Markov chain* $\{\varphi_n^* : n \geq 0\}$ *is the matrix*

$$\Phi = \int_0^\infty \Delta_a e^{-\Delta_a y} \widetilde{P} e^{Uy} \, \mathrm{d}y,$$

defined in (2.26). It is stochastic if $\boldsymbol{\pi}^{\mathrm{T}} \boldsymbol{\mu} \leq 0$ *and substochastic if* $\boldsymbol{\pi}^{\mathrm{T}} \boldsymbol{\mu} > 0$.

Proof. We condition on the level y reached at the first jump time θ_1' and the new phase k at that time:

$$\begin{aligned}
&\mathrm{P}[\varphi_1^* = j | \varphi_0^* = i] \\
&\quad = \int_0^\infty \sum_{k \in \mathcal{E}} \mathrm{P}[Z(\theta_1') \in (y, y + \mathrm{d}y) | Z(0) = 0, \varphi(0) = i] \\
&\qquad\qquad \times \mathrm{P}[\varphi(\theta_1') = k | \varphi(0) = i] \, \mathrm{P}[\varphi(\theta_1^*) = j | Z(\theta_1') = y, \varphi(\theta_1') = k] \\
&\quad = \int_0^\infty \sum_{k \in \mathcal{E}} a_i e^{-a_i y} \widetilde{P}_{ik} (e^{Uy})_{kj} \, \mathrm{d}y,
\end{aligned}$$

where the vector $\boldsymbol{a}$ and the matrix $\widetilde{P}$ are defined in (2.18) and (1.30), respectively. Indeed,

- the transition probability to y at time θ_1' is $a_i e^{-a_i y} \, \mathrm{d}y$ by Lemma 2.6, because θ_1' is the length of the interval of uninterrupted sojourn in phase i, and $Z(\theta_1') = X(\theta_1') - m(\theta_1')$,
- the next factor in the integral is $\widetilde{P}_{ik}$ since $\widetilde{P}$ is the transition matrix of the jump chain for the phase process,
- the last factor is equal to

$$\mathrm{P}[\tau_{-y} < \infty, \varphi(\tau_{-y}) = j | X(0) = y, \varphi(0) = k]$$

  because in the interval (θ_1', θ_1^*), the process $Z(t)$ evolves exactly like $X(t)$.

This proves that $\mathrm{P}[\varphi_1^* = j | \varphi_0^* = i] = \Phi_{ij}$.

If $\boldsymbol{\pi}^{\mathrm{T}} \boldsymbol{\mu} \leq 0$, then e^{Ux} is a stochastic matrix, so that

$$\Phi \mathbf{1} = \int_0^\infty \Delta_a e^{-\Delta_a y} \widetilde{P} \, \mathrm{d}y \, \mathbf{1} = \int_0^\infty \Delta_a e^{-\Delta_a y} \, \mathrm{d}y \, \mathbf{1} = \mathbf{1}.$$

If $\boldsymbol{\pi}^{\mathrm{T}} \boldsymbol{\mu} > 0$, then $e^{Ux} \mathbf{1} < \mathbf{1}$ and $\Phi \mathbf{1} < \mathbf{1}$ by extension. □

We have seen that Φ is a solution of the Riccati equation (2.27). By (2.25), we may write that Φ is also a solution of the system

$$\Delta_a \Phi - \Phi U = \Delta_a \widetilde{P}.$$

As $\boldsymbol{a} > 0$ and the eigenvalues of U are all in $\mathbb{C}_{\leq}$, this is a nonsingular Sylvester equation by Lemma 2.18.

In the next step, we determine the expected local time for the regulated process. To that end, we replace in the definition (4.1) the MMBM process $\{X\}$ with the regulated process $\{Z\}$ and we use the subscript "r" to emphasise that we refer to the regulated process; in particular, $L_{\mathrm{r},y}[x,t]$ is the expected local time for $\{Z\}$ at level x, starting from level y, during the interval of time $[0,t]$.

The matrix

$$M_{\mathrm{r}}(x) = \mathrm{E}[\int_0^{\theta_1^*} \mathbb{1}\{Z(s) \leq x\}; \varphi(s)\,\mathrm{d}s]$$

is the matrix of expected occupation time for $\{Z\}$ in $[0,x]$ during an interval of time between two regeneration epochs. By the occupation density formula (4.3), we have

$$M_{\mathrm{r}}(x) = \int_0^x L_{\mathrm{r},0}[v, \theta_1^*]\,\mathrm{d}v. \tag{5.5}$$

We saw in Remark 4.10 that $(e^{Rx})_{ij}$ is the expected local time for $\{X\}$ in (x,j) per unit of expected local time in $(0,i)$. Furthermore, we have already observed that the process $\{Z\}$ behaves exactly like $\{X\}$ when visiting a strictly positive level, thus,

$$L_{\mathrm{r},0}[x, \theta_1^*]_{ij} = L_{\mathrm{r},0}[0, \theta_1^*]_{ii} (e^{Rx})_{ij} \qquad \text{for } x > 0, \tag{5.6}$$

and $M_{\mathrm{r}}(x) = \Delta_{\mathrm{r},L} \int_0^x e^{Rv}\,\mathrm{d}v$, where

$$\Delta_{\mathrm{r},L} = \mathrm{diag}(L_{\mathrm{r},0}[0, \theta_1^*]_{ii} : i \in \mathcal{E}).$$

If $\boldsymbol{\pi}^{\mathrm{T}} \boldsymbol{\mu} < 0$, the matrix R is nonsingular by Theorem 4.21 and

$$M_{\mathrm{r}}(x) = \Delta_{\mathrm{r},L}(-R^{-1})(I - e^{Rx}) \qquad \text{for } x > 0. \tag{5.7}$$

If $\boldsymbol{\pi}^{\mathrm{T}} \boldsymbol{\mu} \geq 0$, the expression for the integral of e^{Rv} is a bit more involved, and we defer its analysis to Sect. 5.2.

The only visits to level 0 during $[0, \theta_1^*]$ occur before the first change of phase, and so the right-hand side of (5.7) will be fully determined, once we shall know the local time at 0 for a regulated Brownian motion during an exponential interval. This is the object of the next theorem.

Theorem 5.2. *Assume that $\{Z\}$ is a Brownian motion regulated at 0, with parameters μ and σ^2, and that T is an exponentially distributed random variable with parameter λ. The expected local time at 0 during the interval of time $[0,T]$ is*

$$L_{\text{r},0}[0,T] = \frac{2}{\mu + \sqrt{\mu^2 + 2\lambda\sigma^2}} = \frac{2}{\alpha_+(\mu,\sigma,\lambda)\,\sigma^2}, \tag{5.8}$$

with $\alpha_+(\mu,\sigma,\lambda)$ given in (2.14).

Proof. Our proof makes use of the fast-switching process defined in Sect. 1.11. We construct a two-phase Fluid flow $\{X_\varepsilon(t), \varphi_\varepsilon(t) : t \geq 0\}$, on the state space $\{1, 2\}$, with generator

$$Q_\varepsilon = \begin{bmatrix} -1/\varepsilon^2 & 1/\varepsilon^2 \\ 1/\varepsilon^2 & -1/\varepsilon^2 \end{bmatrix}, \tag{5.9}$$

and drift parameters $\boldsymbol{d}^{\text{T}} = \begin{bmatrix} \mu + \sigma/\varepsilon & \mu - \sigma/\varepsilon \end{bmatrix}$. For ε small enough, $d_1 = \mu + \sigma/\varepsilon$ is positive, while $d_2 = \mu - \sigma/\varepsilon$ is negative.

We denote by N_ε the matrix of expected number of crossings of level 0 during $[0, T]$, meaning that $N_{\varepsilon;11}$ is the expected number of times the process *leaves* level 0 in the "up" phase, given that $\varphi(0) = 1$, and $N_{\varepsilon;12}$ is the expected number of times the process *enters* level 0 in the "down" phase. The entries $N_{\varepsilon;21}$ and $N_{\varepsilon;22}$ are similarly defined, with $\varphi(0) = 2$.

Finally, we define Ψ_ε to be the conditional probability, given that the process starts in $(0, 1)$, that it returns to level 0 (of necessity in phase 2) before time T; clearly, $0 < \Psi_\varepsilon < 1$. By Theorem 2.19, it is the solution in $(0, 1)$ of the scalar quadratic equation

$$d_1^{-1}\frac{1}{\varepsilon^2} - d_1^{-1}(\frac{1}{\varepsilon^2} + \lambda)\Psi_\varepsilon - \Psi_\varepsilon|d_2^{-1}|(\frac{1}{\varepsilon^2} + \lambda) + \Psi_\varepsilon|d_2^{-1}|\frac{1}{\varepsilon^2}\Psi_\varepsilon = 0,$$

which we may rewrite after simple manipulations as

$$(\sigma - \varepsilon\mu) - 2\sigma(1 + \varepsilon^2)\Psi_\varepsilon + (\sigma + \varepsilon\mu)\Psi_\varepsilon^2 = 0. \tag{5.10}$$

For ε sufficiently small, Ψ_ε is an analytic function of ε, and we may write it as

$$\Psi_\varepsilon = \Psi_0 + \varepsilon\Psi_1 + O(\varepsilon^2).$$

By equating the coefficients of ε^0, ε^1 and ε^2 in (5.10), we find that $\Psi_0 = 1$ and that Ψ_1 is a root of the quadratic equation

$$-2\sigma\lambda + 2\mu x + \sigma x^2 = 0.$$

There are two roots, one positive, one negative, and Ψ_1 is the negative root since $\Psi_\varepsilon = 1 + \varepsilon\Psi_1 + O(\varepsilon^2)$ has to be strictly less than 1. Thus,

$$\Psi_\varepsilon = 1 - \varepsilon\frac{\mu + \sqrt{\mu^2 + 2\lambda\sigma^2}}{\sigma} + O(\varepsilon^2) = 1 - \varepsilon\alpha_+(\mu,\sigma,\lambda)\sigma + O(\varepsilon^2). \tag{5.11}$$

We use the expression (4.4) for the local time at level 0, the initial phase $\varphi(0)$ is chosen with the stationary distribution of Q_ε, and the expected local time at 0 is

$$L_\varepsilon[0,T] = \varepsilon \begin{bmatrix} 0.5 \; 0.5 \end{bmatrix} N_\varepsilon \begin{bmatrix} (\sigma + \varepsilon\mu)^{-1} \\ (\sigma - \varepsilon\mu)^{-1} \end{bmatrix}.$$

To determine N_ε, we condition on the first epoch when the fast-switching process leaves level 0. If $\varphi(0) = 1$, the fluid queue leaves level 0 immediately and

$$N_{\varepsilon;11} = 1 + \Psi_\varepsilon N_{\varepsilon;21}, \qquad N_{\varepsilon;12} = \Psi_\varepsilon(1 + N_{\varepsilon;22}).$$

If $\varphi(0) = 2$, the process leaves level 0 if $\{\varphi_\varepsilon\}$ jumps to phase 2 (and the fluid starts going up) before time T. The probability of that event is

$$(\frac{1}{\varepsilon^2} + \lambda)^{-1}\frac{1}{\varepsilon^2} = (1 + \varepsilon^2\lambda)^{-1},$$

and

$$N_{\varepsilon;21} = (1 + \varepsilon^2\lambda)^{-1}(1 + \Psi_\varepsilon N_{\varepsilon;21}), \qquad N_{\varepsilon;22} = (1 + \varepsilon^2\lambda)^{-1}\Psi_\varepsilon(1 + N_{\varepsilon;22}).$$

We replace Ψ_ε by its expression in (5.11) and we find, after belaboured but simple manipulations, that

$$N_\varepsilon = (\varepsilon\alpha_+(\mu, \sigma, \lambda)\sigma)^{-1} \begin{bmatrix} 1 & 1 \\ 1 & 1 \end{bmatrix} + O(\varepsilon),$$

so that $L_\varepsilon[0,T] = 2(\alpha_+(\mu,\sigma,\lambda)\sigma^2)^{-1} + O(\varepsilon)$. The proof is completed by taking the limit as $\varepsilon \longrightarrow 0$. □

Remark 5.3. The expression (5.8) for $L_{\mathrm{r},0}[0,T]$ makes sense from a physical point of view. By its definition (5.1), $Z(t)$ is equal to 0 when $X(t)$ is equal to its running minimum $m_X(t)$. In other words, the set of epochs when the regulated process is equal to 0 is the set of points of increase of $|m_X(t)|$, and $L_{\mathrm{r},0}[0,T]$ has to be proportional to the expectation of $|m_X(t)|$. By Lemma 2.6, that is proportional to $(\alpha_+(m,\sigma,\lambda))^{-1}$. From Eq. (4.73), we interpret the coefficient $2/\sigma^2$ as a conversion factor to associate a difference of expected local time to a difference of level. We shall find this conversion factor again in various expressions.

We have now all the ingredients necessary to determine the stationary distribution of the regulated MMBM, provided that the stationary drift is negative.

Theorem 5.4. *Assume that $\boldsymbol{\pi}^{\mathrm{T}}\boldsymbol{\mu} < 0$. The stationary distribution function of the MMBM regulated at 0 is*

$$\boldsymbol{G}^{\mathrm{T}}(x) = \boldsymbol{\pi}^{\mathrm{T}}(I - e^{Rx}), \qquad \textit{for } x \ge 0, \tag{5.12}$$

and its density is $\boldsymbol{g}^{\mathrm{T}}(x) = \boldsymbol{\pi}^{\mathrm{T}}(-R)e^{Rx}$.

Proof. By Çinlar [49, Section 10.4, Proposition 4.9],

$$\boldsymbol{G}^{\mathrm{T}}(x) = c\boldsymbol{\rho}^{\mathrm{T}}M_{\mathrm{r}}(x), \tag{5.13}$$

where $\boldsymbol{\rho}$ is the stationary probability vector of the Markov chain $\{\varphi_n^*\}$, and c, equal to $(\boldsymbol{\rho}^{\mathrm{T}} M_{\mathrm{r}}(\infty)\mathbf{1})^{-1}$, is a normalising constant that we shall deal with at the end of the proof.

By Theorem 5.2 and the definition (2.18) of the vector $\boldsymbol{b}$, $\Delta_{\mathrm{r},L} = 2(\Delta_b \Delta_\sigma^2)^{-1}$, and by (5.7) the matrix of expected occupation times in $[0, x]$ is given by

$$M_{\mathrm{r}}(x) = 2(\Delta_b \Delta_\sigma^2)^{-1}(-R)^{-1}(I - e^{Rx}), \tag{5.14}$$

so that

$$\boldsymbol{G}^{\mathrm{T}}(x) = \boldsymbol{\xi}^{\mathrm{T}}(I - e^{Rx})$$

for some vector $\boldsymbol{\xi}$. We shall show algebraically that $\boldsymbol{\xi} = \boldsymbol{\pi}$, but there is a shorter argument, which we give here.

The eigenvalues of R are all in $\mathbb{C}_<$ and so e^{Rx} tends to 0 as x tends to ∞. Therefore, $\lim_{x\to\infty} \boldsymbol{G}^{\mathrm{T}}(x) = \boldsymbol{\xi}^{\mathrm{T}}$. That limit is also, by definition of $\boldsymbol{G}$, the stationary distribution of $\{\varphi\}$. Since the regulation at level 0 modifies the behaviour of X but not of φ, the stationary distribution of φ is not modified by the regulation, and $\boldsymbol{\xi} = \boldsymbol{\pi}$. This concludes our proof. □

Remark 5.5. We give now the algebraic argument mentioned at the end of the proof above, we shall refer to it again in Sect. 5.4. The symbol $\propto$ means "is proportional to".

We start from (5.13). By Lemma 5.1, $\boldsymbol{\rho}$ is the stationary probability vector associated with Φ, hence $\boldsymbol{\rho}^{\mathrm{T}} = \boldsymbol{\rho}^{\mathrm{T}}\Phi = \boldsymbol{\rho}^{\mathrm{T}}(I + \Delta_b^{-1}U)$ by (2.25), which we may write as $\boldsymbol{\rho}^{\mathrm{T}}\Delta_b^{-1}U = 0$.

Therefore,

$$\boldsymbol{\rho}^{\mathrm{T}}\Delta_b^{-1} \propto \boldsymbol{u}^{\mathrm{T}}, \tag{5.15}$$

where $\boldsymbol{u}$ is the stationary probability vector of the generator U. We combine (5.13) and (5.14) and find that

$$\boldsymbol{G}^{\mathrm{T}}(x) \propto \boldsymbol{u}^{\mathrm{T}}\Delta_\sigma^{-2}(-R)^{-1}(I - e^{Rx}), \tag{5.16}$$

or $\boldsymbol{\xi}^{\mathrm{T}} \propto \boldsymbol{u}^{\mathrm{T}}\Delta_\sigma^{-2}(-R)^{-1}$. We show after a few substitutions that

$$\boldsymbol{\xi}^{\mathrm{T}} R \Delta_\sigma^2 \propto \boldsymbol{u}^{\mathrm{T}},$$

or

$$\boldsymbol{\xi}^{\mathrm{T}}(\Delta_\sigma^2 U + 2\Delta_\mu) \propto \boldsymbol{u}^{\mathrm{T}} \qquad \text{by (4.44)},$$

so that

$$\boldsymbol{\xi}^{\mathrm{T}}(\Delta_\sigma^2 U + 2\Delta_\mu)U = \mathbf{0},$$

or

$$\boldsymbol{\xi}^{\mathrm{T}} Q = \mathbf{0}$$

since U is a solution of the quadratic equation (2.23). This means that $\boldsymbol{\xi} \propto \boldsymbol{\pi}$ and that $\boldsymbol{G}^{\mathrm{T}}(x) \propto \boldsymbol{\pi}^{\mathrm{T}}(I - e^{Rx})$. Since $\boldsymbol{G}^{\mathrm{T}}(\infty)\mathbf{1} = 1$, Eq. (5.12) is proved.

We shall give with the next lemma an expression for the stationary distribution that is directly connected with the Skorokhod regulator $S(t) = |m_X(t)|$. We define the sub-regulators

$$S_i(t) = \int_0^t \mathbb{1}\{\varphi(s) = i\}\, \mathrm{d}S(s), \qquad i \in \mathcal{E}. \tag{5.17}$$

By extension to the characteristics given in Parenthesis 1.6, the function $S_i(t)$ increases when (and only when) $Z(t) = 0$ and $\varphi(t) = i$.

Lemma 5.6. *Assume that $\boldsymbol{\pi}^{\mathrm{T}}\boldsymbol{\mu} < 0$. The ratio $S_i(t)/t$, $i \in \mathcal{E}$, converges almost surely to a constant as $t \to \infty$, and we write*

$$\ell_i = \lim_{t\to\infty} S_i(t)/t. \tag{5.18}$$

The vector $\boldsymbol{\ell}$ is proportional to the vector $\boldsymbol{u}$, and it is normalised as $\boldsymbol{\ell}^{\mathrm{T}}\mathbf{1} = -\boldsymbol{\pi}^{\mathrm{T}}\boldsymbol{\mu}$. The stationary distribution function of the MMBM regulated at 0 is

$$\boldsymbol{G}^{\mathrm{T}}(x) = c_\ell\, \boldsymbol{\ell}^{\mathrm{T}}\Delta_\sigma^{-2}R^{-1}(I - e^{Rx}), \qquad \text{for } x \geq 0, \tag{5.19}$$

where $c_\ell = (\boldsymbol{\ell}^{\mathrm{T}}\Delta_\sigma^{-2}R^{-1}\mathbf{1})^{-1}$ is the normalising constant. The density function is

$$\boldsymbol{g}^{\mathrm{T}}(x) = c_\ell\, \boldsymbol{\ell}^{\mathrm{T}}\Delta_\sigma^{-2}e^{Rx}.$$

Proof. Starting from the regeneration epochs $\{\theta_n^*\}$ and the process $\{\varphi_n^*\}$ defined at the beginning of this section, we focus on a phase i, arbitrary but fixed for the time being, and we define the sequence $\{\theta_n^\dagger\}$ of epochs when $\{Z, \varphi\}$ returns to $(0, i)$ after at least one change of phase, and possibly after returns to 0 in another phase:

$$\begin{aligned} \theta_1^\dagger &= \inf\{\theta_\nu^* : \varphi_\nu^* = i\}, \\ \theta_{n+1}^\dagger &= \inf\{\theta_\nu^* > \theta_n^\dagger : \varphi_\nu^* = i\}, \qquad n \geq 1. \end{aligned}$$

The differences $\delta_n^\dagger = \theta_{n+1}^\dagger - \theta_n^\dagger$ are i.i.d. and define a renewal process $\{N_i(t) : t \geq 0\}$ where

$$N_i(t) = \sup\{n : \theta_n^\dagger \leq t\}$$

is the number of visits of $\{\varphi_n^*\}$ to phase i during $(0, t)$. It is well known (see Resnick [132, Theorem 3.3.2] for instance) that

$$\lim_{t\to\infty} \frac{N_i(t)}{t} = \frac{1}{M_i^\dagger} \qquad \text{a.s.,}$$

where $M_i^\dagger = \mathrm{E}[\theta_{n+1}^\dagger - \theta_n^\dagger]$ is the expected length of the interval of time between two such visits.

Next, we analyse the increase

$$S_n^\dagger = S_i(\theta_{n+1}^\dagger) - S_i(\theta_n^\dagger) = \int_{\theta_n^\dagger}^{\theta_{n+1}^\dagger} \mathbb{1}\{\varphi(s) = i\}\, \mathrm{d}S(s), \qquad \text{for } n \geq 1,$$

of the sub-regulator $S_i(\cdot)$ during the nth renewal interval. Define $\theta_n^\dagger + T_n$ to be the epoch of the first jump to a phase different from i. As the MMBM is never at level 0 in phase i during $(\theta_n^\dagger + T_n, \theta_{n+1}^\dagger)$,

$$S_n^\dagger = \int_{\theta_n^\dagger}^{\theta_n^\dagger + T_n} \mathbb{1}\{\varphi(s) = i\}\, \mathrm{d}S(s), \qquad \text{for } n \geq 1,$$

and it has the same distribution as $|m_B(T)|$, where B is a Brownian motion with parameters (μ_i, σ_i), and T is exponential with parameter $|Q_{ii}|$. Thus, by Lemma 2.6, the $S_n^\dagger$s are i.i.d., exponentially distributed with parameter b_i defined in (2.18), and

$$\lim_{t\to\infty} \frac{1}{t} \sum_{1 \leq \nu \leq N_i(t)} S_\nu^\dagger = \frac{1}{b_i\, M_i^\dagger} \qquad \text{a.s.}$$

(see Resnick [132, Proposition 3.4.1]). We conclude by the usual sandwich argument that

$$\lim_{t\to\infty} S_i(t)/t = 1/(b_i M_i^\dagger)$$

with probability 1. This proves the first statement, with

$$\ell_i = 1/(b_i M_i^\dagger), \qquad \text{for } i \in \mathcal{E}.$$

Now, denote by $f_j = \mathrm{E}_j[\theta_{n+1}^* - \theta_n^*]$ the expected interval between two regeneration epochs for j arbitrary in $\mathcal{E}$. By [132, Proposition 2.12.2 and the remark after Proposition 2.12.4], we have

$$\rho_i M_i^\dagger = \sum_{j\in\mathcal{E}} \rho_j f_j = \mathrm{E}_\rho[\theta_1^*].$$

Therefore,

$$\lim_{t\to\infty} \frac{1}{t} S_i(t) = \rho_i b_i^{-1}/\mathrm{E}_\rho[\theta_1^*] \qquad \text{a.s.}$$

and $\boldsymbol{\ell}^{\mathrm{T}} \propto \boldsymbol{\rho}^{\mathrm{T}} \Delta_b^{-1}$. We conclude from (5.15) that $\boldsymbol{\ell}$ is proportional to $\boldsymbol{u}$.

Finally, we follow Asmussen and Kella [16, Eq. 4.2] to prove as follows that $\boldsymbol{\ell}^{\mathrm{T}}\mathbf{1} = -\boldsymbol{\pi}^{\mathrm{T}}\boldsymbol{\mu}$:

$$\lim_{t\to\infty} Z(t)/t = \lim_{t\to\infty} (X(t) + S(t))/t = \lim_{t\to\infty} (X(t) + \sum_{i\in\mathcal{E}} S_i(t))/t = \boldsymbol{\pi}^{\mathrm{T}}\boldsymbol{\mu} + \boldsymbol{\ell}^{\mathrm{T}}\mathbf{1}.$$

As $\boldsymbol{\pi}^{\mathrm{T}}\boldsymbol{\mu} < 0$, $Z(t)/t$ tends to 0 with probability 1, as t grows to infinity, and this concludes our proof. □

Remark 5.7. Different expressions for $\boldsymbol{G}(x)$ are to be found in the literature. We have given three so far, in Theorem 5.4, Lemma 5.6 and Eq. (5.16), and we shall mention a few others in Sect. 5.5, obtained by following different approaches. In each case, the matrix $I - e^{Rx}$ is pre-multiplied by a vector which depends on some behaviour of the process at the boundary. The expressions (5.12) and (5.19) crucially depend on the fact that the evolution of the phase is independent of the fluid level. Indeed, both $\boldsymbol{\ell}$ and $\boldsymbol{\pi}$ are defined by the unrestricted MMBM as

(a) the vector $\boldsymbol{\pi}$ is the stationary marginal distribution of the phase when its evolution is governed by the matrix Q and is not modified in any way,
(b) the vector $\boldsymbol{\ell}$ is defined in (5.18) as the rate of increase of $|m_X(\cdot)|$ in the absence of any barrier.

We conclude this section with a few words about the decay rate of the stationary distribution. Define η to be the spectral radius of the matrix e^R. If $\boldsymbol{\pi}^{\mathrm{T}}\boldsymbol{\mu} < 0$, then $\eta = e^{-\lambda_{m+1}} < 1$, where $-\lambda_{m+1} < 0$ is the eigenvalue of R closest to 0 (see the Wiener-Hopf factorisation theorem 2.11, and Eq. (4.67)). Define $\boldsymbol{v}_R$ and $\boldsymbol{u}_R$ as the right and left eigenvectors of R for the eigenvalue $-\lambda_{m+1}$, normalised by $\boldsymbol{u}_R^{\mathrm{T}}\mathbf{1} = \boldsymbol{u}_R^{\mathrm{T}}\boldsymbol{v}_R = 1$. We have

$$e^{Rx} = \eta^x \boldsymbol{v}_R \cdot \boldsymbol{u}_R^{\mathrm{T}} + o(\eta^x),$$

asymptotically as $x \to \infty$. The tail of the stationary distribution of $\{Z\}$ is

$$\begin{aligned}\lim_{t\to\infty} \mathrm{P}[Z(t) > x] &= \boldsymbol{\pi}^{\mathrm{T}} e^{Rx}\mathbf{1} \\ &= (\boldsymbol{\pi}^{\mathrm{T}}\boldsymbol{v}_R)\eta^x + o(\eta^x),\end{aligned}$$

the stationary density is

$$\boldsymbol{g}^{\mathrm{T}}\mathbf{1} = (\boldsymbol{\pi}^{\mathrm{T}}\boldsymbol{v}_R)|\log\eta|\,\eta^x + o(\eta^x),$$

and the ratio of the time spent above level x to the time spent in a neighbourhood of x is $|\log\eta|^{-1}$, asymptotically as x tends to ∞.

Thus, η may be used as an indicator of the stickiness of high levels: The MMBM may have a small probability of reaching a high level x between two returns to level 0, but if η is very close to 1, the process will spend an enormous amount of time above x, once it gets there. On the other hand, if η is much smaller than 1, this means that excursions to high levels are short-lived and that the process recovers fast from visits to high levels, when they do occur.

Neuts [117], in the context of QBDs, calls η the *caudal characteristic* and suggests to use it as a tool to compare the tail dynamic behaviour of different queueing systems.

Example 5.8. The model in this example is inspired from [106, Example 9.2.1], and it represents the workload of a noisy PH/M/1 queue: Customers arrive according to a renewal process and bring an exponentially distributed amount of work, with mean $\mathrm{E}[S]$. The distribution of the interval of time between two successive arrivals is the mixture

$$F(\cdot) = pE_{m,\nu}(\cdot) + (1-p)E_{n,\nu}(\cdot),$$

where we denote as $E_{k,\nu}(\cdot)$ the Erlang distribution with parameters k and ν. The sets of parameters in this example are given in the table below. The parameter ν is chosen such that the mean of the PH distribution $F(\cdot)$ is equal to 1 in all three cases, and we give in the last column the variance of the interval between two arrivals.

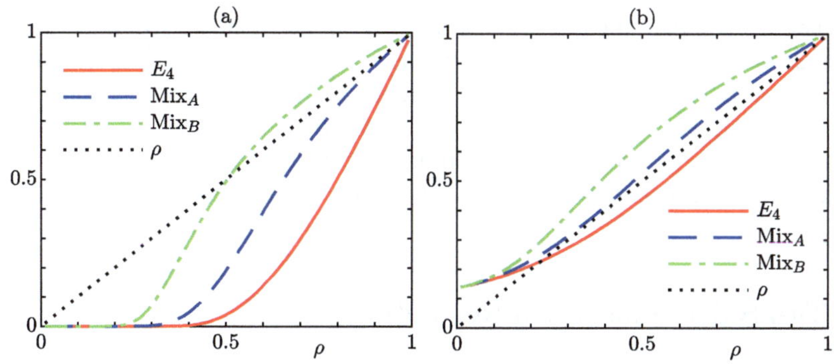

Fig. 5.1 Decay rate η for the workload in a noisy PH/M/1 queue for different distributions of the interarrival times and different volatility. The parameters σ_i, $i = 1 \ldots m$, are equal to 0.1 in (**a**) on the left, they are equal to 1 in (**b**) on the right

Label	p	m	n	ν	Var.
E_4	0	4	4	4	0.25
Mix_A	0.1	25	4	6.1	1.23
Mix_B	0.15	50	4	10.9	2.36

The generator for the phase process is the m by m matrix

$$Q = \begin{bmatrix} -\nu & \nu & & & & \\ & \ddots & \ddots & & & \\ & & -\nu & \nu & & \\ & & & -\nu & \nu & \\ & & & & \ddots & \ddots \\ p\nu & & \ldots & (1-p)\nu & \ldots & -\nu \end{bmatrix}$$

(on the last row, the entry $(1-p)\nu$ is in column $m - n + 1$); the drift vector is

$$\boldsymbol{\mu}^{\mathrm{T}} = \begin{bmatrix} -1 & \ldots & -1 & -1 + \nu \mathrm{E}[S] \end{bmatrix},$$

indicating that the workload decreases at the constant rate of 1 per unit of time, except during the last phase, which corresponds to the arrival of a new customer.

We show in Fig. 5.1 the caudal characteristic η as a function of $\mathrm{E}[S]$, note that the maximal value is 1 as the arrival rate itself is equal to 1. The volatility parameter is the same for all phases in both figures, with $\sigma_i = 0.1$ in Fig. 5.1a on the left and $\sigma_i = 1$ in Fig. 5.1b on the right.

Our first observation is that, just like the QBD example in [106], the caudal characteristic η increases with the variance of the interval between two arrivals. In other words, the system needs more time to recover from an excursion to high levels if the interarrival distribution has a greater variance.

When we compare the plots on the left and on the right, we see that η increases if the Brownian volatility increases, this effect being significant for small values of $\mathrm{E}[S]$, that is, for small values of the system load. This confirms the conclusion in Karandikar and Kulkarni [95] where it is shown that congestion is underestimated when the volatility is neglected. In addition, we notice that there is less difference between the caudal characteristic curves when the σ_is are greater.

5.2 Two-Sided Skorokhod Regulation

The process in this section is regulated at level 0 and level b, for some finite $b > 0$, and is denoted as $\{Y(t), \varphi(t) : t \geq 0\}$ so as to distinguish it from the MMBM $\{X\}$ and from the one-sided regulated process $\{Z\}$ of Sect. 5.1. It is defined as

$$Y(t) = X(t) + S_0(t) - S_b(t),$$

where $S_0(t)$ and $S_b(t)$ are the lower and upper regulators. In technical terms, the Skorokhod regulators are such that

1. $S_0(\cdot)$ and $S_b(\cdot)$ are increasing and continuous, with $S_0(0) = S_b(0) = 0$,
2. $Y(t) \in [0, b]$ for all $t \geq 0$, and
3. $S_0(t)$ and $S_b(t)$ increase only when $Y(t) = 0$ and $Y(t) = b$, respectively.

To complete the picture, we give the following expressions for the regulators (Harrison [80, Section 2.4]):

$$\begin{aligned} S_0(t) &= \max\{0, |X(s) - S_b(s)| : 0 \leq s \leq t\}, \\ S_b(t) &= \max\{0, X(s) + S_0(s) - b : 0 \leq s \leq t\}. \end{aligned}$$

To determine the stationary distribution function $\boldsymbol{G}(x)$ of $\{Y, \varphi\}$, with

$$G_i(x) = \lim_{t \to \infty} \mathrm{P}[Y(t) \leq x, \varphi(t) = i],$$

we choose as semi-renewal epochs the first visits alternatively to level 0 and to level b:

$$\widetilde{\theta}_0 = \inf\{t \geq 0 : Y(t) = 0\},$$

$$\widetilde{\theta}_{2n+1} = \inf\{t > \widetilde{\theta}_{2n} : Y(t) = b\}, \tag{5.20}$$

$$\widetilde{\theta}_{2n+2} = \inf\{t > \widetilde{\theta}_{2n+1} : Y(t) = 0\}, \tag{5.21}$$

for $n \geq 0$. As $\{Y\}$ is forced to remain between 0 and b always, the $\widetilde{\theta}_n$s are finite for all n.

The set of states at regeneration epochs is $\widetilde{\mathcal{E}} = \{0, b\} \times \mathcal{E}$ and we define $(\widetilde{y}_n, \widetilde{\varphi}_n) = (Y(\widetilde{\theta}_n), \varphi(\widetilde{\theta}_n))$. The process $\{\widetilde{y}_n, \widetilde{\varphi}_n : n \geq 0\}$ is a Markov chain with transition probability matrix

$$\widetilde{\Phi} = \begin{bmatrix} 0 & H^{(b)} \\ \widetilde{H}^{(b)} & 0 \end{bmatrix}, \tag{5.22}$$

and the matrices $H^{(b)}$ and $\widetilde{H}^{(b)}$ of transition probability from the boundary 0 to b and vice versa are analysed in Sects. 3.4 and 3.5. The stationary probability vector of $\widetilde{\Phi}$ is $\widetilde{\boldsymbol{\rho}} = \begin{bmatrix} \widetilde{\boldsymbol{\rho}}_0 \\ \widetilde{\boldsymbol{\rho}}_b \end{bmatrix}$, with

$$\widetilde{\boldsymbol{\rho}}_0^{\mathrm{T}} = \widetilde{\boldsymbol{\rho}}_0^{\mathrm{T}} H^{(b)} \widetilde{H}^{(b)}, \qquad \widetilde{\boldsymbol{\rho}}_b^{\mathrm{T}} = \widetilde{\boldsymbol{\rho}}_b^{\mathrm{T}} \widetilde{H}^{(b)} H^{(b)},$$

normalised by $\widetilde{\boldsymbol{\rho}}_0^{\mathrm{T}}\mathbf{1} = \widetilde{\boldsymbol{\rho}}_b^{\mathrm{T}}\mathbf{1} = 0.5$.

To determine the expected occupation time of $\{Y\}$ during a regeneration interval is more involved than for the process $\{Z\}$ with only one boundary, and we shall proceed in three steps.

First Step. We start with the unrestricted process $\{X\}$ and we determine the expected local time at level x, $0 < x < b$, per unit of local time at either level 0 or level b. Define t^* as the first return time to the level $X(0)$ after the first jump time of $\{\varphi\}$:

$$t^* = \inf\{t > t' : X(t) = X(0)\},$$

where $t' = \inf\{t > 0 : \varphi(t) \neq \varphi(0)\}$.[1] Define δ^* as the first passage time to b if $X(0) = 0$, and the first passage time to 0 if $X(0) = b$. Next, define the matrices

$$\widetilde{L}_0(x) = L_0[x, t^* \wedge \delta^*], \qquad \widetilde{L}_b(x) = L_b[x, t^* \wedge \delta^*], \tag{5.23}$$

for $0 < x < b$. These are the expected local time at level x, starting from level 0 or level b, respectively, until either

- the MMBM returns to the starting level after a change of phase, or
- the MMBM reaches the alternate level.

Finally, we define the expected local times $\widetilde{N}_0(x)$ and $\widetilde{N}_b(x)$ at level x, in the time interval $(0, t^* \wedge \delta^*)$, per unit of local time at level 0 or level b, respectively, as

$$(\widetilde{N}_0(x))_{ij} = \frac{(\widetilde{L}_0(x))_{ij}}{(L_0[0, t^*])_{ii}} \quad \text{and} \quad (\widetilde{N}_b(x))_{ij} = \frac{\widetilde{L}_b(x))_{ij}}{(L_b[b, t^*])_{ii}}. \tag{5.24}$$

Note the similarity with the definition (4.23); the difference with e^{Rx} and $e^{\widehat{R}x}$ is that we use in the numerators the interval of time $[0, t^* \wedge \delta^*]$ instead of $[0, t^*]$. By the definition (4.11) of Δ_d, these equations may be written as

$$\widetilde{N}_0(x) = \Delta_d^{-1}\widetilde{L}_0(x) \quad \text{and} \quad \widetilde{N}_b(x) = \Delta_d^{-1}\widetilde{L}_b(x). \tag{5.25}$$

Lemma 5.9. *The matrices* $\widetilde{N}_0(x)$ *and* $\widetilde{N}_b(x)$ *defined in (5.24) are solutions of the system*

[1] These have the same distribution as t_1' and t_1^* in (4.6, 4.7), but we do not need to specify the index in this section.

$$\begin{bmatrix} I & e^{Rb} \\ e^{\widehat{R}b} & I \end{bmatrix} \begin{bmatrix} \widetilde{N}_0(x) \\ \widetilde{N}_b(x) \end{bmatrix} = \begin{bmatrix} e^{Rx} \\ e^{\widehat{R}(b-x)} \end{bmatrix}, \tag{5.26}$$

for $0 < x < b$.

Proof. Assume that $X(0) = 0$. We decompose $L_0[x, t^*]$ as the sum of the expected local time before $t^* \wedge \delta^*$ and of the local time in $(\delta^*, t^*]$, in case $\delta^* < t^*$:

$$\begin{aligned} L_0[x, t^*] &= L_0[x, t^* \wedge \delta^*] + L_0[x, (t^* \wedge \delta^*, t^*]] \\ &= \widetilde{L}_0(x) + L_0[b, t^*]\widetilde{N}_b(x), \end{aligned} \tag{5.27}$$

the second term being justified as follows. Take any three phases i, j and ν in $\mathcal{E}$:

- $(L_0[b, t^*])_{i\nu}$ is the expected local time of $\{X\}$ in (b, ν), starting from $(0, i)$, until the process returns to level 0 after a change of phase. This local time is equal to 0 if $t^* < \delta^*$.
- $(\widetilde{N}_b(x))_{\nu j}$ is the expected local time of $\{X\}$ in (x, j), starting from (b, ν), per unit of local time in (b, ν), until the process returns to level b or visits 0.

The same argument gives us

$$L_b[x, t^*] = \widetilde{L}_b(x) + L_b[b, t^*]\widetilde{N}_0(x). \tag{5.28}$$

We pre-multiply both sides of (5.27) by Δ_d^{-1} and obtain from (4.23) that

$$e^{Rx} = \widetilde{N}_0(x) + e^{Rb}\widetilde{N}_b(x),$$

which is the first block of equations in (5.26). Next, we use the fact that

$$L_b[x, t^*] = L_0[x - b, t^*] = \Delta_d \, e^{\widehat{R}(b-x)},$$

for $0 < x < b$, and obtain

$$e^{\widehat{R}(b-x)} = \widetilde{N}_b(x) + e^{\widehat{R}b}\widetilde{N}_0(x)$$

from (5.28). This concludes our proof. □

We have seen in Sect. 4.4 that the matrices U and $\widehat{R}$ on the one hand, $\widehat{U}$ and R on the other hand, are similar. In consequence, the matrices

$$\begin{bmatrix} I & e^{Rb} \\ e^{\widehat{R}b} & I \end{bmatrix} \quad \text{and} \quad \begin{bmatrix} I & e^{\widehat{U}b} \\ e^{Ub} & I \end{bmatrix}$$

are similar, and by the same analysis which follows Lemma 3.2, we conclude that the system (5.26) is nonsingular if $\boldsymbol{\pi}^{\mathrm{T}}\boldsymbol{\mu} \neq 0$ and singular if $\boldsymbol{\pi}^{\mathrm{T}}\boldsymbol{\mu} = 0$. In the first case, the matrices $\widetilde{N}_0(x)$ and $\widetilde{N}_b(x)$ are completely characterised.

Corollary 5.10. *If* $\boldsymbol{\pi}^{\mathrm{T}}\boldsymbol{\mu} \neq 0$, *then*

$$\widetilde{N}_0(x) = (e^{-Rb} - e^{\widehat{R}b})^{-1}(e^{-R(b-x)} - e^{\widehat{R}(b-x)}), \tag{5.29}$$

$$\widetilde{N}_b(x) = (e^{-\widehat{R}b} - e^{Rb})^{-1}(e^{-\widehat{R}x} - e^{Rx}), \tag{5.30}$$

for $0 < x < b$. □

Proof. We may repeat the argument in the proof of Theorem 3.3, but we shall instead use the time-reversed MMBM defined in Sect. 4.6, with phase transition generator $\widetilde{Q} = \Delta_\pi^{-1} Q^{\mathrm{T}} \Delta_\pi$; we use the subscript "tr" for matrices pertaining to the time-reversed process.

We replace in (5.26) the matrices R and $\widehat{R}$ with the expressions given in Lemma 4.26, so that the system becomes

$$\begin{bmatrix} I & e^{\widehat{U}_{\mathrm{tr}}^{\mathrm{T}} b} \\ e^{U_{\mathrm{tr}}^{\mathrm{T}} b} & I \end{bmatrix} \begin{bmatrix} \Delta_\pi & \\ & \Delta_\pi \end{bmatrix} \begin{bmatrix} \widetilde{N}_0(x) \\ \widetilde{N}_b(x) \end{bmatrix} = \begin{bmatrix} e^{\widehat{U}_{\mathrm{tr}}^{\mathrm{T}} x} \\ e^{U_{\mathrm{tr}}^{\mathrm{T}}(b-x)} \end{bmatrix} \Delta_\pi .$$

Next, we transpose both sides and find that

$$\Delta_\pi^{-1} \begin{bmatrix} \widetilde{N}_0^{\mathrm{T}}(x) \; \widetilde{N}_b^{\mathrm{T}}(x) \end{bmatrix} \begin{bmatrix} \Delta_\pi & \\ & \Delta_\pi \end{bmatrix} \begin{bmatrix} I & e^{U_{\mathrm{tr}} b} \\ e^{\widehat{U}_{\mathrm{tr}} b} & I \end{bmatrix} = \begin{bmatrix} e^{\widehat{U}_{\mathrm{tr}} x} \; e^{U_{\mathrm{tr}}(b-x)} \end{bmatrix}.$$

This may be re-arranged as

$$\Delta_\pi^{-1} \begin{bmatrix} \widetilde{N}_b^{\mathrm{T}}(x) \; \widetilde{N}_0^{\mathrm{T}}(x) \end{bmatrix} \begin{bmatrix} \Delta_\pi & \\ & \Delta_\pi \end{bmatrix} (I + \mathcal{P}_{\mathrm{tr}}^{(b)}) = \mathcal{U}_{\mathrm{tr}}^{(b)}, \tag{5.31}$$

where the matrices $\mathcal{P}$ and $\mathcal{U}$ are defined in (3.2)—notice that $\widetilde{N}_b^{\mathrm{T}}(x)$ now appears to the left of $\widetilde{N}_0^{\mathrm{T}}(x)$) in the left-hand side.

For $\boldsymbol{\pi}^{\mathrm{T}} \boldsymbol{\mu} \neq 0$, the system is nonsingular, its solution is unique, and this implies by Theorem 3.2 that

$$\Delta_\pi^{-1} \widetilde{N}_0^{\mathrm{T}}(x) \Delta_\pi = \widehat{\Psi}_{\mathrm{tr}}^{(b-x,x)} \qquad \text{and} \qquad \Delta_\pi^{-1} \widetilde{N}_b^{\mathrm{T}}(x) \Delta_\pi = \Psi_{\mathrm{tr}}^{(b-x,x)}, \tag{5.32}$$

two equations which are equivalent to (5.29, 5.30). □

If $\boldsymbol{\pi}^{\mathrm{T}} \boldsymbol{\mu} = 0$, we need an additional equation, and it is provided by Lemma 5.11, in the proof of which we shall need (5.32). To that end, we proceed as in the proof of Theorem 4.9: We introduce an artificial killing mechanism with rate α, independent of the MMBM. For this modified process, the system (5.31) is nonsingular, its solution is unique, and we obtain (5.32) after letting α tend to 0.

Lemma 5.11. *If* $\boldsymbol{\pi}^{\mathrm{T}} \boldsymbol{\mu} = 0$*, then the matrices* $\widetilde{N}_0$ *and* $\widetilde{N}_b$ *are such that*

$$\widetilde{\boldsymbol{\beta}}^{\mathrm{T}} \begin{bmatrix} \widetilde{N}_0(x) \\ \widetilde{N}_b(x) \end{bmatrix} = \widetilde{\boldsymbol{h}}^{\mathrm{T}}, \qquad \text{for } 0 < x < b, \tag{5.33}$$

where $\widetilde{\boldsymbol{h}}^{\mathrm{T}} = \boldsymbol{\pi}^{\mathrm{T}} \Delta_\mu (-Q)^{\#}$ *and*

$$\widetilde{\boldsymbol{\beta}}^{\mathrm{T}} = \left[x\boldsymbol{\pi}^{\mathrm{T}} + \widetilde{\boldsymbol{h}}^{\mathrm{T}} \;\; -(b-x)\boldsymbol{\pi}^{\mathrm{T}} + \widetilde{\boldsymbol{h}}^{\mathrm{T}} \right].$$

The solution of (5.26, 5.33) is

$$\begin{bmatrix} \widetilde{N}_0(x) \\ \widetilde{N}_b(x) \end{bmatrix} = \begin{bmatrix} I & e^{Rb} \\ e^{\widehat{R}b} & I \end{bmatrix}^{\#} \begin{bmatrix} e^{Rx} \\ e^{\widehat{R}(b-x)} \end{bmatrix} + \boldsymbol{v}_R \cdot \boldsymbol{y}_R^{\mathrm{T}}, \tag{5.34}$$

where $\boldsymbol{v}_R$ is the right eigenvector of the matrix $\begin{bmatrix} I & e^{Rb} \\ e^{\widehat{R}b} & I \end{bmatrix}$, associated to the eigenvalue 0, and

$$\boldsymbol{y}_R^{\mathrm{T}} = (\widetilde{\boldsymbol{\beta}}^{\mathrm{T}} \boldsymbol{v}_R)^{-1} (\widetilde{\boldsymbol{h}}^{\mathrm{T}} - \widetilde{\boldsymbol{\beta}}^{\mathrm{T}} \begin{bmatrix} I & e^{Rb} \\ e^{\widehat{R}b} & I \end{bmatrix}^{\#} \begin{bmatrix} e^{Rx} \\ e^{\widehat{R}(b-x)} \end{bmatrix}).$$

Proof. We apply Lemma 3.7 to the time-reversed process and, using (5.32), the additional Eq. (3.9) becomes

$$\left[\Delta_\pi^{-1} \widetilde{N}_b^{\mathrm{T}}(x) \Delta_\pi \;\; \Delta_\pi^{-1} \widetilde{N}_0^{\mathrm{T}}(x) \Delta_\pi \right] \begin{bmatrix} -(b-x)\mathbf{1} + \boldsymbol{h}_{\mathrm{tr}} \\ x\mathbf{1} + \boldsymbol{h}_{\mathrm{tr}} \end{bmatrix} = \boldsymbol{h}_{\mathrm{tr}}$$

with $\boldsymbol{h}_{\mathrm{tr}} = (-\widetilde{Q})^{\#}\boldsymbol{\mu}$. To complete the proof of (5.33), we transpose both sides, define $\widetilde{\boldsymbol{\beta}} = \Delta_\pi \boldsymbol{\beta}_{\mathrm{tr}}$ and $\widetilde{\boldsymbol{h}} = \Delta_\pi \boldsymbol{h}_{\mathrm{tr}}$, and perform some simple algebraic manipulations.

Equation (5.34) follows from (1.32) in the same manner as (3.11) follows from (1.33). □

Second Step. Next, we determine the expected local time for the regulated process at the boundaries during each regeneration interval. We readily see that we only need to deal with one boundary at a time: If $Y(0) = 0$, a regeneration occurs as soon as $Y(t)$ is equal to b, there is no time spent at level b until then, and

$$L_{r,0}[0, (\widetilde{\theta}_n, \widetilde{\theta}_{n+1})] = L_{\mathrm{r},0}[0, \delta^*],$$

where the right-hand side is determined in Lemma 5.12. Conversely, if $Y(0)$ is equal to b, the regeneration occurs when $Y(t)$ is equal to 0 for the first time, and

$$L_{r,b}[b, (\widetilde{\theta}_n, \widetilde{\theta}_{n+1})] = L_{\mathrm{r},b}[b, \delta^*].$$

Lemma 5.12. *For the MMBM $\{Y, \varphi\}$ regulated on two boundaries, the conditional expected local time at the boundaries during a regeneration interval, given the process starts at level 0, is*

$$L_{\mathrm{r},0}[0, \delta^*] = 2(-L^{(b)})^{-1} \Delta_\sigma^{-2}. \tag{5.35}$$

If it starts at level b, then

$$L_{\mathrm{r},b}[b, \delta^*] = 2(-\widehat{D}^{(b)})^{-1} \Delta_\sigma^{-2}. \tag{5.36}$$

The matrices $L^{(b)}$ and $\widehat{D}^{(b)}$ are given in (3.39, 3.43) and (3.49, 3.52), respectively.

Proof. We give details for $L_{r,0}[0, \delta^*]$ only, the argument for $L_{r,b}[b, \delta^*]$ being nearly identical.

We condition on the first passage through some intermediary level y, $0 < y < b$, and write that

$$L_{r,0}[0, \delta^*] = L_{r,0}[0, \delta_y] + H^{(y)}\Psi^{(y,b-y)}L_{r,0}[0, \delta^*],$$

where $H^{(y)}$ is the transition probability matrix from the regulated level 0 to level $y > 0$, its expression is given in Theorem 3.14. The left-hand side is independent on y, and so

$$\frac{\partial}{\partial y}L_{r,0}[0, \delta_y] + (\frac{\partial}{\partial y}H^{(y)})\Psi^{(y,b-y)}L_{r,0}[0, \delta^*] + H^{(y)}(\frac{\partial}{\partial y}\Psi^{(y,b-y)})L_{r,0}[0, \delta^*] = 0,$$

for any $y > 0$. We take the limit as $y \to 0$, use (3.30, 3.34), and find that

$$\frac{\partial}{\partial y}L_{r,0}[0, \delta_y]\bigg|_{y=0} + L^{(b)}L_{r,0}[0, \delta^*] = 0. \tag{5.37}$$

Finally, we repeat the argument in Theorem 4.22 and write $L_{r,0}[0, \delta_y]$ as

$$L_{r,0}[0, \delta_y] = \int_0^y H^{(v)}(2\Delta_\sigma^{-2})\widehat{R}_{r,v}(0)\,\mathrm{d}v,$$

where $\widehat{R}_{r,v}(0)$ is the expected local time at the regulated level 0, starting from level v, per unit of local time at level v. From this equation, we obtain that

$$\begin{aligned}\frac{\partial}{\partial y}L_{r,0}[0, \delta_y]\bigg|_{y=0} &= H^{(0)}(2\Delta_\sigma^{-2})\widehat{R}_{r,0}(0)\\ &= 2\Delta_\sigma^{-2}\end{aligned}$$

which, together with (5.37), proves (5.35)—remember that $L^{(b)}$ is nonsingular by Theorem 3.14. □

Last Step. The matrices of expected occupation in $[0, x]$ during an interval between two regeneration epochs are

$$\widetilde{M}_0(x) = \mathrm{E}[\int_{\widetilde{\theta}_0}^{\widetilde{\theta}_1} \mathbb{1}\{Y(s) \le x\}; \varphi(s)\,\mathrm{d}s] = L_{r,0}[0, \delta^*]\int_0^x \widetilde{N}_0(v)\,\mathrm{d}v \tag{5.38}$$

and

$$\widetilde{M}_b(x) = \mathrm{E}[\int_{\widetilde{\theta}_1}^{\widetilde{\theta}_2} \mathbb{1}\{Y(s) \le x\}; \varphi(s)\mathrm{d}s] = L_{r,b}[b, \delta^*]\int_0^x \widetilde{N}_b(v)\,\mathrm{d}v \tag{5.39}$$

for $0 \le x \le b$ and i, j in $\mathcal{E}$.

The stationary distribution

$$\widetilde{\boldsymbol{G}}^{\mathrm{T}}(x) = \widetilde{\boldsymbol{\rho}}^{\mathrm{T}} \begin{bmatrix} \widetilde{M}_0(x) \\ \widetilde{M}_b(x) \end{bmatrix} \qquad \text{for } 0 \le x \le b \tag{5.40}$$

is now fully determined: $\widetilde{\boldsymbol{\rho}}$ is the stationary distribution of the phase at regeneration epochs, it is the stationary probability vector associated to the matrix $\widetilde{\Phi}$ in (5.22), the matrices $\widetilde{M}_0(x)$ and $\widetilde{M}_b(x)$ are given above in (5.38, 5.39), and their constituent factors are given in Lemmas 5.9, 5.11 and 5.12. We wish, however, to give $\widetilde{\boldsymbol{G}}(\cdot)$ a more explicit expression, and we show in Theorem 5.13 that the density is a linear combination of two matrix-exponential functions when $\boldsymbol{\pi}^{\mathrm{T}}\boldsymbol{\mu} \neq 0$.

It is obvious that we need to deal with integrals of e^{Rv} and of $e^{\widehat{R}v}$, and we know from Lemma 4.21 that either R or $\widehat{R}$, or both, has an eigenvalue 0. To simplify the resulting equations, we introduce the notation

$$\mathcal{G}(K, x) = \int_0^x e^{Ku} \, \mathrm{d}u.$$

If all eigenvalues of K are different from zero, then

$$\mathcal{G}(K, x) = (-K)^{-1}(I - e^{Kx}). \tag{5.41}$$

If one eigenvalue is equal to 0 and the others are different from 0, then

$$\mathcal{G}(K, x) = (-K)^{\#}(I - e^{Kx}) + x\, \boldsymbol{v}_K \boldsymbol{u}_K^{\mathrm{T}}, \tag{5.42}$$

where $\boldsymbol{v}_K$ and $\boldsymbol{u}_K$ are the normalised right and left eigenvectors of K for the eigenvalue 0, normalised so that $\boldsymbol{u}_K^{\mathrm{T}}\boldsymbol{v}_K = 1$. To verify this last equation, we observe that both sides are equal to 0 for $x = 0$, and have the same derivatives with respect to x.

Theorem 5.13. *Assume that $\boldsymbol{\pi}^{\mathrm{T}}\boldsymbol{\mu} \neq 0$. The stationary distribution function of the Markov-modulated Brownian motion regulated at both 0 and b is*

$$\widetilde{\boldsymbol{G}}^{\mathrm{T}}(x) = \widetilde{c}\,\widetilde{\boldsymbol{\gamma}}^{\mathrm{T}} \begin{bmatrix} 2\Delta_\sigma^{-2} & 0 \\ 0 & 2\Delta_\sigma^{-2} \end{bmatrix} \begin{bmatrix} I & e^{Rb} \\ e^{\widehat{R}b} & I \end{bmatrix}^{-1} \begin{bmatrix} \mathcal{G}(R, x) \\ e^{\widehat{R}(b-x)}\mathcal{G}(\widehat{R}, x) \end{bmatrix}, \tag{5.43}$$

for $0 \le x \le b$, where

$\widetilde{c}$ is a normalisation scalar such that $\widetilde{\boldsymbol{G}}^{\mathrm{T}}(b)\mathbf{1} = 1$,
$\widetilde{\gamma}$ is the stationary probability vector of the generator defined in (3.37):

$$\widetilde{\gamma}^{\mathrm{T}} \begin{bmatrix} L^{(b)} & \widehat{L}^{(b)} \\ D^{(b)} & \widehat{D}^{(b)} \end{bmatrix} = \mathbf{0}, \qquad \widetilde{\gamma}^{\mathrm{T}}\mathbf{1} = 1. \tag{5.44}$$

The stationary density function is

$$\widetilde{\boldsymbol{g}}^{\mathsf{T}}(x) = \widetilde{c}\,\widetilde{\boldsymbol{\gamma}}^{\mathsf{T}} \begin{bmatrix} 2\Delta_\sigma^{-2} & 0 \\ 0 & 2\Delta_\sigma^{-2} \end{bmatrix} \begin{bmatrix} I & e^{Rb} \\ e^{\widehat{R}b} & I \end{bmatrix}^{-1} \begin{bmatrix} e^{Rx} \\ e^{\widehat{R}(b-x)} \end{bmatrix}. \tag{5.45}$$

Proof. As we assume that $\boldsymbol{\pi}^{\mathsf{T}}\boldsymbol{\mu} \neq 0$, we know that the system (5.26) is nonsingular, and we may write that

$$\begin{bmatrix} \widetilde{N}_0(x) \\ \widetilde{N}_b(x) \end{bmatrix} = \begin{bmatrix} I & e^{Rb} \\ e^{\widehat{R}b} & I \end{bmatrix}^{-1} \begin{bmatrix} e^{Rx} \\ e^{\widehat{R}(b-x)} \end{bmatrix}. \tag{5.46}$$

Next, we bring together Lemma 5.12 and (5.38, 5.39, 5.40), so that

$$\begin{aligned} \widetilde{\boldsymbol{G}}^{\mathsf{T}}(x) = \widetilde{\boldsymbol{\rho}}^{\mathsf{T}} & \begin{bmatrix} (-L^{(b)})^{-1} & 0 \\ 0 & (-\widehat{D}^{(b)})^{-1} \end{bmatrix} \\ & \times \begin{bmatrix} 2\Delta_\sigma^{-2} & 0 \\ 0 & 2\Delta_\sigma^{-2} \end{bmatrix} \begin{bmatrix} I & e^{Rb} \\ e^{\widehat{R}b} & I \end{bmatrix}^{-1} \begin{bmatrix} \int_0^x e^{Rv}\,\mathrm{d}v \\ e^{\widehat{R}(b-x)} \int_0^x e^{\widehat{R}v}\,\mathrm{d}v \end{bmatrix}. \end{aligned}$$

By definition,

$$\widetilde{\boldsymbol{\rho}}^{\mathsf{T}} \begin{bmatrix} -I & H^{(b)} \\ \widetilde{H}^{(b)} & -I \end{bmatrix} = 0$$

or, by the expressions (3.29, 3.46) for the exit probability matrices,

$$\widetilde{\boldsymbol{\rho}}^{\mathsf{T}} \begin{bmatrix} (-L^{(b)})^{-1} & 0 \\ 0 & (-\widehat{D}^{(b)})^{-1} \end{bmatrix} \begin{bmatrix} L^{(b)} & \widehat{L}^{(b)} \\ D^{(b)} & \widehat{D}^{(b)} \end{bmatrix} = \mathbf{0}.$$

Therefore,

$$\widetilde{\boldsymbol{\rho}}^{\mathsf{T}} \begin{bmatrix} (-L^{(b)})^{-1} & 0 \\ 0 & (-\widehat{D}^{(b)})^{-1} \end{bmatrix} \propto \widetilde{\boldsymbol{\gamma}}^{\mathsf{T}}, \tag{5.47}$$

and this completes the proof. □

If $\boldsymbol{\pi}^{\mathsf{T}}\boldsymbol{\mu} = 0$, we should use (5.34) instead of (5.46). We do not write up the details, however, because the vector $\boldsymbol{y}_R$ in (5.34) is a function of x, its derivative is easy to obtain, but the resulting expression for the stationary density $\widetilde{\boldsymbol{g}}(x)$ is not enlightening.

Like in Lemma 5.6, we may connect the stationary distribution to the Skorokhod regulators S_0 and S_b defined at the beginning of the section. Here, the sub-regulators are

$$\begin{aligned} S_{0,i}(t) &= \int_0^t \mathbb{1}\{\varphi(s) = i\}\,\mathrm{d}S_0(s), \\ S_{b,i}(t) &= \int_0^t \mathbb{1}\{\varphi(s) = i\}\,\mathrm{d}S_b(s), \end{aligned}$$

for $i \in \mathcal{E}$.

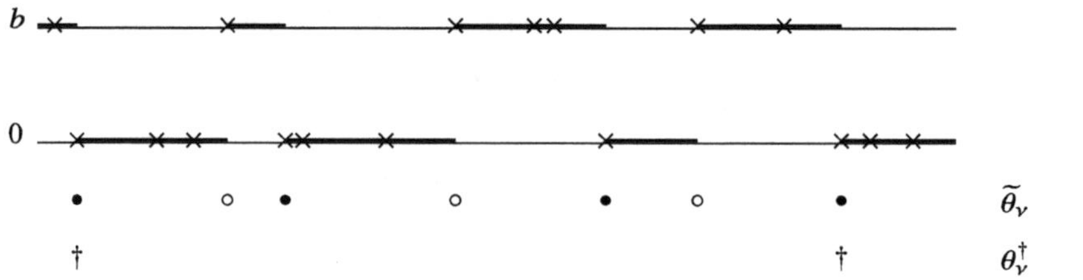

Fig. 5.2 Three interleaved sequences of passage times to the boundaries. The thick lines at level 0 and level b mark intervals between regenerations

Lemma 5.14. *The ratios $S_{0,i}(t)/t$ and $S_{b,i}(t)/t$ converge almost surely to constants as t tends to infinity, and we write*

$$\ell_{0,i} = \lim_{t\to\infty} \frac{1}{t} S_{0,i}(t) \qquad \text{and} \qquad \ell_{b,i} = \lim_{t\to\infty} \frac{1}{t} S_{b,i}(t),$$

for $i \in \mathcal{E}$.

The vector $\begin{bmatrix} \boldsymbol{\ell}_0^{\mathrm{T}} & \boldsymbol{\ell}_b^{\mathrm{T}} \end{bmatrix}$ is proportional to $\widetilde{\boldsymbol{\gamma}}^{\mathrm{T}} \begin{bmatrix} \Delta_\sigma^{-2} & \\ & \Delta_{\sigma}^{-2}, \end{bmatrix}$ where $\widetilde{\boldsymbol{\gamma}}$ is defined in (5.44). It is normalised by $\boldsymbol{\ell}^{\mathrm{T}}\mathbf{1} = -\boldsymbol{\pi}^{\mathrm{T}}\boldsymbol{\mu}$.

Proof. The proof is closely adapted from Lemma 5.6 and we give the main steps only.

Let i be an arbitrary but fixed phase in $\mathcal{E}$ and define the renewal sequence $\{\theta^\dagger_n : n \geq 1\}$ of epochs when $\{Y, \varphi\}$ returns to $(0, i)$ immediately after a visit to b. Precisely, $\{\theta^\dagger_n\}$ is embedded in the semi-renewal process $\{\widetilde{\theta}_n\}$ defined in (5.20, 5.21) in such a way that

$$\begin{aligned} \theta^\dagger_1 &= \inf\{\widetilde{\theta}_\nu : \widetilde{y}_\nu = 0, \widetilde{\varphi}_\nu = i\}, \\ \theta^\dagger_{n+1} &= \inf\{\widetilde{\theta}_\nu > \theta^\dagger_n : \widetilde{y}_\nu = 0, \widetilde{\varphi}_\nu = i\}, \qquad \text{for } n \geq 1. \end{aligned}$$

We work with three interleaved time sequences and we draw the picture in Fig. 5.2 to make the argument easier to follow.The lines across the page represent the two boundaries, 0 and b. The symbols $\times$ mark epochs of visit to the boundaries, similar to the θ^*_ns of Sect. 5.1 in that there has to be at least one change of phase if the visit is a return to the same boundary. The epochs marked by a circle are the regeneration epochs $\widetilde{\theta}_n$s of (5.20, 5.21), the circles are black if the index n is even, meaning a transition to level 0, they are white if the index is odd, for a return to level b; the states at these epochs form the Markov chain with transition matrix $\widetilde{\Phi}$. The epochs marked with a † are the $\theta^\dagger_n$s defined above, when $\{Y\}$ moves to level 0 in the fixed phase i, following a visit to level b.

We show that

$$\lim_{t\to\infty} \frac{1}{t} S_{0,i}(t) = \frac{1}{M^\dagger_{0,i}} \Theta_{0,i} \qquad \text{a.s.,} \tag{5.48}$$

where

$$M^{\dagger}_{0,i} = \mathrm{E}[\theta^{\dagger}_{n+1} - \theta^{\dagger}_{n}]$$

is the expected time between two renewals (two successive † in Fig. 5.2) and

$$\Theta_{0,i} = (L_0[0, \theta^{\dagger}_1])_{ii}$$

is the expected local time in $(0, i)$ during any of the intervals $[\theta^{\dagger}_n, \theta^{\dagger}_{n+1})$.

Note that there may be several visits to $(0, i)$ during such an interval, at times marked with a ×, provided that they do not belong to the sequence $\{\widetilde{\theta}_{2n}\}$ (marked with a black circle). Therefore,

$$\Theta_{0,i} = \sum_{j \in \mathcal{E}} N_{0,j} (L_{\mathrm{r},0}[0, \delta^*])_{ji},$$

where

- $N_{0,j}$ is the expected number of visits to $(0, j)$ at epochs in $\{\widetilde{\theta}_{2n}\}$ between two visits to $(0, i)$, and
- $(L_{\mathrm{r},0}[0, \delta^*])_{ji}$ is the expected local time of $\{Y\}$ in $(0, i)$ during an interval $[\widetilde{\theta}_{2n}, \widetilde{\theta}_{2n+1})$, given that the phase at time $\widetilde{\theta}_{2n}$ is j.

By Resnick [132, Proposition 2.12.2], $N_{0,j} = \widetilde{\rho}_{0,j} / \widetilde{\rho}_{0,i}$, and we have given $L_{\mathrm{r},0}[0, \delta^*]$ in (5.35), so that

$$\Theta_{0,i} = \frac{2}{\widetilde{\rho}_{0,i}} (\widetilde{\boldsymbol{\rho}}_0^{\mathrm{T}} (-L^{(b)})^{-1} \Delta_{\sigma}^{-2})_i. \tag{5.49}$$

Now, define the vector $\boldsymbol{f} = \begin{bmatrix} \boldsymbol{f}_0 \\ \boldsymbol{f}_b \end{bmatrix}$ with

$$f_{0,j} = \mathrm{E}_j[\widetilde{\theta}_{2n+1} - \widetilde{\theta}_{2n}] \qquad f_{b,j} = \mathrm{E}_j[\widetilde{\theta}_{2n+2} - \widetilde{\theta}_{2n+1}].$$

We have

$$\begin{aligned} M^{\dagger}_{0,i} &= \sum_{j \in \mathcal{E}} (N_{0,j} f_{0,j} + N_{b,j} f_{b,j}) \\ &= \frac{1}{\widetilde{\rho}_{0,i}} \widetilde{\boldsymbol{\rho}}^{\mathrm{T}} \boldsymbol{f}. \end{aligned} \tag{5.50}$$

We bring together (5.48, 5.49, 5.50), and we conclude that

$$\boldsymbol{\ell}_0 = \frac{2}{\widetilde{\boldsymbol{\rho}}^{\mathrm{T}} \boldsymbol{f}} \widetilde{\boldsymbol{\rho}}_0^{\mathrm{T}} (-L^{(b)})^{-1} \Delta_{\sigma}^{-2}.$$

By a similar argument,

$$\boldsymbol{\ell}_b = \frac{2}{\widetilde{\boldsymbol{\rho}}^{\mathrm{T}} \boldsymbol{f}} \widetilde{\boldsymbol{\rho}}_b^{\mathrm{T}} (-\widehat{D}^{(b)})^{-1} \Delta_{\sigma}^{-2}.$$

We bring the two equations together and we get

$$\begin{aligned}\boldsymbol{\ell}^{\mathsf T} &\propto \widetilde{\boldsymbol{\rho}}^{\mathsf T}\begin{bmatrix}(-L^{(b)})^{-1} & 0\\ 0 & (-\widehat{D}^{(b)})^{-1}\end{bmatrix}\begin{bmatrix}\Delta_\sigma^{-2} & 0\\ 0 & \Delta_\sigma^{-2}\end{bmatrix}\\ &\propto \widetilde{\boldsymbol{\gamma}}^{\mathsf T}\begin{bmatrix}\Delta_\sigma^{-2} & 0\\ 0 & \Delta_\sigma^{-2}\end{bmatrix}\end{aligned}$$

by (5.47). We show by the same argument as in Lemma 5.6 that $\boldsymbol{\ell}^{\mathsf T}\mathbf{1} = -\boldsymbol{\pi}^{\mathsf T}\boldsymbol{\mu}$ and so conclude our proof. □

If $\{Y\}$ represents the evolution of a finite buffer in some flow system, the first passage times τ_b and τ_0 are of independent interest. In the first case, τ_b is the time when the buffer would begin to overflow, a state of affairs that may call for emergency measures. Conversely, τ_0 is the time when the buffer dries out, which might be a nuisance as well. We give simple expressions in the next corollary for their expected values, in the case when $\boldsymbol{\pi}^{\mathsf T}\boldsymbol{\mu} \neq 0$.

Corollary 5.15. *Consider the MMBM process $\{Y, \varphi\}$ regulated at levels 0 and b. Define $\mathrm{E}_{\mathrm{r},0}[\tau_b]$ to be the expected time to overflow starting from level 0, and $\mathrm{E}_{\mathrm{r},b}[\tau_0]$ to be the expected time to dry out, starting from level b.*

Denote by $\boldsymbol{\alpha}$ the initial probability vector of the phase. We have

$$\mathrm{E}_{\mathrm{r},0}[\tau_b] = \boldsymbol{\alpha}^{\mathsf T}\widetilde{M}_0(b)\mathbf{1} \qquad \text{and} \qquad \mathrm{E}_{\mathrm{r},b}[\tau_0] = \boldsymbol{\alpha}^{\mathsf T}\widetilde{M}_b(b)\mathbf{1}, \tag{5.51}$$

where the matrix functions $\widetilde{M}_0(\cdot)$ and $\widetilde{M}_b(\cdot)$ are given in (5.38, 5.39).

If $\boldsymbol{\pi}^{\mathsf T}\boldsymbol{\mu} \neq 0$, then

$$\widetilde{M}_0(b) = (-L^{(b)})^{-1}(2\Delta_\sigma^{-2})(I - e^{Rb}e^{\widehat{R}b})^{-1}(\mathcal{G}(R,b) - e^{Rb}\mathcal{G}(\widehat{R},b)), \tag{5.52}$$

$$\widetilde{M}_b(b) = (-\widehat{D}^{(b)})^{-1}(2\Delta_\sigma^{-2})(I - e^{\widehat{R}b}e^{Rb})^{-1}(\mathcal{G}(\widehat{R},b) - e^{\widehat{R}b}\mathcal{G}(R,b)). \tag{5.53}$$

Proof. Equation (5.51) is obvious since the matrices $\widetilde{M}_0(b)$ and $\widetilde{M}_b(b)$ are the expected occupation times in the whole interval $[0, b]$ between two regeneration points.

If $\boldsymbol{\pi}^{\mathsf T}\boldsymbol{\mu} \neq 0$, we use (5.46) and the integral equations (5.38, 5.39) to write

$$\begin{bmatrix}\widetilde{M}_0(b)\\ \widetilde{M}_b(b)\end{bmatrix} = \begin{bmatrix}L_{\mathrm{r},0}[0,\delta^*] & 0\\ 0 & L_{\mathrm{r},b}[b,\delta^*]\end{bmatrix}\begin{bmatrix}I & e^{Rb}\\ e^{\widehat{R}b} & I\end{bmatrix}^{-1}\begin{bmatrix}\mathcal{G}(R,b)\\ \mathcal{G}(\widehat{R},b)\end{bmatrix}.$$

We prove by direct verification that

$$\begin{bmatrix}I & e^{Rb}\\ e^{\widehat{R}b} & I\end{bmatrix}^{-1} = \begin{bmatrix}(I - e^{Rb}e^{\widehat{R}b})^{-1} & 0\\ 0 & (I - e^{\widehat{R}b}e^{Rb})^{-1}\end{bmatrix}\begin{bmatrix}I & -e^{Rb}\\ -e^{\widehat{R}b} & I\end{bmatrix}$$

and the remainder of the proof follows immediately from Lemma 5.12. □

Example 5.16. We conclude this section with an illustration based on the system analysed in Bean et al. [29]. The regulated MMBM $\{Y, \varphi\}$ in this example is a model for a hydropower generation system. The component Y represents the level of deterioration of the equipment, 0 being the state of a newly operating machine and $Y = b$ indicating that it has to be removed and replaced. We take $b = 1$. There are six phases:

1. *On-design:* Energy is produced at a normal rate.
2. *Off-design:* At times of lower demand, energy is produced at a lower rate and wear is greater.
3. *Start:* Production is being started.
4. *Stop:* Production is being stopped.
5. *Idle:* The generator is actually idle.
6. *Maintenance:* The equipment is undergoing repairs.

The adjacency matrix of the transition graph is

	1	*2*	*3*	*4*	*5*	*6*
1		1		1		
2	1			1		
3	1	1		1		
4			1		1	
5			1			1
6					1	

and the transition rates matrix is taken from [29]—the missing entries are equal to 0:

$$Q = \begin{bmatrix} -83.9 & 40.7 & & 43.2 & & \\ 180.2 & -262.9 & & 82.7 & & \\ 1085.5 & 314.9 & -1440 & 39.6 & & \\ & & 17.6 & -1440 & 1422.4 & \\ & & 62.3 & & -64.8 & 2.5 \\ & & & & 39.4 & -39.4 \end{bmatrix}.$$

The time unit is one month, so that the *Start* and the *Stop* phases last 30 minutes on average, etc. We assume that $Y(0) = 0$ and that the initial phase distribution is $\alpha = \begin{bmatrix} 1 \; 0 \dots 0 \end{bmatrix}^{\mathrm{T}}$, so that at time 0 a newly operating machine starts production at the normal rate.

The rate at which wear increases is

$$\mu = \begin{bmatrix} 0.004 & 0.017 & 0.020 & 0.020 & 0.001 & -0.010 \end{bmatrix}^{\mathrm{T}},$$

it is greatest when the equipment is running in the *off-design* phase, it is small but positive in the *idle* state, and it is negative when in *maintenance*, meaning that the equipment recovers from some of its deterioration. The value of μ_1 is such that the generator might run for 250 months if it were continuously run *on-design*.

The model in [29] is a Fluid flow system with $\sigma_i = 0$ for all i. We use four sets of strictly positive σ_is to observe the effect of the Brownian volatility on the expected time until the equipment has to be replaced. The four sets are given in the next table, we use three values for individual parameters: The smallest is $v_\ell = 0.01$, the largest is $v_m = 0.1$, and the intermediary is $w = 0.05$.

Case	σ					
	1	2	3	4	5	6
A	v_ℓ	v_ℓ	v_ℓ	v_ℓ	v_ℓ	v_ℓ
B	w	w	v_ℓ	v_ℓ	v_ℓ	v_ℓ
C	v_ℓ	v_ℓ	v_ℓ	v_ℓ	v_ℓ	v_m
D	w	w	v_ℓ	v_ℓ	v_ℓ	v_m

In Case A, all volatility parameters are small, and we expect that performances are not much different from what they would be if $\boldsymbol{\sigma}$ were actually equal to 0. We assume that environmental conditions may influence the rate of deterioration, and for this we set $\sigma_1 = \sigma_2 = 0.05$ in case B. Furthermore, it is likely that the term "maintenance" covers different activities, from periodical tune-ups to major repairs, and may improve to varying degrees the state of the machine, and to reflect this, we set $\sigma_6 = 0.1$ in Case D. Case C, in between Cases B and D, is there to illustrate the effect of increasing σ_6 without increasing σ_1 and σ_2.

The expected time to failure is reported in [29] to be equal to 243.96 when $\boldsymbol{\sigma} = 0$: here, the computed values are

Case	A	B	C	D
$\mathrm{E}_{\mathrm{r},0}[\tau_1]$	240.78	204.07	232.84	196.40

confirming our expectation that Case A would be close to the Fluid flow model.

The components of the vector $\boldsymbol{\alpha}^{\mathrm{T}}\widetilde{M}_0(1)$ are the expected time spent in each phase until failure of the machine. We only give the *proportion* of time, as it so happens that this is the same in all four cases, up to four decimal places. Defining $p_i = (\boldsymbol{\alpha}^{\mathrm{T}}\widetilde{M}_0(1))_i/\mathrm{E}_{\mathrm{r},0}[\tau_1]$, we have

$$\boldsymbol{p} = \begin{bmatrix} 0.4279 & 0.0883 & 0.0184 & 0.0184 & 0.4204 & 0.0267 \end{bmatrix}^{\mathrm{T}}.$$

The system spends most of the time in phases 1 and 5, producing energy at the normal rate, or being idle. The generator spends slightly less than 10% of the time in phase 2, producing at a lower rate, but as we show below, this small amount of time has a significant effect on the wear of the machine.

We show in Fig. 5.3 the expected contribution C_i of individual phases to the total degradation of the power generator: These contributions are defined as

$$C_i = (\boldsymbol{\alpha}^{\mathrm{T}}\widetilde{M}_0(1))_i\,\mu_i, \qquad \text{for } i \text{ in } \mathcal{E},$$

that is, C_i is the expected wear during the expected total sojourn time in phase i. The first observation is that C_2 is much larger than one would expect from the small proportion p_2 of time spent in the *off-design* state. Also, the contributions do depend to some extent on the volatility $\boldsymbol{\sigma}$, with $\boldsymbol{C}$ being more sensitive to σ_1 and σ_2 than to σ_6. A last observation is that the total expected degradation $\boldsymbol{C}\mathbf{1}$ is strictly less

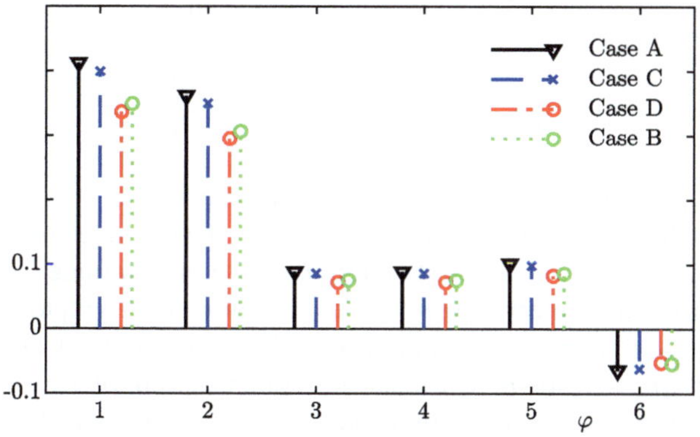

Fig. 5.3 Example 5.16. Expected individual contribution C_i of each phase to the total degradation. Cases are plotted in the order A, C, D, B from left to right

than the total actual degradation, which is 1 by design of the example. In Case A for instance, $\boldsymbol{C}1$ is equal to 0.9879, and it is even smaller in the other cases.

5.3 Change of Regime at Level 0

We argued in Parenthesis 1.6 that the Skorokhod regulation is not always the adequate model when the level is constrained by one or two boundaries. Systems in real life are designed with feedback loops: A finite buffer does not repeatedly overflow without its input being throttled, water conservation measures are taken before reservoirs get thoroughly dry, and so on. This is our reason for being interested in stochastic processes with reactive boundaries, that is, processes that change behaviour upon hitting some boundary. We give two examples, here and in the next section.

Assume that visits to level zero are critical epochs and that some procedure needs to be implemented until the fluid has moved sufficiently far away from the boundary. In traditional queueing theory, such considerations have led to the analysis of so-called queues with server vacation, where the server is not allowed to remain idle and is temporarily assigned to other tasks. A simple example we have in mind is that of a buffer in a communication network. When the level gets down to zero, an additional source may be allowed to use it temporarily.

The detailed specifications of the model in this section are as follows: Under normal circumstances, the MMBM is characterised by the usual parameters $\boldsymbol{\mu}$, $\boldsymbol{\sigma}$ and Q, on the state space $\mathcal{E}$. When the level drops down to 0, there is an instantaneous change to the alternate set of parameters $\boldsymbol{\mu}'$, $\boldsymbol{\sigma}'$ and Q', on the state space $\mathcal{E}'$, and the parameters remain so until the level reaches $b > 0$, at which time the parameters return to their original setting until the next visit to level 0. Thus, time is marked by the epochs

$$\overline{\theta}_0 = \inf\{t > 0 : Z(t) = 0\},$$
$$\theta'_n = \inf\{t > \overline{\theta}_n : Z(t) = b\},$$
$$\overline{\theta}_{n+1} = \inf\{t > \theta'_n : Z(t) = 0\},$$

for $n \geq 0$, and $\{\overline{\theta}_n\}$s is the sequence of regeneration epochs.

Like in Sect. 5.1, we assume that $\boldsymbol{\pi}^{\mathrm{T}}\boldsymbol{\mu} < 0$, so that the regulated process repeatedly returns to level 0, and $\overline{\theta}_n$, θ'_n are finite for all n. Since the alternate set of parameters only apply in the finite interval $[0, b]$, we might allow the asymptotic rate $\boldsymbol{\pi}'^{\mathrm{T}}\boldsymbol{\mu}'$ to take any value, where $\boldsymbol{\pi}'$ is the stationary probability vector of Q'. However, we assume that $\boldsymbol{\pi}'^{\mathrm{T}}\boldsymbol{\mu}' > 0$ for it will make for somewhat simpler expressions in the sequel. Furthermore, this assumption makes sense since the objective is to have the MMBM move away from the boundary 0. We assume that either $\sigma'_i > 0$ for all $i \in \mathcal{E}'$ (Case 1) or that $\sigma'_i = 0$, $\mu'_i > 0$ for all i (Case 2): In both cases, the process does not remain in level 0 for any positive interval of time. We shall use the notation U', $\widehat{U}'$, etc., for matrices and vectors pertaining to the alternate set of parameters.

We also specify two transition matrices, P_0 and P_b: For i in $\mathcal{E}$ and j in $\mathcal{E}'$, $(P_0)_{ij}$ is the probability that at a renewal epoch $\overline{\theta}_n$, phase j is selected, given that the process was in phase i upon reaching level 0, and $(P_b)_{ji}$ is the probability that the phase changes from j to i when the level returns to b at time θ'_n.

Before getting into the detailed analysis, we observe that $U + \widehat{U}$ and $U' + \widehat{U}'$ are both nonsingular: the former because $\boldsymbol{\pi}^{\mathrm{T}}\boldsymbol{\mu} < 0$, so that $\widehat{U}$ is a nonsingular generator, and the latter because $\boldsymbol{\pi}'^{\mathrm{T}}\boldsymbol{\mu}' > 0$, so that U' is a nonsingular generator.

Case 1 $\sigma'_i > 0$ for all i in $\mathcal{E}'$.

We immediately recognise that the transition matrix for $\{\varphi(\overline{\theta}_n) : n \geq 0\}$ is

$$\overline{\Phi} = H' \, P_b \, e^{Ub} \, P_0, \tag{5.54}$$

where H' gives the conditional distribution of the phase in $\mathcal{E}'$, upon reaching level b at the end of the interval $[\overline{\theta}_n, \theta'_n)$, given the phase at time $\overline{\theta}_n$ after hitting the regulated level 0. That matrix is given by the right-hand side of (3.40), where U and $\widehat{U}$ are replaced with the generators U' and $\widehat{U}'$. By definition, the factor P_b gives the distribution of the new phase in $\mathcal{E}$ at time θ'_n; the third factor takes us to the end of the interval $[\theta'_n, \overline{\theta}_{n+1})$, and we conclude with the change of phase at time $\overline{\theta}_{n+1}$, ready to start anew with a phase in $\mathcal{E}'$.

To determine the expectedoccupation time $\overline{M}(x)$ in $[0, x]$ during an interval of time between regenerations, we analyse separately the occupation times before and after the first passage to level b:

$$(\overline{M}_1)_{ij}(x) = \mathrm{E}[\int_{\overline{\theta}_n}^{\theta'_n} \mathbb{1}\{Z(t) \leq x, \varphi(t) = j\} \, \mathrm{d}t \mid \varphi(\overline{\theta}_n) = i], \qquad i, j \in \mathcal{E}',$$

$$(\overline{M}_2)_{ij}(x) = \mathrm{E}[\int_{\theta'_n}^{\overline{\theta}_{n+1}} \mathbb{1}\{Z(t) \leq x, \varphi(t) = j\} \, \mathrm{d}t \mid \varphi(\theta'_n) = i], \qquad i, j \in \mathcal{E},$$

and to begin with, we determine the local times

$$\overline{m}_1(x) = \frac{\partial}{\partial x}\overline{M}_1(x) \qquad \text{and} \qquad \overline{m}_2(x) = \frac{\partial}{\partial x}\overline{M}_2(x).$$

During $[\overline{\theta}_n, \theta'_n)$, the process behaves exactly like the process $\{Y\}$ of Sect. 5.2 during the interval $[0, \delta^*)$: Both processes feel the effect of the regulation at level 0 and neither has to react to the control at level b. Therefore, we may immediately write that

$$\overline{m}_1(x) = L'_{\mathrm{r},0}[0, \delta^*]\widetilde{N}'_0(x), \tag{5.55}$$

for $0 < x < b$, where $L'_{\mathrm{r},0}[0, \delta^*]$ is the local time at level 0, and $\widetilde{N}'_0(x)$ is the local time at level x per unit of the local time at 0, during the interval $[\overline{\theta}_n, \theta'_n)$. By Lemmas 5.12 and 3.16,

$$\begin{aligned} L'_{\mathrm{r},0}[0, \delta^*] &= -(L'^{(b)})^{-1}(2\Delta_{\sigma'}^{-2}) \\ &= -(I - e^{\widehat{U}'b}e^{U'b})(U' + \widehat{U}'e^{\widehat{U}'b}e^{U'b})^{-1}(2\Delta_{\sigma'}^{-2}) \end{aligned}$$

since we assume that $\boldsymbol{\pi}'^{\mathrm{T}}\boldsymbol{\mu}' > 0$, and

$$\widetilde{N}'_0(x) = (e^{-R'b} - e^{\widehat{R}'b})^{-1}(e^{-R'(b-x)} - e^{\widehat{R}'(b-x)})$$

by Corollary 5.10.

During the second part $[\theta'_n, \overline{\theta}_{n+1})$ of the regenerative interval, as the process starts in level b and is not subjected to the constraint of any boundary until its return to 0, we have $\overline{m}_2(x) = L_b[x, \tau_0]$.

If $x > b$, then

$$\overline{m}_2(x) = L_b[b, \tau_0]\, e^{R(x-b)} \tag{5.56}$$

as $e^{R(x-b)}$ is the expected local time in level $x > b$ per unit of local time at b, and $L_b[b, \tau_0]$ is the expected local time at level b, starting from level b, before the first visit to level 0.

If $x < b$, then

$$\overline{m}_2(x) = L_b[b, \tau_0]\, \widetilde{N}_b(x) \tag{5.57}$$

as $\widetilde{N}_b(x)$, analysed in Sect. 5.2 and given in (5.30), is the expected local time at level $x < b$, per unit of the local time at level b, before the process hits level 0.

Obviously,

$$\begin{aligned} L_b[b, \tau_0] &= L_0[0, \tau_{-b}], \\ &= -2(I - e^{Ub}e^{\widehat{U}b})(U + \widehat{U})^{-1}\Delta_\sigma^{-2} \end{aligned}$$

by Theorem 4.4 and Corollary 4.5, and so $\overline{m}_2(x)$ is fully determined for all $x > 0$.

We collect in Theorem 5.17 the various bits and pieces obtained above, and we do not give a detailed expression for the stationary distribution function, as it is quite cumbersome and not enlightening.

Theorem 5.17. *Assume that* $\boldsymbol{\pi}^{\mathrm{T}}\boldsymbol{\mu} < 0$, *that* $\sigma'_i > 0$ *for all* i *in* $\mathcal{E}'$, *and that* $\boldsymbol{\pi}'^{\mathrm{T}}\boldsymbol{\mu}' > 0$. *The stationary distribution function of the Markov-modulated Brownian motion with change of regime at level 0 is given as*

$$\overline{\boldsymbol{G}}(x) = \overline{c}\,\overline{\boldsymbol{\rho}}^{\mathrm{T}}\,(\int_0^x \overline{m}_1(v)\,\mathrm{d}v + H'P_b \int_0^x \overline{m}_2(v)\,\mathrm{d}v), \tag{5.58}$$

for $x \geq 0$, *where*

$\overline{\boldsymbol{\rho}}$ *is the stationary probability vector of* $\overline{\Phi}$,
$\overline{m}_1(x)$ *is given by (5.55) for* $0 < x < b$ *and is equal to 0 for* $x > b$,
$\overline{m}_2(x)$ *is given by (5.56, 5.57).*

The normalisation constant $\overline{c}$ *is*

$$\overline{c} = (\mathrm{E}'_{\mathrm{r},0}[\tau_b] + \mathrm{E}_b[\tau_0])^{-1}. \tag{5.59}$$

The expected time $\mathrm{E}'_{\mathrm{r},0}[\tau_b]$ *to reach level* b *is given in Lemma 5.15, using* $\overline{\boldsymbol{\rho}}$ *as initial probability vector and the alternate parameters* $\boldsymbol{\mu}'$, $\boldsymbol{\sigma}'$ *and* Q'. *The expected time* $\mathrm{E}_b[\tau_0]$ *to return to level 0 is given by (2.86) with initial probability vector* $\overline{\boldsymbol{\rho}}^{\mathrm{T}}H'P_b$.

Proof. The vector $\overline{\boldsymbol{\rho}}$ is the probability distribution of the phase at the beginning of a regeneration interval and $\overline{\boldsymbol{\rho}}^{\mathrm{T}}H'P_b$ is its distribution at the time the process reaches level b. This, together with the definition of $\overline{m}_1(x)$ and $\overline{m}_2(x)$, justifies (5.58).

The normalisation constant $\overline{c}$ may be written as

$$(\overline{c})^{-1} = \overline{\boldsymbol{\rho}}^{\mathrm{T}}\,(\int_0^b \overline{m}_1(v)\,\mathrm{d}v + H'P_b \int_0^\infty \overline{m}_2(v)\,\mathrm{d}v)\,\mathbf{1}.$$

Equation (5.59) is equivalent and reads better. □

Case 2 $\mu'_i > 0$, $\sigma'_i = 0$ for all i in $\mathcal{E}'$.

As the evolution of the MMBM does not change during the second part $[\theta'_n, \overline{\theta}_{n+1})$ of the regeneration interval, we only need to deal with the beginning. During $[\overline{\theta}_n, \theta'_n)$, as $\boldsymbol{\mu}' > \mathbf{0}$ and $\boldsymbol{\sigma}' = 0$, the fluid increases monotonically with time. We know from the discussion at the beginning of Sect. 2.4 that the rates $|\mu'_i|$, i in $\mathcal{E}'$, act as conversion rates between time and level, and so, the generator of the ladder process $\{\varphi(\tau_x) : x \geq 0\}$ is in this case

$$\widehat{U}' = \Delta_{\mu'}^{-1} Q',$$

and the phase transition probability matrix from $\overline{\theta}_n$ to θ'_n is

$$H' = e^{\widehat{U}'b}.$$

Furthermore, each level $v \leq x$ being visited once and only once during the interval $[0, \tau_x]$, we have by (4.4) that the local time is given by

$$\ell_j(v, \tau_x) = \mathbb{1}\{\varphi(\tau_v) = j\}/\mu'_j,$$

and the matrix $\overline{M}_1(x)$ of expected occupation time of $[0, x]$ during the first part of a regeneration interval is

$$\overline{M}_1(x) = \int_0^x e^{\widehat{U}'v} \Delta_{\mu'}^{-1} \, \mathrm{d}v$$
$$= ((-\widehat{U}')^{\#}(I - e^{\widehat{U}'x}) + x\mathbf{1} \cdot \boldsymbol{w}'^{\mathrm{T}})\Delta_{\mu'}^{-1} \quad (5.60)$$

by (5.42), where $\boldsymbol{w}'^{\mathrm{T}} = (\boldsymbol{\pi}'^{\mathrm{T}}\boldsymbol{\mu}')^{-1}\boldsymbol{\pi}'^{\mathrm{T}}\Delta_{\mu'}$ is the left eigenvector of $\widehat{U}'$ for the eigenvalue 0.

We give the next theorem without proof. The main difference with Theorem 5.17 is the much simpler expression for $\overline{M}_1(x)$ in the formulation of (5.61).

Theorem 5.18. *Assume that $\boldsymbol{\pi}^{\mathrm{T}}\boldsymbol{\mu} < 0$, that $\mu'_i > 0$, and that $\sigma'_i = 0$ for all i in $\mathcal{E}'$. The stationary distribution function of the Markov-modulated Brownian motion with change of regime at level 0 is given as*

$$\overline{\boldsymbol{G}}(x) = \overline{c}\,\overline{\boldsymbol{\rho}}^{\mathrm{T}} (\overline{M}_1(x) + H' P_b \int_0^x \overline{m}_2(v) \, \mathrm{d}v), \quad (5.61)$$

for $x \geq 0$, where

$\overline{\boldsymbol{\rho}}$ is the stationary probability vector of $\overline{\Phi}$,
$\overline{M}_1(x)$ is given in (5.60) for $0 < x < b$, and is equal to $\overline{M}_1(b)$ for $x > b$,
$\overline{m}_2(x)$ is given by (5.56 5.57),

and the normalisation constant $\overline{c}$ is

$$\overline{c} = (\mathrm{E}'_{\mathrm{r},0}[\tau_b] + \mathrm{E}_b[\tau_0])^{-1}. \quad (5.62)$$

The expected time to reach level b is $\mathrm{E}'_{\mathrm{r},0}[\tau_b] = \overline{\boldsymbol{\rho}}\overline{M}_1(b)\mathbf{1}$. The expected time $\mathrm{E}_b[\tau_0]$ to return to level 0 is given by (2.86) with initial probability vector $\overline{\boldsymbol{\rho}}^{\mathrm{T}} H' P_b$. □

Example 5.19. We illustrate in Fig. 5.4 the effect of changing the regime at level 0 in Case 2. The process $\{X, \varphi\}$ is similar to Example 2.28: The phase cycles through the values 1 to m, with $m = 6$, at a constant switching rate equal to 0.1. The parameters μ_i are all equal to -0.5, the parameters σ_i are equal to 1 for $i \neq 3$, and σ_3 is equal to 4. Thus, the MMBM behaves very much like a Brownian motion BM(-0.5, 1), except that the volatility increases whenever the process is in phase 3.

The alternate parameters are

$$Q' = \begin{bmatrix} -0.5 & 0.5 \\ 0.5 & -0.5 \end{bmatrix}, \qquad \mu' = \begin{bmatrix} 5 \\ 1 \end{bmatrix}$$

and $\boldsymbol{\sigma}' = \mathbf{0}$. The transition matrices are

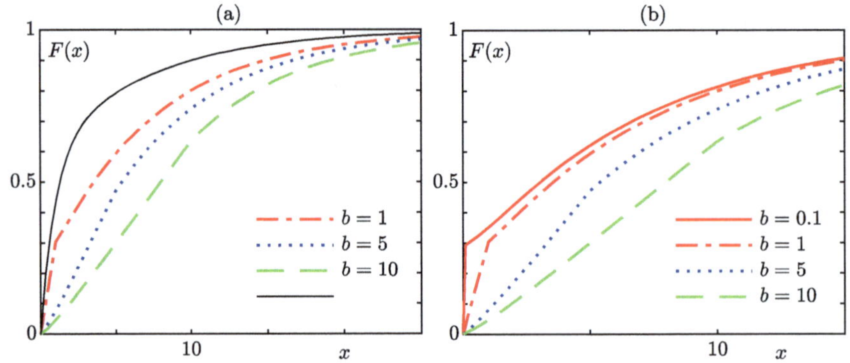

Fig. 5.4 Distribution functions for a system with change of regime at level 0 for different values of b, equal to 0.1 (plain red line), 1 (dot-dotted red line), 5 (dotted blue line), and 10 (dashed green line). In (**a**) on the left, the plain black line is the distribution function of the process without change of regime. In (**b**) on the right, the range for x is restricted to a smaller interval, so as to better show the limiting behaviour as b gets smaller

b	0.1	1	5	10
$\mathrm{E}'_{\mathrm{r},0}[\tau_b]$	0.082	0.712	2.464	4.171
$\overline{c}^{-1}$	0.282	2.712	12.464	24.171

Table 5.1 Expected time to reach b at the beginning of a regeneration interval, and expected length of the inter-regeneration interval for varying b

$$P_b = \begin{bmatrix} 0 & 0 & 1 & 0 & 0 & 0 \\ 0 & 0 & 1 & 0 & 0 & 0 \end{bmatrix} \qquad \text{and} \qquad P_0 = \begin{bmatrix} 0.8 & 0.2 \\ 1 & 0 \\ 0.2 & 0.8 \\ 1 & 0 \\ 0.2 & 0.8 \\ 0 & 1 \end{bmatrix},$$

so that upon hitting level b at time θ'_n, the volatility parameter immediately takes its higher value σ_3.

Not surprisingly, we see that the occupancy of the buffer increases with b. Still, the *tail* of the distribution is not getting heavier, as the tail of $\overline{m}_2(x)$ is determined by the same matrix e^{Rx} as the Skorokhod-regulated process of Sect. 5.1, independently of the change of regime.

One interesting feature of Case 2 is that the limiting distribution, as b tends to 0, has a positive mass at 0, even though $\overline{\boldsymbol{G}}(0) = 0$ for all $b > 0$. We make this visible in Fig. 5.4b on the right by adding $b = 0.1$ as a fourth value, and by restricting the range of x to the interval $[0, 15]$. This limiting behaviour is proved formally in Latouche and Simon [107], the argument going as follows: As b gets smaller, the expected time to reach level b gets smaller also, and the expected inter-regeneration time $\overline{c}^{-1}$ gets smaller as well. This is illustrated in Table 5.1. Eventually, the expression (5.61) for

$\overline{G}(x)$ tends to the indeterminate form 0/0, and we apply L'Hôpital's rule, to obtain the limiting distribution. We write $G(x)\mathbf{1} = T_b(x)/T_b(\infty)$, with

$$T_b(x) = \bar{\rho}^{\mathrm{T}}(\overline{M}_1(b) + H'P_b \int_0^x \overline{m}_2(v)\,\mathrm{d}v)\,\mathbf{1}$$

for any given x. As b tends to 0, H' and e^{Ub} both converge to the identity matrix, so that $\overline{\rho}$ tends to $\overline{\rho}_0$, the stationary probability vector of P_bP_0. Next, one easily shows that

$$\frac{\partial}{\partial b}\overline{M}_1(b)|_{b=0} = \Delta_{\mu'}^{-1}$$

and, after simple manipulations, that

$$\lim_{b\to 0}\frac{\partial}{\partial b}\int_0^x \overline{m}_2(v)\,\mathrm{d}v) = 2\Delta_{\sigma}^{-2}(-R)^{-1}(I - e^{Rx}).$$

Finally, $\lim_{b\to 0} G(x)\mathbf{1} = T_0'(x)/T_0'(\infty)$ with

$$T_0'(x) = \overline{\rho}_0^{\mathrm{T}}(\Delta_{\mu'}^{-1} + 2\Delta_{\sigma}^{-2}(-R)^{-1}(I - e^{Rx}))\mathbf{1},$$

and we see that $T_0'(0) > 0$.

5.4 Sticky Boundary at Level 0

We analyse in this section a family of regulated Brownian processes that behave as usual away from the boundary but are slowed down at level 0, in such a way that, without actually staying at level 0 for any interval of time, nevertheless they have a strictly positive expected occupation time there. Our system with a sticky boundary is inspired from Harrison and Lemoine [81], and we start with a description of their model.

Take a Brownian motion $\{X\}$ with parameters $\mu < 0$ and $\sigma^2 > 0$ and assume that $X(0) = 0$. Next, define the same regulated process $Z(t) = X(t) - m_X(t)$ as in (5.1), where $m_X(t) = \inf\{X(s) : 0 \le s \le t\}$ is the running minimum, and $|m_X(\cdot)|$ is the Skorokhod regulator. Define

$$V(t) = t + |m_X(t)|/\omega, \qquad \omega > 0,$$

together with the function $\Gamma(\cdot)$ such that $V(\Gamma(s)) = s$ for all $s \ge 0$. Finally, define the new process $\{Y(t) : t \ge 0\}$ with

$$Y(t) = Z(\Gamma(t)). \tag{5.63}$$

The regulator $\{|m_X|\}$ is nonnegative, non-decreasing, and continuous, and it remains constant when $Z(t) > 0$. Therefore, $V(t)$ is continuous and strictly increasing, and so is $\Gamma(t)$. We refer to Γ as the new clock imposed on $\{Z\}$.

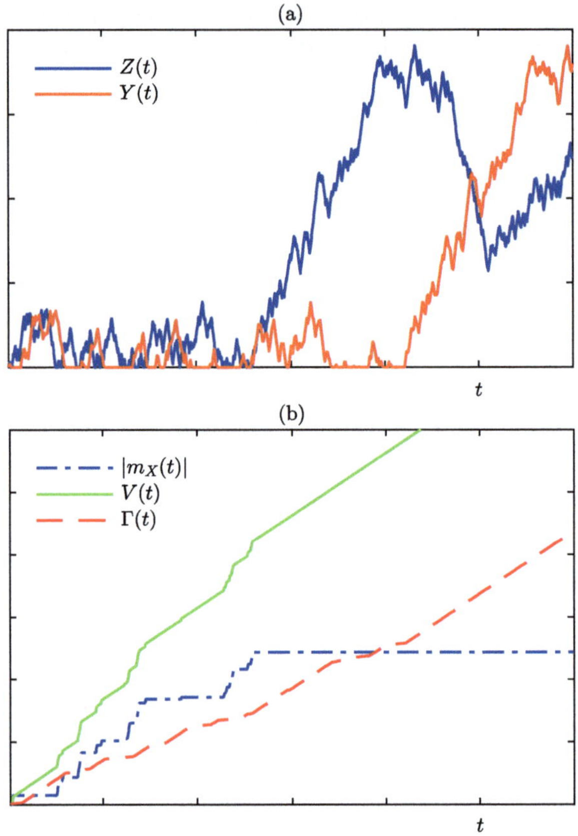

Fig. 5.5 Illustration for the use of the new clock $\{\Gamma\}$. In (**a**), a trajectory $Z(t)$ of the regulated Brownian motion is drawn with a plain blue line and the transformed trajectory $Y(t)$ with a plain red line. In (**b**), the regulator $|m_X(t)|$ is shown as a dot-dashed blue line, the function $V(t)$ in plain green, and the new clock $\Gamma(t)$ in dashed red. The parameters are $\mu = -1$, $\sigma^2 = 1$, and $\omega = 1.5$

As $V(t) \geq t$, therefore $\Gamma(t) \leq t$, and the level of $\{Y\}$ at any time t_0 is a copy of the level of $\{Z\}$ at the earlier time $\Gamma(t_0)$. Conversely, the level of $\{Z\}$ at any time s_0 is a copy of the level of $\{Y\}$ at the later time $V(s_0)$.

Consider an interval of time during which $Z(t)$ is constantly strictly positive. During such an interval, the regulator remains constant, $V(t)$ is equal to t plus some constant quantity, and the trajectory of $\{Z\}$ is translated without change. We illustrate this in Fig. 5.5.

Harrison and Lemoine [81] prove that $\{Y(t) : t \geq 0\}$ is a sticky Brownian motion with parameters μ, σ^2 and ω, with stationary distribution function

$$\lim_{t\to\infty} \mathrm{P}[Y(t) \leq x] = \frac{|\mu|}{|\mu| + \omega} + \frac{\omega}{|\mu| + \omega}(1 - e^{2\mu x/\sigma^2})$$

for $x \geq 0$. Notice that the stationary distribution has a strictly positive probability mass at 0.

Getting back to the MMBM $\{X, \varphi\}$, we define as follows a Markov-modulated Brownian motion with sticky boundary at 0: Similarly to (5.17) for the one-sided regulated process, we decompose $|m_X(t)|$ as the sum of m sub-regulators

$$S_i(t) = \int_0^t \mathbb{1}\{\varphi(s) = i\} \, |m_X(\mathrm{d}s)|, \qquad i \in \mathcal{E}.$$

It is clear that $S_i(t)$ increases only at such times t when both $Z(t) = 0$ and $\varphi(t) = i$. Next, we define

$$V(t) = t + \sum_{i \in \mathcal{E}} S_i(t)/\omega_i,$$

where $\omega_i > 0$ for all i; we define the new clock $\Gamma(t)$ such that $V(\Gamma(s)) = s$ for all $s \geq 0$. The parameters ω_i need not be equal, so that we allow the clock $\Gamma(t)$ to slow down at different rates for different phases. The process analysed in this section is $\{Y(t), \kappa(t) : t \geq 0\}$, with

$$Y(t) = Z(\Gamma(t)) \qquad \text{and} \qquad \kappa(t) = \varphi(\Gamma(t)). \tag{5.64}$$

Like in Sect. 5.1, we assume that $\boldsymbol{\pi}^{\mathrm{T}}\boldsymbol{\mu} < 0$, and we take as regeneration times the sequence $\{\xi_n^* : n \geq 0\}$ of return times to 0 after at least one change of phase:

$$\begin{aligned}
\xi_0^* &= \inf\{t \geq 0 : Y(t) = 0\}, \\
\xi_n' &= \inf\{t > \xi_{n-1}^* : \kappa(t) \neq \kappa(\xi_{n-1}^*)\}, \\
\xi_n^* &= \inf\{t > \xi_n' : Y(t) = 0\}
\end{aligned}$$

for $n \geq 1$. The epochs ξ_n^* are finite for all n since we assume that $\boldsymbol{\pi}^{\mathrm{T}}\boldsymbol{\mu} < 0$.

Equation (5.64) defines a pathwise transformation, the regulators S_i remain constant when $Z(t) > 0$ and, like in the case of a simple Brownian motion, the sample path of Z is modified solely through the insertion of additional time at level 0. In consequence,

$$\xi_n^* = V(\theta_n^*) \qquad \text{and} \qquad \xi_n' = V(\theta_n'), \qquad\qquad n \geq 0,$$

with the sequences $\{\theta_n^*\}$ and $\{\theta_n'\}$ defined in (5.2–5.4) for the Skorokhod-defined process.

The sequence of phases visited by $\{\kappa\}$ is the same as $\{\varphi\}$: The phase remains constant over the intervals (ξ_{n-1}^*, ξ_n'), by definition, and it is copied without change from $\{\varphi\}$ over $[\xi_n', \xi_n^*]$. For that reason, the transition matrix for the Markov chain $\{\kappa(\xi_n^*) : n \geq 0\}$ is the matrix Φ in (2.26), and its stationary probability vector is $\boldsymbol{\rho}_{\mathrm{s}}$, such that

$$\boldsymbol{\rho}_{\mathrm{s}}^{\mathrm{T}}(I - \Phi) = \mathbf{0}, \qquad \boldsymbol{\rho}_{\mathrm{s}}^{\mathrm{T}}\mathbf{1} = 1.$$

It is the same as the vector $\boldsymbol{\rho}$ defined in Theorem 5.4 and analysed in Remark 5.5, but we use the subscript "s" (for *sticky*) to identify the elements of the stationary distribution in this section.

The next lemma will be used to determine the stationary distribution of our sticky MMBM.

Lemma 5.20. *Assume that $\{Y(t) : t \geq 0\}$ is a sticky Brownian motion with parameters $\mu < 0$, $\sigma^2 > 0$ and $\omega > 0$ defined by (5.63) where $\{Z\}$ is a Skorokhod-regulated Brownian motion with parameters μ and σ^2. Assume that T is a random variable, independent of the regulated Brownian motion, exponentially distributed with parameter λ.*

Define the expected occupation time of $\{Y\}$ at level 0 during $[0, V(T)]$ as

$$\Lambda[0,T] = \mathrm{E}[\int_0^{V(T)} \mathbb{1}\{Y(v) = 0\} \, \mathrm{d}v \mid Y(0) = 0].$$

We have

$$\Lambda[0,T] = (b\,\omega)^{-1},$$

where $b = \alpha_+(\mu, \sigma, \lambda)$ is defined in (2.18).

Proof. By construction, the trajectory $\{Z(s) : 0 \leq s \leq T\}$ is modified solely through the insertion of additional time at level 0, and the evolution of $\{Z\}$ above 0 is reproduced, without change, as part of the trajectory $\{Y(t) : 0 \leq t \leq V(T)\}$. Therefore,

$$\int_0^{V(T)} \mathbb{1}\{Y(v) > 0\} \, \mathrm{d}v = \int_0^{T} \mathbb{1}\{Z(v) > 0\} \, \mathrm{d}v = T,$$

the last equality being justified because the set of points $\{s : Z(s) = 0\}$ has Lebesgue measure 0, as we saw in Sect. 1.4. Thus,

$$\int_0^{V(T)} \mathbb{1}\{Y(v) = 0\} \, \mathrm{d}v = V(T) - T = |m_X(T)|/\omega.$$

By Lemma 2.6, $|m_X(T)|$ is exponentially distributed with parameter $\alpha_+(\mu, \sigma, \lambda)$, and this concludes the proof. □

To determine the expected time of the Markov-modulated process $\{Y, \kappa\}$ at level 0, assume that $\kappa(\xi_n^*) = i$ for some arbitrary n. In that case, $\varphi(\theta_n^*) = i$, and $\theta_n' - \theta_n^*$ is exponentially distributed with parameter $|Q_{ii}|$. The sub-regulators $S_j(\cdot)$ with $j \neq i$ remain constant during the interval of time $[\theta_n^*, \theta_n']$ and therefore

$$\begin{aligned}
\xi_n' - \xi_n^* &= \theta_n' - \theta_n^* + (S_i(\theta_n^*) - S_i(\theta_n'))/\omega_i \\
&\stackrel{\mathrm{d}}{=} \theta_n' - \theta_n^* + (S_i(\theta_n^* - \theta_n'))/\omega_i \\
&\stackrel{\mathrm{d}}{=} V(\theta_n' - \theta_n^*).
\end{aligned}$$

By Lemma 5.20, the expected time at level 0 during $(\xi_n' - \xi_n^*)$ is

$$M_s(0) = (\Delta_b \Delta_\omega)^{-1},$$

where $\boldsymbol{b}$ is defined in (2.18) and $\boldsymbol{\omega} = \left[\omega_i : i \in \mathcal{E}\right]^{\mathrm{T}}$.

Finally, the expected local time of $\{Y, \kappa\}$ in $x > 0$ during $[\theta_n^*, \theta_{n+1}^*)$ is identical to that of the Skorokhod-regulated process of Sect. 5.1 since their sample paths are the same. Therefore,

$$L_{\mathrm{s},0}[x, \theta_1^*] = L_{\mathrm{r},0}[x, \theta_1^*] = 2(\Delta_b \Delta_\sigma^2)^{-1} e^{Rx}, \qquad \text{for } x > 0,$$

by (5.6) and Theorem 5.2.

Theorem 5.21. *Consider an MMBM $\{Y, \kappa\}$ with sticky boundary at 0, with parameters $\boldsymbol{\mu}$, $\boldsymbol{\sigma}^2$, ω and Q, and assume that $\boldsymbol{\pi}^{\mathrm{T}}\boldsymbol{\mu} < 0$. Its stationary distribution function is given by*

$$\boldsymbol{G}_s^{\mathrm{T}}(x) = c_s \boldsymbol{u}^{\mathrm{T}}(\Delta_\omega^{-1} + 2\Delta_\sigma^{-2}(-R)^{-1}(I - e^{Rx})), \tag{5.65}$$

where $\boldsymbol{u}$ is the eigenvector of U for the eigenvalue 0, and

$$c_s = (\boldsymbol{u}^{\mathrm{T}}(\Delta_\omega^{-1} + 2\Delta_\sigma^{-2}(-R)^{-1})\mathbf{1})^{-1}$$

is the normalisation constant.

Proof. It immediately follows from our analysis that

$$\begin{aligned}\boldsymbol{G}_s^{\mathrm{T}}(x) &= c_1 \boldsymbol{\rho}_s^{\mathrm{T}}(M_s(0) + \int_0^x L_{\mathrm{s},0}[v, \theta_1^*]\, \mathrm{d}v) \\ &= c_1 \boldsymbol{\rho}_s^{\mathrm{T}} \Delta_b^{-1}(\Delta_\omega^{-1} + 2\Delta_\sigma^{-2} \int_0^x e^{Rv}\, \mathrm{d}v),\end{aligned}$$

for some scalar c_1, where $\boldsymbol{\rho}_s = \boldsymbol{\rho}$ is the stationary probability vector of Φ. As $\boldsymbol{\pi}^{\mathrm{T}}\boldsymbol{\mu} < 0$, the matrix R is nonsingular, and $\int_0^x e^{Rv}\, \mathrm{d}v$ is equal to $R^{-1}(I - e^{Rx})$. Finally, we have seen in (5.15) that $\boldsymbol{u}^{\mathrm{T}}$ is proportional to $\boldsymbol{\rho}^{\mathrm{T}}\Delta_b^{-1}$. This concludes the proof. □

If we compare $\boldsymbol{G}_s^{\mathrm{T}}(x)$ to the expression (5.16) of the stationary distribution of the one-sided Skorokhod-regulated MMBM, we see that the only difference resides in the introduction of the probability mass vector $c_s \boldsymbol{u}^{\mathrm{T}}\Delta_\omega^{-1}$ at $x = 0$.

5.5 Notes

The Markov-regenerative approach that I have systematically followed here was originally introduced for Fluid flows in Ramaswami [130], and its versatility was subsequently demonstrated by numerous extensions of the basic model: finite buffer models (da Silva Soares and Latouche [54]), piecewise level-dependent parametrisation (Bean and O'Reilly [24], Horváth and Telek [85], Deiana et al. [59]), phase belonging to a general state space (Bean and O'Reilly [26]), to cite but a few examples. A useful feature of that approach is to separate the analysis at the boundary from the analysis in the remainder of the state space, as I demonstrate in Sects. 5.3 and 5.4, two sections where much of the material from Sects. 5.1 and 5.2 has easily been recycled.

I think it is fair to say that most of the results in this chapter have been published before, sometimes in multiple different versions, but that the arguments which I use here are new. To make this possible, I needed properties which, to the best of my knowledge, are original: I am referring to Lemma 5.9 and its Corollary 5.10, also Lemma 5.11 about $\boldsymbol{\pi}^{\mathrm{T}}\boldsymbol{\mu} = 0$, and Lemma 5.12 about the expected local time at the boundary of the 2-regulated process.

The vectors $\boldsymbol{\ell}$ in Lemmas 5.6 and 5.14 are borrowed from Asmussen and Kella [16], and from D'Auria et al. [56], the characterisation I give here is new.

Several authors have used the time-reversed MMBM of Sect. 4.6 to obtain the stationary distribution of the Skorokhod-regulated processes: Asmussen [13], Rogers [133], Breuer [44], Ivanovs [87], as well as D'Auria et al. [57]. This has led to the production of different expressions for the stationary 2-regulated process, like I already mentioned for 1-regulated MMBMs in Remark 5.7. Asmussen and Kella [16] use martingales, this is also described in Asmussen [15].

In Latouche and Nguyen [100–103], and in Latouche and Simon [107], the analysis is built upon the fast-switching fluid approximation described in Sect. 1.11. On the plus side, the argument from Fluid flows is easy to follow, but on the minus side, working with that approach requires rather cumbersome notation. Anyway, I am convinced that the present arguments are better.

As I mentioned in a few places already, it is often necessary to rely on the evolution of the process over *intervals of time*, when making conditional probability arguments. This may be done in different ways. Asmussen [13] uses Lemma 2.6 and so do we in (5.2–5.4) when we insert θ's between successive θ^*s. Harrison [80] uses as regeneration points the epochs of return to level 0 separated by a visit to some arbitrary level $b > 0$. In [103], we use an artificial timer, exponentially distributed, to make sure that renewal epochs are separated by strictly positive intervals of time; this, however, has introduced an extraneous parameter which obscures the interpretation to some of the results. Finally, such intervals of time occur in a natural manner for some models: For processes with a lower *and* an upper bound, one may alternate between visits to the lower and higher boundaries, as I do in Sect. 5.2, another example is the model in Simon [138], where catastrophic returns to 0 are followed by an interval of recovery.

When $\boldsymbol{\pi}^{\mathrm{T}}\boldsymbol{\mu} = 0$, it would be nice if we had a physical justification for the additional Eq. (5.33), in the spirit of the additional equation (3.9), but this has eluded me so far. Another question concerns the existence, or not, of an invariant measure in case $\boldsymbol{\pi}^{\mathrm{T}}\boldsymbol{\mu} > 0$. It has been answered in the positive for QBDs in Latouche et al. [109], and for Fluid queues in Van Lierde et al. [140]. I imagine that this would also be the case for MMBMs.

Chapter 6
Algorithms

⋄ *M-matrices* ⋄ *Functional iteration* ⋄ *Physical justification* ⋄
Linear convergence ⋄ *Newton's iteration* ⋄ *Quadratic convergence*
⋄ *Component-wise accuracy* ⋄

This is where we make good on the implied promise made in the foreword that the analysis of MMBMs from the stochastic process viewpoint would go hand in hand with the elaboration of provably efficient numerical procedures. We present two iterative algorithms for the actual computation of the matrix U, both based on the fixed point equation of Theorem 2.8. To prove convergence for the first algorithm, we give a physical, probabilistic interpretation of the successive iterations. We use Newton's method to set up the second, quadratically convergent algorithm, its analysis relies mostly on linear algebraic arguments.

It is obvious that equivalent algorithms may be designed for the computation of $\widehat{U}$, and it is also true, although perhaps not quite as obvious, that similar procedures may be adapted to the computation of R and $\widehat{R}$. We briefly return to this claim in Sect. 6.4.

Parenthesis 6.1. Nonsingular M-Matrices

We refer in this chapter to properties of nonsingular M-matrices, and we give here a short summary. For further details, we should cite Berman and Plemmons [31] and Varga [141, Chapter 3].

M-matrices occur in many areas, including partial differential equations, economics, operations research, statistics and, most interestingly for us, Markov processes. Different fields of study emphasise different defining features of M-matrices, Plemmons [128] gives a survey of 40 equivalent characterisations, and we present a small sample in Theorem 6.3.

The story starts with *Z-matrices*.

© The Author(s), under exclusive license to Springer Nature Switzerland AG 2026

G. Latouche, *Markov-Modulated Brownian Motion*, Springer Series in Operations Research and Financial Engineering, https://doi.org/10.1007/978-3-031-98576-8_6

Definition 6.2. *A square matrix A is a Z-matrix if $A_{ij} \le 0$ for all $i \neq j$.*

It is an M-matrix if, in addition, it may be written as $A = sI - B$, where $B \ge 0$ and $s \ge \mathrm{sp}(B)$. It is nonsingular if $s > \mathrm{sp}(B)$.

M-matrices are ubiquitous in Markov chains analysis: $I - P$ is an M-matrix, where P is the transition matrix of a discrete-time Markov chain, it is singular if P is stochastic, nonsingular if P is substochastic. Similarly, $-Q$ is an M-matrix, where Q is the generator of a continuous-time Markov chain. The following is a selection from [128, Theorem 1] of a few characterisations of nonsingular M-matrices.

Theorem 6.3. *Assume that A is a Z-matrix. It is a nonsingular M-matrix if and only if any of the following conditions hold:*

1. *All eigenvalues of A are in $\mathbb{C}_{>}$.*
2. *A^{-1} exists and $A^{-1} \ge 0$.*
3. *(Regular splitting) There exist M, N such that $A = M - N$, $M^{-1} \ge 0$, $N \ge 0$, and $\mathrm{sp}(M^{-1}N) < 1$.*
4. *(Regular splitting, second version) For every M, N, such that $A = M - N$, $M^{-1} \ge 0$, $N \ge 0$, one has $\mathrm{sp}(M^{-1}N) < 1$.*
5. *(Diagonal dominance) There exists $\boldsymbol{v} \ge \mathbf{0}$ such that $A\boldsymbol{v} > \mathbf{0}$.*
6. *(Diagonal dominance, second version) There exists $\boldsymbol{v} \ge \mathbf{0}$ such that $A\boldsymbol{v} \ge \mathbf{0}$ and $\sum_{1 \le j \le i} A_{ij} v_j > 0$ for all i.* □

We shall also use Bini et al. [40, Theorem A.20]:

Theorem 6.4. *Assume that A and B are M-matrices. The sum $I \otimes A + B \otimes I$ is an M-matrix, it is nonsingular if either A or B is nonsingular.* □

A consequence of this theorem is that, assuming that A and B are M-matrices, the Sylvester equation $AX + XB = C$ is nonsingular if either A or B is nonsingular. Indeed, that equation is identical to

$$(I \otimes A + B^{\mathrm{T}} \otimes I)\mathrm{vec}(X) = \mathrm{vec}(C),$$

as we saw in (2.46), where $\mathrm{vec}(M)$ is the vectorisation of the matrix M, formed by stacking its columns.

△

6.1 Functional Iteration

A standard method to numerically evaluate the matrix U is to start from one fixed point equation and proceed by functional iteration. We use (2.19) and define the recurrence

$$\begin{aligned} U_{n+1} &= -\Delta_b + \Delta_b \int_0^\infty \Delta_a e^{-\Delta_a y} \widetilde{P} e^{U_n y} \, \mathrm{d}y, \qquad (6.1)\\ &= -\Delta_b + \Delta_b \widetilde{P} + \Delta_a^{-1}(U_{n+1} + \Delta_b)U_n, \qquad n \ge 0, \end{aligned}$$

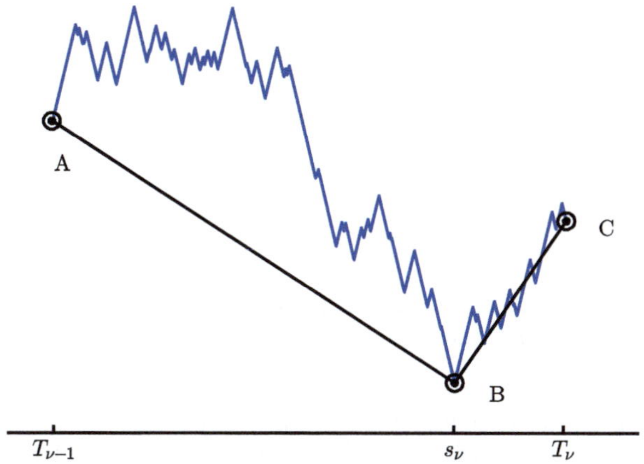

Fig. 6.1 Three-point schematic representation of a Brownian motion during the exponential interval of sojourn in a phase: A marks the starting point $X(T_{\nu-1})$, B the minimum point reached in $(T_{\nu-1}, T_\nu)$, and C the final point $X(T_\nu)$

after integration by part. We pre-multiply by Δ_a and we conclude that U_{n+1} is a solution of the linear system

$$U_{n+1}U_n - \Delta_a U_{n+1} + \Delta_b U_n + 2\Delta_\sigma^{-2} Q = 0. \tag{6.2}$$

We prove in Theorem 6.5 that the sequence $\{U_n : n \geq 1\}$ of solutions to (6.2), starting with $U_0 = 0$, *monotonically* converges to U, meaning that $e^{U_n x} \leq e^{U_{n+1} x}$ for all $n \geq 1$ and all $x > 0$. The proof is based on the evolution of a virtual stack $\zeta(t)$ that we associate with the MMBM, and that we describe now.

Starting with $T_0 = 0$, we define the sequence $\{T_\nu\}$ of jump epochs for the phase process:

$$T_\nu = \inf\{t > T_{\nu-1} : \varphi(t) \neq \varphi(T_{\nu-1})\} \qquad \text{for } \nu \geq 1.$$

Next, we define the sequence $\{m_\nu\}$ of minimum values of the MMBM over the intervals during which the phase remains constant:

$$m_\nu = \inf\{X(t) : T_{\nu-1} < t < T_\nu\} \qquad \text{for } \nu \geq 1.$$

We also define a sequence $\{s_\nu\}$ of epochs when the MMBM reaches the local minimum m_ν:

$$s_\nu = \inf\{t : T_{\nu-1} < t < T_\nu, X(t) = m_\nu\} \qquad \text{for } \nu \geq 1.$$

The points A, B and C in Fig. 6.1 represent the three points $(T_{\nu-1}, X(T_{\nu-1}))$, (s_ν, m_ν) and $(T_\nu, X(T_\nu))$ for some unspecified ν. The two line segments will serve as a schematic representation of the MMBM trajectory during the interval $[T_{\nu-1}, T_\nu)$.

We associate with each s_ν a time s'_ν at which the MMBM returns to level m_ν:

$$s'_\nu = \inf\{t > T_\nu : X(t) = m_\nu\},$$

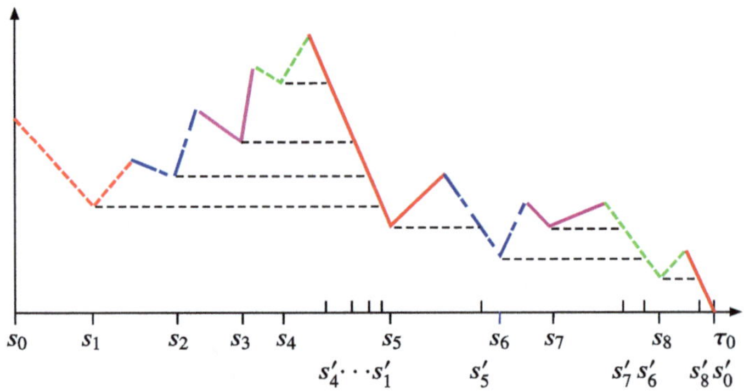

Fig. 6.2 Physical interpretation of the functional iteration algorithm. The level value is pushed on the stack at time s_i, $i = 0, 1, \ldots, 8$. The epochs s'_1 to s'_8 of removal of the top are marked but not all labelled, they are ordered as $s'_4 < s'_3 < s'_2 < s'_1 < s'_5 < s'_7 < s'_6 < s'_8 < s'_0$. The maximum size of the stack is 5, reached at time s_4

and $s'_\nu = \infty$ if $X(t) > m_\nu$ for all $t > s_\nu$, that is, if the process drifts to $+\infty$ after having gone all the way down to m_ν. If $s'_\nu = \infty$ for some ν_0, then s_ν and s'_ν are equal to ∞ for any $\nu \geq \nu_0$. The finite epochs s_ν and s'_ν are defined without ambiguity as they are the first visits to m_ν, after $T_{\nu-1}$ and after T_ν, respectively.

We represent in Fig. 6.2 a sample path over$[0, \tau_0]$. In that figure, there are nine intervals over which the phase remains constant, and we give a three-point schema of the MMBM evolution over each. To make this easier to follow, we use different line styles for different intervals $[T_{\nu-1}, T_\nu)$, and we mark the epochs $s_1, s_2, \ldots$ of local minimum.

The stack ζ increases at times s_ν and decreases at times s'_ν: At each epoch s_ν, we push the value m_ν on top of the stack, thereby increasing its size $|\zeta|$ by one unit. At time s'_ν, we remove the top of ζ, thereby reducing its size by one unit. The horizontal dashed intervals in Fig. 6.2 indicate when the push/pop actions take place: The left extremities of these intervals mark the epochs s_ν when a new value is pushed, the right extremities mark the time points s'_ν where the corresponding value is popped.

At time 0, we push an artificial value $m_0 = 0$ as initial state for the stack and so $|\zeta(0)| = 1$. Next, the values m_1, m_2, m_3, m_4 are pushed in succession, the value m_4 at the top is removed at time s'_4, then m_3 is removed at time s'_3, and so on. Notice that the stack size is equal to 1 whenever a new running minimum is reached. Eventually, we remove the last item on the stack when the MMBM reaches level 0.

Define the phase transition matrix from level $x > 0$ to level 0, under the constraint that the stack size is at most equal to n, and denote it as $H_n(x)$:

$$(H_n(x))_{ij} = \mathrm{P}_{x,i}[\tau_0 < \infty, \varphi(\tau_0) = j, \max_{0 \leq t \leq \tau_0} |\zeta(t)| \leq n],$$

for $n \geq 1$. Because we push m_0 right at the beginning of the trajectory, ζ may not grow by more than $n - 1$ units during the open interval $(0, \tau_0)$.

Theorem 6.5. *There exists a sequence of matrices U_n, $n \geq 1$, such that*

$$(e^{U_n x})_{i,j} = \mathrm{P}_{x,i}[\tau_0 < \infty, \max_{0 \leq t \leq \tau_0} |\zeta(t)| \leq n, \varphi(\tau_0) = j]. \tag{6.3}$$

The sequence $\{e^{U_n x}\}$ converges monotonically to e^{Ux} for any $x > 0$, and U_n converges to U, as n grows to infinity.

The matrices U_n are generators, and U_{n+1} is the unique solutions of the Sylvester equation (6.2) for $n \geq 1$, with $U_1 = -\Delta_b$.

Proof. Obviously, the matrices $H_n(x)$, $n \geq 1$, form an increasing sequence since the constraint on $|\zeta|$ is progressively weakened as n grows to infinity, until it disappears altogether, and $\lim_{n\to\infty} H_n(x) = e^{Ux}$ for any $x > 0$.

Furthermore, the matrices satisfy the semi-group property

$$H_n(x + y) = H_n(x)H_n(y)$$

for any fixed n, because the stack has returned to its initial state at the end of the interval of time from level $x + y$ to level y, and it is allowed to grow up to size n again during the interval from level y to level 0. Therefore, there exists a generator U_n such that $H_n(x) = e^{U_n x}$.

We adapt the proof of (2.22) and obtain

$$e^{U_{n+1}x} = e^{-\Delta_b x} + \int_0^x \int_0^\infty V_{ab}(y, v)\, e^{U_n y} e^{U_{n+1}(x-v)}\, dv\, dy, \tag{6.4}$$

for $n \geq 1$. To see this, consider T_1, the first jump time of the phase process. The first term in the right-hand side of (6.4) is the probability that the running minimum in $(0, T_1)$ is smaller than $-x$, so that $\tau_0 < T_1$ and the stack size has not grown beyond its initial value $|\zeta(0)| = 1$. The expression under the two integrals is the product of probabilities that

- the running minimum at time T_1 is $x - v > 0$, and it is pushed on ζ, which is now of size 2;
- the process is at level $x - v + y$ at time T_1;
- denoting as s'_1 the first return time to level $x - v$, the stack may not grow by more than $n - 1$ units in (T_1, s'_1), and at time s'_1, the stack has been cleaned up, except for the initial value $m_0 = 0$;
- in the interval (s'_1, τ_0), the stack is again allowed to grow by n units on its path from level $x - v$ to 0.

For $n = 1$, the stack is not allowed to increase at all during $(0, \tau_0)$ as its initial size is already equal to 1. Thus, $e^{U_1 x} = e^{-\Delta_b x}$ and $U_1 = -\Delta_b$. We take in (6.4) the derivative with respect to x evaluated at $x = 0$ and find that

$$U_{n+1} = -\Delta_b + \int_0^\infty V_{ab}(y, 0) e^{U_n y}\, dy, \tag{6.5}$$

identical to (6.1), which proves that U_{n+1} is a solution of (6.2).

The matrix U_n is a generator and so $-U_n$ is an M-matrix, furthermore, the diagonal matrix Δ_a having strictly positive diagonal elements is a nonsingular M-matrix. Therefore, the Sylvester equation (6.2) is nonsingular by Theorem 6.4, and U_{n+1} is its unique solution. □

Algorithm 6.1 Functional iteration

Input: The MMBM parameters $\boldsymbol{\mu}, \boldsymbol{\sigma}$, Q, a tolerance $\epsilon > 0$
Output: An approximation U to the solution of (2.23)
1: Set $\boldsymbol{\lambda} = [|Q_{ii}| : i \in \mathcal{E}]$
2: Set $\Delta_a = \text{diag}(\alpha_+(\boldsymbol{\mu}, \boldsymbol{\sigma}, \boldsymbol{\lambda}))$, $\Delta_b = \text{diag}(\alpha_-(\boldsymbol{\mu}, \boldsymbol{\sigma}, \boldsymbol{\lambda}))$, $\Delta_\sigma = \text{diag}(\boldsymbol{\sigma})$
3: Set $U_0 = \Delta_b$
4: Set $err = \|U_0\|_\infty$ and set $k = 0$
5: **while** $err > \epsilon$ **do**
6: Compute U_{k+1}, the unique solution of the Sylvester system
$U_{k+1}U_k - \Delta_a U_{k+1} + \Delta_b U_k - 2\Delta_\sigma^{-2} Q = 0$
7: Compute $err = \|U_{k+1} - U_k\|_\infty$ and set $k = k + 1$
8: **end while**
9: Set $U = U_k$

We give in Algorithm 6.1 a pseudo-code for the functional iteration algorithm, and we analyse now its convergence property. Proceeding like we did in (2.25) when writing U as $-\Delta_b + \Delta_b \Phi$, we rewrite (6.5) as $U_n = -\Delta_b + \Delta_b \Phi_n$, where

$$\Phi_n = \int_0^\infty \Delta_a e^{-\Delta_a y} \widetilde{P} e^{U_{n-1} y} \, dy,$$

and we observe that $\Phi_n \le \Phi$ since $e^{U_n y} \le e^{U y}$ for all $n \ge 1$ and all y. Therefore, $U - U_n = \Delta_b(\Phi - \Phi_n)$ is nonnegative, and the difference $E_n = U - U_n$ may be used as a measure of the error at step n.

From (2.15, 2.23, 6.2), we find that

$$\Delta_a E_{n+1} - E_{n+1} U_n = (U + \Delta_b) E_n,$$

an equation equivalent to

$$(I \otimes \Delta_a)\boldsymbol{e}_{n+1} - (U_n^{\mathrm{T}} \otimes I)\boldsymbol{e}_{n+1} = (I \otimes (U + \Delta_b))\boldsymbol{e}_n,$$

where $\boldsymbol{e}_n = \text{vec}(E_n)$. Since $U_n \le U$, this implies that

$$(I \otimes \Delta_a)\boldsymbol{e}_{n+1} - (U^{\mathrm{T}} \otimes I)\boldsymbol{e}_{n+1} \le (I \otimes (U + \Delta_b))\boldsymbol{e}_n.$$

We observed in the proof of Theorem 6.5 that Δ_a is a nonsingular M-matrix, and so $I \otimes \Delta_a - U \otimes I$ is itself a nonsingular M-matrix by Theorem 6.4. Its inverse is a nonnegative matrix by Condition 2 in Theorem 6.3, and so $\boldsymbol{e}_{n+1} \le R_{\text{FI}} \boldsymbol{e}_n$, where

$$R_{\text{FI}} = (I \otimes \Delta_a - U^{\mathrm{T}} \otimes I)^{-1}(I \otimes (U + \Delta_b)). \tag{6.6}$$

We conclude that $E_n = O(r^n)$ asymptotically as $n \to \infty$, where r is the spectral radius of R_{FI}. We determine in the next lemma the link between r and the stationary drift $\boldsymbol{\pi}^{\mathrm{T}}\boldsymbol{\mu}$.

Lemma 6.6. *If $\boldsymbol{\pi}^{\mathrm{T}}\boldsymbol{\mu} \neq 0$, then the spectral radius r of the matrix R_{FI} is strictly less than 1, and the functional iteration algorithm is linearly convergent.*

If $\boldsymbol{\pi}^{\mathrm{T}}\boldsymbol{\mu} = 0$, then $r = 1$, and convergence is sublinear.

Proof. To simplify the presentation, we temporarily define

$$\begin{aligned} M_1 &= I \otimes \Delta_a - U^{\mathrm{T}} \otimes I \\ M_2 &= I \otimes (U + \Delta_b), \end{aligned}$$

so that $R_{\text{FI}} = M_1^{-1}M_2$. Also, we define

$$\begin{aligned} M^* &= M_2 - M_1 \\ &= U^{\mathrm{T}} \otimes I + I \otimes (U + 2\Delta_\sigma^{-2}\Delta_\mu) \qquad \text{by (2.15)} \\ &= U^{\mathrm{T}} \otimes I + I \otimes (\Delta_\sigma^{-2}R\Delta_\sigma^2) \end{aligned}$$

by Theorem 4.14, and $-M^*$ is an M-matrix by Theorem 6.4.

Now, assume that $\boldsymbol{\pi}^{\mathrm{T}}\boldsymbol{\mu} \neq 0$. By Theorem 2.11 and Lemma 4.21, either U or R is nonsingular, and so M^* is nonsingular as well. It follows, therefore, that $M_1 - M_2$ is a regular splitting of $-M^*$ and that $\text{sp}(M_1^{-1}M_2) < 1$ by Condition 4 of Theorem 6.3. This proves the first statement of the lemma.

On the other hand, if $\boldsymbol{\pi}^{\mathrm{T}}\boldsymbol{\mu} = 0$, then both U and R are singular, and we need a different argument. Define $\boldsymbol{u}_U > \mathbf{0}$ to be the left eigenvector associated to the eigenvalue 0 of U, $\boldsymbol{v}_R > \mathbf{0}$ to be the right eigenvector associated to the eigenvalue 0 of R, and define $\boldsymbol{v}^* = \boldsymbol{u}_U \otimes \Delta_\sigma^{-2}\boldsymbol{v}_R > \mathbf{0}$.

We easily verify that $M^*\boldsymbol{v}^* = \mathbf{0}$ and, in consequence, that $M_1^{-1}M_2\boldsymbol{v}^* = \boldsymbol{v}^*$ so that $\text{sp}(M_1^{-1}M_2) = 1$. This concludes the proof. □

Remark 6.7. If $\boldsymbol{\pi}^{\mathrm{T}}\boldsymbol{\mu} < 0$, one may improve the convergence rate of the functional iteration algorithm by choosing U_1 such that $e^{U_1 x}$ is a stochastic matrix. This is proved in Meini [113] for QBDs, and the argument should hold for MMBMs as well.

Example 6.8. We illustrate in Fig. 6.3 the linear convergence of $\{U_n\}$ to U. This example is a small system with generator

$$Q = \begin{bmatrix} -3 & 1 & 2 \\ 1 & -1 & 0 \\ 1 & 1 & -2 \end{bmatrix},$$

and parameters

$$\boldsymbol{\mu} = \gamma \begin{bmatrix} 0.01 & -2 & 1 \end{bmatrix} \qquad \text{and} \qquad \boldsymbol{\sigma} = \begin{bmatrix} 0.2 & 0.5 & 1 \end{bmatrix}.$$

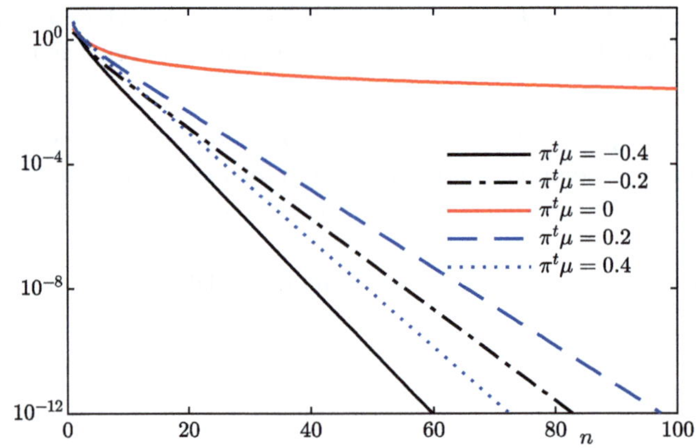

Fig. 6.3 Example 6.8. Inf norm $||E_n||_\infty$ of the error, in function of the number of iterations of the functional iteration algorithm, for different values of $\boldsymbol{\pi}^{\mathrm{T}}\boldsymbol{\mu}$

The scalar coefficient γ is there so that we may give any chosen value to $\boldsymbol{\pi}^{\mathrm{T}}\boldsymbol{\mu}$. We give the semilog plot of $||E_n||_\infty$ for $n = 1, \ldots, 100$ and for five values of $\boldsymbol{\pi}^{\mathrm{T}}\boldsymbol{\mu}$, given in the table below together with the corresponding values of the decay rate r.

$\boldsymbol{\pi}^{\mathrm{T}}\boldsymbol{\mu}$	−0.4	−0.2	0	0.2	0.4
r	0.6231	0.7149	1	0.7494	0.6728

In computing the successive errors $\mathrm{E}_n = U - U_n$, we use a value for U obtained from a quadratically convergent algorithm.

The slowly decreasing plain red line in Fig. 6.3 corresponds to $\boldsymbol{\pi}^{\mathrm{T}}\boldsymbol{\mu} = 0$, the decay rate r is equal to 1 and the convergence is clearly sublinear. The other four curves are obviously asymptotically linear. The rate r is slightly smaller when the drift is negative ($\boldsymbol{\pi}^{\mathrm{T}}\boldsymbol{\mu} = -0.4$ and -0.2, as opposed to 0.2 and 0.4) and r is smaller as well when the drift takes values further away from 0 ($\boldsymbol{\pi}^{\mathrm{T}}\boldsymbol{\mu} = -0.4$ and 0.4, compared to -0.2 and 0.2).

Example 6.9. Here, the process is similar to those in Examples 2.28 and 5.19. The phase cycles through the values 1 to m, with $m = 6$, at a constant jump rate equal to 0.1. The parameters σ_i are equal to 1 for $i \neq 3$ and $\sigma_3 = 4$. We use two different vectors $\boldsymbol{\mu}$:

A $\boldsymbol{\mu}_A = \begin{bmatrix}1 \; 0.5 \; -2 \; -0.5 \; 0.5 \; 0.5\end{bmatrix}^{\mathrm{T}} + \gamma_A \mathbf{1}$, we have $\boldsymbol{\pi}^{\mathrm{T}}\boldsymbol{\mu}_A = \gamma_A$ and we use γ_A to control the value of the stationary drift.

B $\boldsymbol{\mu}_B = \gamma_B \begin{bmatrix}1 \; 0.5 \; -2 \; -0.5 \; 0.5 \; 1\end{bmatrix}^{\mathrm{T}}$, where γ_B is used to control the stationary drift $\boldsymbol{\pi}^{\mathrm{T}}\boldsymbol{\mu}_B$, equal here to $\gamma_B/12$.

The plot in Fig. 6.4 shows the decay rate as a function of $\boldsymbol{\pi}^{\mathrm{T}}\boldsymbol{\mu}$. The plain blue line is for Case A, the dashed red line is for Case B. Both are equal to 1 when $\boldsymbol{\pi}^{\mathrm{T}}\boldsymbol{\mu} = 0$, and decrease as $\boldsymbol{\pi}^{\mathrm{T}}\boldsymbol{\mu}$ gets further away from 0. The shapes of the two lines show

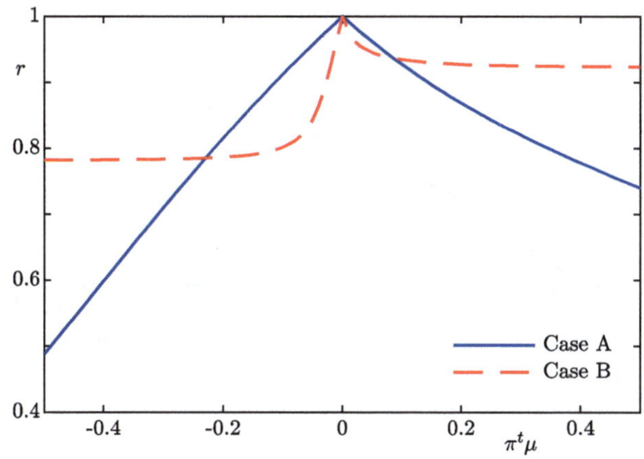

Fig. 6.4 Example 6.9. Plot of the decay rate r of functional iteration as a function of $\boldsymbol{\pi}^{\mathrm{T}}\boldsymbol{\mu}$. The plain line is for Case A, the dashed line is for Case B

different profiles: In Case B, r depends mostly on the sign of the stationary drift, and variations are concentrated in a small interval around $\boldsymbol{\pi}^{\mathrm{T}}\boldsymbol{\mu} = 0$. Like in Example 6.8, the decay rate is somewhat smaller if the drift is negative rather than positive. We give in the table below a few typical values for the decay rates r_A in Case A and r_B in Case B.

$\boldsymbol{\pi}^{\mathrm{T}}\boldsymbol{\mu}$	−0.5	−0.3	−0.1	0.1	0.3	0.5
r_A	0.4879	0.7093	0.9123	0.9268	0.8198	0.7393
r_B	0.7818	0.7834	0.8015	0.9347	0.9252	0.9238

Example 6.10. We show here the effect of the volatility parameters on the rate of convergence of U_n to U. This example is taken from Karandikar and Kulkarni [95] and is similar to Example 3.18: m sources generate fluid to a multiplexer in an `on-off` manner, with the same parameters and independently of each other. During the `on` intervals, each source behaves like a Brownian motion with parameters μ_A/m and σ_A^2/m, and during the `off` intervals, the parameters are μ_B/m and σ_B^2/m, with $\mu_A > 0 > \mu_B$. When there are i sources in the `on` state, the level in the buffer behaves as a Brownian motion with parameters $\mu_i = (i\mu_A + (m-i)\mu_B)/m$ and $\sigma_i^2 = (i\sigma_A^2 + (m-i)\sigma_B^2)/m$. Individual `on` and `off` intervals are exponentially distributed with parameters ω_A and ω_B, respectively.

We plot in Fig. 6.5 the decay rate r as a function of σ_A for $m = 5$, 10, 15 and 20. The other parameters are $\mu_A = 1$, $\mu_B = -2$, $\sigma_B^2 = 1$, and $\omega_A = \omega_B = 1$, so that $\boldsymbol{\pi}^{\mathrm{T}}\boldsymbol{\mu}$ has the same value −0.5 in the four cases. One observes that the decay rate gets closer to 1, and convergence is consequently slower, if m increases or if the variance increases. We let σ_A only vary for this particular model, similar observations were made in other examples.

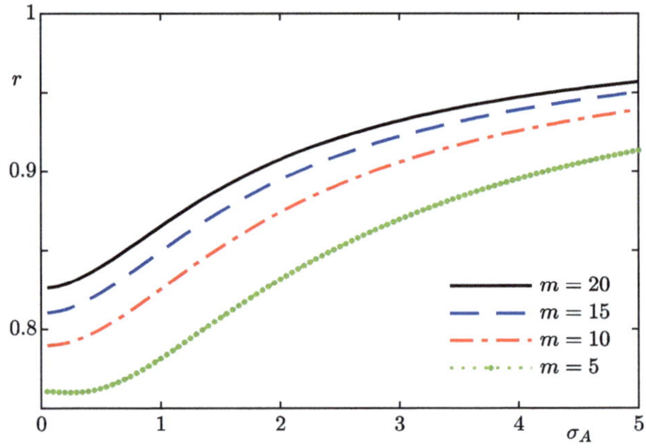

Fig. 6.5 Example 6.10. Decay rate r of the functional iteration algorithm as a function of the volatility parameter σ_A

6.2 Newton's Iterations

Newton's method is frequently used as a tool for accelerating the numerical resolution of nonlinear equations. One defines a sequence $\{X_k : k \geq 0\}$ by the recurrence

$$X_{k+1} = X_k - (F'_{X_k})^{-1}[F(X_k)], \qquad k \geq 0, \tag{6.7}$$

with some suitable initial value X_0, where the operator F'_X is the Fréchet derivative of F evaluated at X. For a detailed analysis, we refer to Ortega and Rheinboldt [124, Sections 3.1 and 12.6].

We saw in (2.25, 2.27) that $U = -\Delta_b + \Delta_b \Phi$, where Φ is the minimal nonnegative solution of the Riccati equation $F(X) = 0$ with

$$F(X) = X\Delta_b X - \Delta_a X - X\Delta_b + \Delta_a \widetilde{P}. \tag{6.8}$$

Such equations have been extensively analysed, both from an analytic and from a numerical perspective, in the context of Fluid flows. Two overview references are Bean et al. [28] and Bini et al. [40]. The matrix Φ is the unique stochastic solution if $\boldsymbol{\pi}^{\mathrm{T}}\boldsymbol{\mu} \leq 0$ and it is the unique substochastic solution if $\boldsymbol{\pi}^{\mathrm{T}}\boldsymbol{\mu} > 0$. Therefore, the Riccati equation completely characterises Φ and, by extension, the matrix U.

In the case of (6.8), the Fréchet derivative is given as

$$F'_X[H] = (X\Delta_b - \Delta_a)H + H(\Delta_b X - \Delta_b),$$

and the recurrence (6.7) becomes

$$(\Phi_k \Delta_b - \Delta_a)(\Phi_{k+1} - \Phi_k) + (\Phi_{k+1} - \Phi_k)(\Delta_b \Phi_k - \Delta_b) = -F(\Phi_k),$$

or

$$(\Delta_a - \Phi_k \Delta_b)\Phi_{k+1} + \Phi_{k+1}(\Delta_b - \Delta_b \Phi_k) = \Delta_a \widetilde{P} - \Phi_k \Delta_b \Phi_k, \tag{6.9}$$

for $k \geq 0$. This equation is equivalent to

$$M_k \text{vec}(\Phi_{k+1}) = \text{vec}(\Delta_a \widetilde{P} - \Phi_k \Delta_b \Phi_k) \tag{6.10}$$

with

$$M_k = I \otimes (\Delta_a - \Phi_k \Delta_b) + (\Delta_b - \Delta_b \Phi_k)^{\mathrm{T}} \otimes I. \tag{6.11}$$

We give in Algorithm 6.2 a pseudo-code for Newton's method.

Algorithm 6.2 Newton's algorithm

Input: The MMBM parameters $\boldsymbol{\mu}, \boldsymbol{\sigma}$, Q, a tolerance $\epsilon > 0$
Output: An approximation U to the solution of (2.23)

```
Set λ = [|Q_ii| : i ∈ ℰ]
Set Δ_a = diag(α_+(μ, σ, λ)), Δ_b = diag(α_-(μ, σ, λ)), Δ_σ = diag(σ)
Set Δ_a P̃ = 2Δ_b^{-1} Δ_σ^{-2}(Q + diag(λ))
Set Φ_0 = 0
Set err = 1 and set k = 0
while err > ϵ do
    Compute Φ_{k+1}, the unique solution of the Sylvester system
        (Δ_a − Φ_k Δ_b)Φ_{k+1} + Φ_{k+1}(Δ_b − Δ_b Φ_k) = Δ_a P̃ − Φ_k Δ_b Φ_k
    Compute err = ||Φ_{k+1} − Φ_k||_∞ and set k = k + 1
end while
Set U = −Δ_b + Δ_b Φ_k
```

As mentioned at the end of Parenthesis 2.17, it is more efficient for actual computations to work with the system (6.9), for which $O(m^3)$ algorithms are available, than to solve (6.10) which requires a $O(m^6)$ algorithm, but it will be convenient to use the formulation (6.11) in the proof of Theorem 6.11.

In particular, we prove by induction that M_k is a nonsingular M-matrix for all k; this is intuitively plausible as $\Delta_b - \Delta_b \Phi_k$ approximates $-U$ and $\Delta_a - \Phi_k \Delta_b$ approximates $\Delta_a - \Phi \Delta_b$, which is equal to $-\Delta_b^{-1} \Delta_\sigma^{-2} R \Delta_\sigma^2 \Delta_b$ by (2.15, 4.44). Thus, we expect the matrices M_k to have significant properties similar to the matrix $-M^*$ used in the proof of Lemma 6.6.

Theorem 6.11. *Assume* $\Phi_0 = 0$. *The Newton sequence* $\{\Phi_k : k \geq 0\}$ *in (6.9) is well-defined, it is monotone, with* $\Phi_k \leq \Phi_{k+1} < \Phi$ *for all* $k \geq 0$, *and* $\lim_{k\to\infty} \Phi_k = \Phi$.

Proof. We follow the same steps as in Guo and Laub [76, Theorem 2.1] and Guo and Higham [75, Theorem 2.3], adapted to the equation at hand. The proof makes repeated use of Theorem 6.4.

We saw in Lemma 2.4 that U is irreducible, so that $\Phi > 0$ by the integral equation (2.26), and $\Phi_0 < \Phi$. Furthermore, the matrix $M_0 = I \otimes \Delta_a + \Delta_b \otimes I$ is nonsingular, its inverse is nonnegative and, by (6.10), $\text{vec}(\Phi_1) = M_0^{-1} \text{vec}(\Delta_a \widetilde{P}) \geq \text{vec}\Phi_0 = \mathbf{0}$.

We make the induction assumption that for some $k \geq 0$,

$$\Phi_k < \Phi, \qquad \Phi_k \leq \Phi_{k+1}, \qquad \text{and } M_k \text{ is a nonsingular M-matrix,}$$

and we prove that these statements hold for $k + 1$.

As Φ is a solution of (6.8), we have that

$$\Delta_a \Phi + \Phi \Delta_b = \Delta_a \widetilde{P} + \Phi \Delta_b \Phi.$$

We subtract this from (6.9) and find after some simple manipulations that

$$\begin{aligned}(\Delta_a - &\Phi_k \Delta_b)(\Phi_{k+1} - \Phi) + (\Phi_{k+1} - \Phi)(\Delta_b - \Delta_b \Phi_k) \\ &= -(\Phi - \Phi_k)\Delta_b(\Phi - \Phi_k) < 0,\end{aligned}$$

or $M_k \operatorname{vec}(\Phi_{k+1} - \Phi) < 0$. As M_k is a nonsingular M-matrix, its inverse is positive by Condition 2 of Theorem 6.3, and $\Phi_{k+1} - \Phi \leq 0$. The inequality is strict because $M_k^{-1} \geq 0$ being nonsingular, each of its rows has at least one non-zero entry. We conclude that $\Phi_{k+1} < \Phi$.

Next,

$$\begin{aligned}(\Delta_a - &\Phi_{k+1}\Delta_b)\Phi_{k+1} + \Phi_{k+1}(\Delta_b - \Delta_b \Phi_{k+1}) \\ &= (\Delta_a - \Phi_k \Delta_b)\Phi_{k+1} + \Phi_{k+1}(\Delta_b - \Delta_b \Phi_k) \\ &\quad - (\Phi_{k+1} - \Phi_k)\Delta_b \Phi_{k+1} - \Phi_{k+1}\Delta_b(\Phi_{k+1} - \Phi_k) \\ &= \Delta_a \widetilde{P} - (\Phi_{k+1} - \Phi_k)\Delta_b(\Phi_{k+1} - \Phi_k) - \Phi_{k+1}\Delta_b \Phi_{k+1} \qquad (6.12)\end{aligned}$$

by (6.9). Using $F(\Phi) = 0$, we replace $\Delta_a \widetilde{P}$ with $\Delta_a \Phi + \Phi \Delta_b - \Phi \Delta_b \Phi$, and it follows that

$$\begin{aligned}(\Delta_a - &\Phi_{k+1}\Delta_b)(\Phi_{k+1} - \Phi) + (\Phi_{k+1} - \Phi)(\Delta_b - \Delta_b \Phi_{k+1}) \\ &= -(\Phi_{k+1} - \Phi_k)\Delta_b(\Phi_{k+1} - \Phi_k) - (\Phi_{k+1} - \Phi)\Delta_b(\Phi_{k+1} - \Phi) \\ &< 0,\end{aligned}$$

or

$$M_{k+1} \operatorname{vec}(\Phi - \Phi_{k+1}) > \mathbf{0},$$

so that M_{k+1} is a nonsingular M-matrix by Case 5 of Theorem 6.3.

Finally, we have from (6.9, 6.12) that

$$\begin{aligned}(\Delta_a - &\Phi_{k+1}\Delta_b)(\Phi_{k+1} - \Phi_{k+2}) + (\Phi_{k+1} - \Phi_{k+2})(\Delta_b - \Delta_b \Phi_{k+1}) \\ &= -(\Phi_{k+1} - \Phi_k)\Delta_b(\Phi_{k+1} - \Phi_k) \leq 0\end{aligned}$$

or

$$M_{k+1} \operatorname{vec}(\Phi_{k+1} - \Phi_{k+2}) \leq 0.$$

By Case 2 of Theorem 6.3, $(M_{k+1})^{-1} \geq 0$, therefore, $\Phi_{k+1} \leq \Phi_{k+2}$, and this completes the proof. □

We do not give a detailed analysis of the convergence property of Newton's sequence, but we refer to Guo and Laub [76] for the following conclusions:

- If $\boldsymbol{\pi}^{\mathrm{T}}\boldsymbol{\mu} \neq 0$, then $I \otimes (\Delta_a - \Phi\Delta_b) + (\Delta_b - \Delta_b\Phi)^{\mathrm{T}} \otimes I$ is nonsingular, since R or U has all its eigenvalues in $\mathbb{C}_<$ and the other matrix has one eigenvalue equal to 0, the remaining $m-1$ eigenvalues being all in $\mathbb{C}_<$. By [76, Theorem 2.3] therefore, convergence of Newton's sequence (6.9) is quadratic.
- If $\boldsymbol{\pi}^{\mathrm{T}}\boldsymbol{\mu} = 0$, then convergence may be linear, with rate equal to $1/2$.

Example 6.12. This is the same as Example 6.8, the parameters μ_is are chosen so that the stationary drift takes the values in the set

$$\{-0.1, -0.05, -0.025, -0.0125, -0.00625,\ 0\}.$$

In Fig. 6.6a, we plot the residues $\mathcal{F}(-\Delta_b + \Delta_b\Phi_k)$ for $k = 1$–20. Quadratic convergence is obvious when $\boldsymbol{\pi}^{\mathrm{T}}\boldsymbol{\mu} < 0$, even for values quite close to 0 and, clearly, the convergence is linear for $\boldsymbol{\pi}^{\mathrm{T}}\boldsymbol{\mu}$ exactly 0. We use negative values only for the stationary drift, as we get quite similar pictures for $\boldsymbol{\pi}^{\mathrm{T}}\boldsymbol{\mu} > 0$.

Example 6.13. This is the same as the multiplexer system of Example 6.10. We plot in Fig. 6.6b the residues $\mathcal{F}(U_k)$ and $\mathcal{F}(-\Delta_b + \Delta_b\Phi_k)$ produced by the functional iteration sequence (6.2) and by Newton's sequence (6.9), respectively. The parameters are $m = 10$, $\mu_A = 1$, $\mu_B = -2$, $\sigma_A^2 = 0.1$, $\omega_A = 1$, $\omega_B = 1$, and σ_B takes the values 1, 2 and 4. The red lines, labelled "fi1" to "fi4", are for the functional iteration algorithm, and the black lines, labelled "N1" to "N4", are for Newton's method. Like with Fig. 6.5, we observe again that convergence of functional iteration gets significantly slower as the volatility (σ_B in this case) increases. Convergence of Newton's method, on the contrary, is insensitive to that parameter.

6.3 Component-Wise Accuracy

M-matrices lend themselves beautifully to the design of component-wise accurate algorithms, that is, algorithms such that $\max_{i,j\in\mathcal{E}} |U^*_{ij} - U_{ij}|/|U_{ij}|$ is bounded, where U^* is the computed approximation for U, instead of the usual $\|U^* - U\|/\|U\|$. In short, this means that all elements of U are computed with the same number of correct significant digits, irrespective of their magnitude. In numerical linear algebra, this requires that calculations should be subtraction-free: Computer operations are carried out without ever subtracting two floating-point numbers of the same sign, thereby avoiding the effect of catastrophic cancellation. The GTH algorithm is well known in this class (Grassmann et al. [72], O'Cinneide [123], Alfa et al. [10]). It was originally developed for computing the stationary distribution of finite Markov chains and is currently implemented in packages such as SMCSolver [39] to perform inversion of M-matrices.

To design a subtraction-free algorithm, one represents an M-matrix A with a set $(\boldsymbol{m}, \boldsymbol{v}, \boldsymbol{w})$ of three vectors, $\boldsymbol{m} \geq \mathbf{0}$, $\boldsymbol{v} > \mathbf{0}$, $\boldsymbol{w} \geq \mathbf{0}$, where $\boldsymbol{m}$ is made up of the

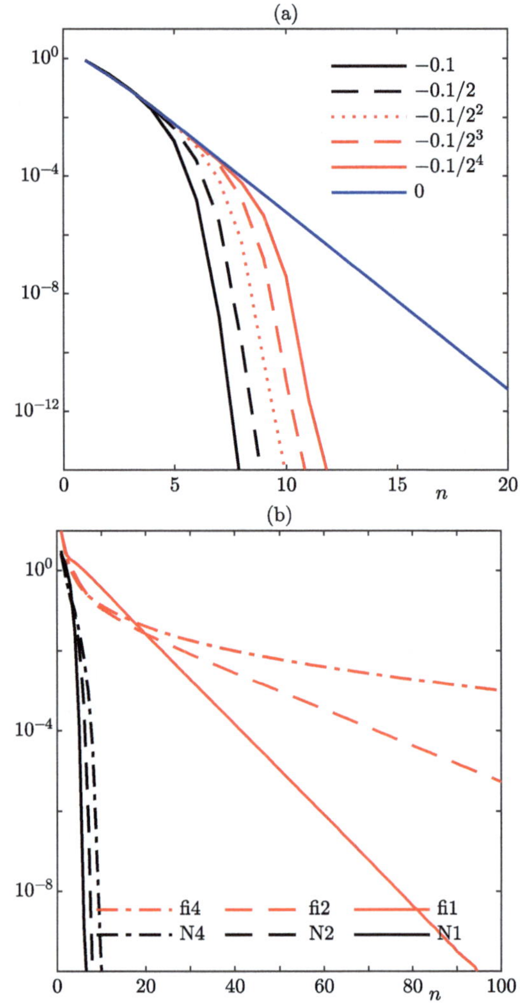

Fig. 6.6 In (**a**) above, we show the residues from Newton's method as a function of the number of iterations for different values of $\boldsymbol{\pi}^{\mathrm{T}}\boldsymbol{\mu}$. In (**b**) below, we compare the residues from Newton's method to the functional iteration algorithm for different volatility values

off-diagonal elements of A, and the vectors $\boldsymbol{v}$ and $\boldsymbol{w}$ are such that $A\boldsymbol{v} = \boldsymbol{w}$. The diagonal elements of A do not appear in the triplet representation, but they may be retrieved from the equation $A\boldsymbol{v} = \boldsymbol{w}$. Guo [74] proves that nonsingular and singular irreducible M-matrices admit of triplet representations, a significant property in our case as the coefficients of the Riccati equation (6.8) are blocks of the matrix

$$A = \begin{bmatrix} -\Delta_b & \Delta_b \\ \Delta_a \widetilde{P} & -\Delta_a \end{bmatrix},$$

an irreducible, singular M-matrix.

Nguyen and Poloni [120] design a component-wise accurate algorithm for the calculation of Φ by combining the triplet representation of A with the structured doubling algorithm of Guo et al. [77], one of the most efficient algorithms known for this class of equations.

6.4 Notes

My intention in this short chapter has been to show the interplay between the analysis of a stochastic process and its trajectories, on the one hand, and the approach from numerical linear algebra on the other hand. The physical interpretation is not absolutely necessary—for instance, the first iterative procedures for the matrix R in QBDs have been proved by purely analytic arguments in Neuts [115]. Still, I do rejoice when I discover the connection between some algorithm and the stochastic process under study. In addition, such understanding often helps to shorten arguments for other models. For its part, numerical linear algebra brings a host of techniques to get more efficient procedures, in terms of speed, stability, or numerical precision, and much has been published in that respect.

The stack I use in the proof of Theorem 6.5 is essentially the same as the one used for Fluid flows in Latouche and Nguyen [104]. For QBDs, the proof of the linear progression algorithm in Latouche and Ramaswami [106, Theorem 8.1.1] is based on the same principle, even though it is not necessary to introduce an explicit stack in that case, the integer-valued level being there to play that role. The physical interpretation of the cyclic reduction algorithm for QBDs (Bini and Meini[34], Bini et al. [38]) and of its close cousin the logarithmic reduction algorithm (Latouche and Ramaswami [105]) is based on the same stack; they make an efficient use of the stack, and they are quadratically convergent when $\boldsymbol{\pi}^{\mathsf{T}}\boldsymbol{\mu} \neq 0$.

I have not given a physical interpretation for Newton's algorithm. There is one for Fluid flows and for QBDs (Bean et al. [27], Bean et al. [30]) and, given the similarities between MMBMs and these two families of processes, I would expect that the interpretations in [27, 30] may be adapted to the Brownian case. Such an interpretation is without doubt much more involved than the one given for functional iteration.

Computational procedures have been developed for the general MMBMs of Sect. 2.5. Nguyen and Poloni [120, 121] start from a quadratic equation, different from (2.83) but with a similar structure, with the intention to use the very efficient cyclic reduction algorithm. That algorithm, however, is designed for matrix-quadratic equations

$$A_{-1} + A_0 X + A_1 X^2 = 0$$

such that the roots of $\det(A_{-1} + A_0 z + A_1 z^2)$ are separated by a circle in $\mathbb{C}$, half of the roots lying inside and half outside, while the roots here are separated by the imaginary axis, as shown in Theorem 2.11. Nguyen and Poloni apply a transformation of $\mathbb{C}$ so that the roots become separated by the unit circle. In addition, the authors make sure that the matrix coefficients A_1 and A_{-1} are nonnegative and that A_0 is the negative of an M-matrix; this allows them to develop a component-wise accurate procedure.

Horváth and Telek [85] determine the stationary distribution of a general MMBM with level-dependent behaviour. Starting from the quadratic equation for the matrix R, the authors transform the coefficients so that the new equation corresponds to a discrete state space QBD. This allows them to directly use the SMCSolver toolbox [37].

Ahn and Ramaswami [8] also take into account the general MMBM. The object there is to construct an algorithm to compute the Laplace-Stieltjes transform (LST) for the distribution of τ_x. The authors start with the fast-switching linear fluid from Sect. 1.11, and they analyse next the LST of the return time to level 0 for the fluid flow approximation, starting from level 0 and avoiding lower levels. In so doing, they obtain a Riccati equation for which they prove that Newton's scheme is quadratically convergent.

I prove in Lemma 6.6, and I show in Fig. 6.6 that both the functional iteration algorithm and Newton's method converge much more slowly when $\boldsymbol{\pi}^{\mathrm{T}}\boldsymbol{\mu} = 0$. This is one more instance of the difficulties one encounters when the MMBM is null recurrent, and it is due to the fact, mentioned in Theorem 2.11, that the eigenvalue 0 has algebraic multiplicity 2. One solution to accelerate the algorithms' convergence is to change the parameters in such a way that the pesky eigenvalue is moved to another location in $\mathbb{C}$ where it does no harm any more. This was suggested in He et al. [83] for the quadratic equation of discrete-time QBDs and later in Guo et al. [78] for the Riccati equation. Bini et al. [41] is a recent paper on this subject.

As I mentioned earlier, one approach to the analysis of MMBMs is based on the spectral decomposition of the Laplace exponent $\mathcal{F}(z)$, or the Wiener-Hopf factorisation in Sects. 2.3 and 4.4. A few authors have used this as the basis for the numerical evaluation of the matrices U or R: Rogers and Shi [134], Karandikar and Kulkarni [95], and D'Auria et al. [55]. However, numerical procedures based on the computation of the Jordan normal form of a matrix are sensitive to numerical errors, and this approach has drawn less interest from researchers in the numerical analysis community.

Bini et al. [35] is a book which covers much of the numerical linear algebra work on the resolution of matrix polynomial equations, the main examples in the present book being (2.23) and (4.45). Algebraic Riccati equations, such as (2.27), and the Wiener-Hopf factorisation (2.29) are treated in depth in Bini et al. [40].

The question of parameter estimation does occur in actual applications, naturally. The estimation-maximisation (EM) algorithm is frequently used to determine the parameters of a Markov-modulated model to fit a set of observations. That is an iterative algorithm to find a (local) maximum of some likelihood function. The so-called E-step consists in using the observed data and a current estimate of the parameters, to calculate the conditional expectation of some statistics, and the M-step consists in getting a new estimate which maximises the obtained likelihood function. A classical reference for the algorithm is Dempster et al. [60], and it was brought to the attention of the Matrix-analytic community in Asmussen et al. [19].

Akbal [9] analyses the efficiency of the EM algorithm to estimate parameters of regime-switching models for applications involving large dimensional data and concludes that it provides satisfactory performance, with respect to accuracy and speed. This conclusion, however, may depend on the algorithm implementation, as mentioned in Ephraim and Roberts [64]. Jamshidian and Jennrich [92] design a modified Newton's method, van Dyk [139] organises the E-step as a nesting of two or more EM algorithms, leading to easier implementation, to faster convergence, and better stability convergence properties, and Bražėnas et al. [42] show how EM is more efficient when tailored to the specific model. Almousa et al. [11] in a recent publication favour a hybrid algorithm to estimate parameters of an MMBM, whereby EM is used to estimate the phase process and a maximisation of the likelihood function is used to estimate mean and variance parameters. A similar recommendation is made in Chevallier and Goutte [50], who use EM to capture the MMBM parameters, followed by the estimation of the jump distributions.

Conclusion

This concludes my presentation of the Matrix-analytic approach to the study of Markov-modulated Brownian motion. I concentrate on the case with $\sigma_i > 0$ for all phases i, and this calls for some comments, beyond my brief justification at the end of Sect. 2.2.

Two key features are the ladder processes of Chap. 2, and the local time analysed in Chap. 4. Much of the presentation in Chaps. 2 and 3 is similar to the analysis by Matrix-analytic methods of first passage times for QBDs and Fluid flows: The arguments used here are pretty much the same, and so are the resulting equations, modulo the differences imposed by specific features governing the stochastic processes, be they discrete state space Markov chains, linear fluid models, and now Brownian trajectories. The matrices U and $\widehat{U}$, omnipresent in the analysis of the ladder processes and the exit probabilities, have been discussed in full generality in Chap. 2, Sect. 2.5. Furthermore, I give pointers in Sect. 6.4 to recent work on computational algorithms. Thus, it seems to me that my restriction to purely Brownian MMBMs is vindicated.

Things are not as simple as far as the matrix R is concerned. For a long time, it looked as if the approach to the matrix R of MMBMs had to be different from those for non-Brownian processes. For the matrix-geometric distribution of QBDs, Neuts gave in 1980 a proof based on the physical interpretation of R. We had to wait for Ramaswami to show in 1999 that in the case of Fluid flows, one may also start from the physical meaning of R. It is only now, with the interpretation given by Ivanovs, that we may treat the three models in a similar manner. Let me give a few indications about the matrix functions $R(x)$ and $\widehat{R}(x)$ when some of the σ_is are equal to 0.

Consider first the case of Fluid flows: $\mathcal{E}_\sigma = \emptyset$, meaning that all σ_is are equal to 0, and both $\mathcal{E}_+$ and $\mathcal{E}_-$ are nonempty. Let us also assume that $\mathcal{E}_0$ is not empty, and denote as $R_{+,\mathcal{E}}(x)$, $R_{-,\mathcal{E}}(x)$, and $R_{0,\mathcal{E}}(x)$ the blocks of rows indexed by an initial phase in $\mathcal{E}_+$, $\mathcal{E}_-$ and $\mathcal{E}_0$, respectively.

If the initial phase is in $\mathcal{E}_+$, one has (see Ahn and Ramaswami [5], or Govorun et al. [70] for instance)

© The Author(s), under exclusive license to Springer Nature Switzerland AG 2026

G. Latouche, *Markov-Modulated Brownian Motion*, Springer Series in Operations Research and Financial Engineering, https://doi.org/10.1007/978-3-031-98576-8

$$R_{+,\mathcal{E}}(x) = C_+ e^{Kx} \begin{bmatrix} C_+^{-1} & \Psi_{+-}|C_-|^{-1} & \Gamma \end{bmatrix},$$

for $x > 0$, where $C = \text{diag}(\mu_i : i \in \mathcal{E}_+ \cup \mathcal{E}_-)$ is the matrix defined in (2.40), Ψ_{+-} is the transition matrix defined in (2.43),

$$K = C_+^{-1}T_{++} + \Psi_{+-}|C_-|^{-1}T_{-+}, \tag{6.13}$$

and

$$\Gamma = (C_+^{-1}Q_{+0} + \Psi_{+-}|C_-|^{-1}Q_{-0})(-Q_{00})^{-1}.$$

The level starts increasing at once if $\varphi(0) = i \in \mathcal{E}_+$ and $(e^{Kx})_{ij}$, for $j \in \mathcal{E}_+$, is the expected number of crossings of (x, j) during the interval of time $(0, t_1)$ until the first return to level 0.

If the initial phase is in $\mathcal{E}_-$, the level becomes negative straightaway, the Fluid flow cannot cross level $x > 0$ before time t_1, and $R_{-,\mathcal{E}}(x) = 0$.

If the initial phase is $i \in \mathcal{E}_0$, the expected local time $L_0[0, t_1])_{ii}$ is equal to the sojourn time $-Q_{ii}^{-1}$ in phase i, and the process may reach level $x > 0$ before time t_1 only if the phase jumps from i to some state in $\mathcal{E}_+$. Therefore,

$$R_{0,\mathcal{E}}(x) = \text{diag}(-Q_{ii} : i \in \mathcal{E}_0)\widetilde{P}_{0+}R_{+,\mathcal{E}}(x),$$

where the jump chain transition matrix $\widetilde{P}$ is defined in (1.30). Eventually, we find that

$$R(x) = \begin{bmatrix} C_+ \\ 0 \\ Q_{0+}C_+ \end{bmatrix} e^{Kx} \begin{bmatrix} C_+^{-1} & \Psi_{+-}|C_-|^{-1} & \Gamma \end{bmatrix}.$$

There is a similar story about the matrix $\widehat{R}$, but I shall not get into that.

I do not know of published results about the matrix functions $R(x)$ and $\widehat{R}(x)$ in the fully general case of Sect. 2.5, with $\sigma_i > 0$ for some phases and $\sigma_i = 0$ for others. If $\boldsymbol{\pi}^{\mathrm{T}}\boldsymbol{\mu} \neq 0$, it seems to me that one may rely on Theorem 4.8 to find a connection between the ladder processes and the expected local time away from level 0, via the equations

$$\widehat{G}(x)L_0[0, \infty] = L_0[0, \infty]R(x), \qquad G(x)L_0[0, \infty] = L_0[0, \infty]\widehat{R}(x),$$

and

$$L_0[0, \infty] = -\begin{bmatrix} \widehat{U}_{\sigma\sigma} + U_{\sigma\sigma} & \widehat{U}_{\sigma+} & U_{\sigma-} & 0 \\ \Psi_{+\sigma} & -I & \Psi_{+-} & 0 \\ \widehat{\Psi}_{-\sigma} & \widehat{\Psi}_{-+} & -I & 0 \\ Q_{0\sigma} & Q_{0+} & Q_{0-} & Q_{00} \end{bmatrix}^{-1} \begin{bmatrix} 2\Delta_\sigma^{-2} & & & \\ & C_+^{-1} & & \\ & & |C_-|^{-1} & \\ & & & I \end{bmatrix}.$$

This equation for $L_0[0, \infty]$ results from Breuer [47, Theorem 1], it is a generalisation of (4.14).

In the special case when both $\mathcal{E}_\sigma$ and $\mathcal{E}_-$ are not empty, while both $\mathcal{E}_+$ and $\mathcal{E}_0$ are empty, it is shown in Ivanovs et al. [91, Section 5.3] that there exists a matrix $\widehat{R}_{\mathrm{dd}}$, indexed by $\mathcal{E}_\sigma \cup \mathcal{E}_-$ such that $\widehat{R}(x) = e^{\widehat{R}_{\mathrm{dd}} x}$, with $\widehat{R}_{\mathrm{dd}}$ being the solution of the matrix-quadratic equation

$$X^2 \begin{bmatrix} \frac{1}{2}\Delta_\sigma^2 & \\ & 0 \end{bmatrix} + X \begin{bmatrix} \Delta_\mu & \\ & C_- \end{bmatrix} + Q = 0.$$

There is, however, no matrix-exponential expression for $R(x)$. The situation is reversed if $\mathcal{E}_+$ is not empty and $\mathcal{E}_- = \mathcal{E}_0 = \emptyset$.

The illustrations I give in this book are mostly about queueing systems (*stochastic flow systems* to borrow Harrison's terminology) because these are the applications I have been most familiar with, but it should be obvious that things do not stop there, and that Markov modulation is a useful modelling tool for applications from finance, risk, and insurance theory. Elliott et al. [63] give a good overview of the literature and applications, Duprey and Klaus [62] argue that Markov switching models do a good job when applied to financial systems, and Weron [144] discusses their strengths and weaknesses for electricity price and production forecasting. These authors generally emphasise the necessity of including jumps in the trajectory of $\{X\}$, and this brings me to Markov-modulated Lévy processes.

Phase-type (PH) random variables may be modelled as the time to absorption of a finite state Markov chain with one absorbing state. In consequence, if the jump distributions of a Markov-modulated Lévy process are of PH type, then one may combine their Markov representation with the set $\mathcal{E}$ of the Markov switching mechanism, and transform the Markov-modulated Lévy process into a general MMBM as discussed in the preceding paragraphs. Asmussen [15], Breuer [44], Deelstra et al. [58], and Jiang and Pistorius [93] are examples of this approach. Among the benefits from using PH distributions, I shall mention that they are dense in the set of all distributions on the real line (Ahn and Ramaswami [6]). One drawback is that their density decays asymptotically geometrically, and there are circumstances when one wishes random variables to be heavy tailed.

When the distribution of jumps is left unspecified, it is often assumed that jumps are restricted to one direction only, all positive as in Ivanovs et al. [91], or all negative as in D'Auria et al. [55]. There is much similarity for these models with Markov chains of the M/G/1-type (jumps in the positive direction) or with GI/M/1-types (negative jumps). Such block-structured Markov chains have been the focus of extensive attention for many years (see Neuts [116, 118] for early work on the topic), and this gives me confidence that Markov-modulated Lévy processes are fertile ground for the application of Matrix-analytic methods.

Guy Latouche
Bruxelles, May 2025

Glossary and Notation

Processes

$\{B(t) : t \in \mathbb{R}^+\}$ written as $\mathcal{B}$ or $\{B\}$ for short: standard Brownian motion.

$\{\varphi(t) : t \in \mathbb{R}^+\}$ written as $\{\varphi\}$ for short: phase process, continuous-time Markov chain with state space $\mathcal{E} = \{1, \ldots, m\}$, $m < \infty$. Important subsets are:

$\mathcal{E}_\sigma = \{i : \sigma_i > 0\}$.
$\mathcal{E}_+ = \{i : \sigma_i = 0, \mu_i > 0\}$.
$\mathcal{E}_- = \{i : \sigma_i = 0, \mu_i < 0\}$.
$\mathcal{E}_0 = \{i : \sigma_i = 0, \mu_i = 0\}$.

$\{X(t), \varphi(t) : t \in \mathbb{R}^+\}$ written as $\{X, \varphi\}$ or MMBM for short: Markov-modulated Brownian motion.

$\{M_X\}$ running maximum.

$\{m_X\}$ running minimum.

$\{Y(t), \varphi(t) : t \in \mathbb{R}^+\}$ written as $\{Y, \varphi\}$ for short: Markov-modulated Brownian motion regulated at levels 0 and b.

In Sect. 5.4, the notation $\{Y, \kappa\}$ is used for the MMBM with sticky boundary.

$\{Z(t), \varphi(t) : t \in \mathbb{R}^+\}$ written as $\{Z, \varphi\}$ for short: Markov-modulated Brownian motion regulated at level 0.
$S(\cdot)$, $S_0(\cdot)$, $S_b(\cdot)$: Skorokhod regulators.
ℓ: $\lim_{t\to\infty} S(t)/t$.

$\{X_\varepsilon(t), \beta_\varepsilon(t), \varphi_\varepsilon(t) : t \geq 0\}$ in Sect. 1.11: family of fast-switching fluid processes, parameterised with $\varepsilon > 0$.
$\{\beta_\varepsilon, \varphi_\varepsilon\}$ is a two-dimensional Markov process on the state space $\{1, 2\} \times \mathcal{E}$.

Random Variables

$\delta_x = \inf\{t > 0 : Z(t) = x\}$ first passage time to level x for the MMBM regulated at 0.

© The Author(s), under exclusive license to Springer Nature Switzerland AG 2026

G. Latouche, *Markov-Modulated Brownian Motion*, Springer Series in Operations Research and Financial Engineering, https://doi.org/10.1007/978-3-031-98576-8

$\boldsymbol{\ell_i(x, t, \omega)}$ local time at x in phase i over the interval of time $[0, t]$.
$L_x[y, t]$: expected local time. $(L_x[y, t])_{ij} = \mathrm{E}_{x,i}[\ell_j(y, t)]$.
$L_{\mathrm{r},x}[y, t]$: expected local time at level y, starting from x during $[0, t]$ for the regulated process $\{Z, \varphi\}$.

$\boldsymbol{\{t_n^*\}}$ defined in (4.7): successive visits to level 0 separated by at least one change of phase.
P^*: transition probability matrix for $\{t_n^*\}$.

$\boldsymbol{\tau_x = \inf\{t > 0 : X(t) = x\}}$ first passage time to level x for the MMBM.

$\boldsymbol{\theta_n, \theta_n'}$ in Sect. 3.1: alternating visits to level 0 and another level x.

$\boldsymbol{\{\theta_n^*\}}$ in Sect. 5.1: Markov-regenerative epochs for the regulated process $\{Z, \varphi\}$.
Φ: transition matrix for $\{\varphi(\theta_n^*)\}$.
ρ: stationary probability vector of Φ.

$\boldsymbol{\theta^*}$ in Sect. 5.2: first return to the starting boundary after the first phase jump.

$\boldsymbol{\{\widetilde{\theta}_n\}}$ in Sect. 5.2: Markov-regenerative epochs for the regulated process $\{Y, \varphi\}$.
$\widetilde{\Phi}$: transition matrix for $\{\varphi(\theta_n)\}$.
$\widetilde{\rho}$: stationary probability vector of $\widetilde{\Phi}$.

$\boldsymbol{\{\overline{\theta}_n\}}$ in Sect. 5.3: Markov-regenerative epochs for the regulated process with change of regime at 0.
$\overline{\Phi}$: transition matrix for $\{\varphi(\widetilde{\theta}_n)\}$.
$\overline{\rho}$: stationary probability vector of $\overline{\Phi}$.

$\boldsymbol{\zeta_n, \eta_n}$ defined in (4.26), 4.27): alternating visits to x and to 0.
ζ_n', $\widehat{\eta}_n$, see (4.58): alternating visits to $-x$ and to 0.

Matrices and Vectors

Vectors are identified by boldface letters, they are column vectors, with the exception of the vector $\mathbf{0}$ with all components equal to zero, which is row or column depending on the circumstances.

Matrices of the same size, as well as vectors, are equipped with a partial order: $A - B < 0$ means that $A_{ij} - B_{ij} < 0$ for all components, the same rule applies to "$>$", "$\leq$", and "$\geq$".

Occasionally, components equal to 0 in a displayed matrix are replaced with a "$\cdot$" or even with a blank space, so that the structure of the non-zero elements is better recognised.

$\boldsymbol{\Delta_x = \mathrm{diag}(x)}$ diagonal matrix with vector $\boldsymbol{x}$ on the diagonal.

$\boldsymbol{E_n}$ difference between U and its approximation U_n.

$\boldsymbol{e_i}$ vector with ith component equal to 1, remaining components equal to 0.

$\mathcal{F}(\cdot)$ matrix Laplace exponent of the MMBM.

$\boldsymbol{G(x)}$ stationary distribution function of the regulated process $\{Z, \varphi\}$.

$\boldsymbol{H(x)}$ transition probability matrix from level 0 to level $-x$, $x > 0$.
$\widehat{H}(x)$: transition probability matrix from level 0 to level x, $x > 0$.
$H_n(x)$: transition probability matrix from level x to level 0 with stack size at most n.

$\boldsymbol{I}$ identity matrix.

$\boldsymbol{H}^{(x)}$ transition probability matrix from 0 to x for the MMBM with lower control at 0.

$\widetilde{H}^{(x)}$ transition probability matrix from x to 0 for the MMBM with upper control at x.

$\boldsymbol{M}^{\#}$ group inverse of the matrix M, if it exists.

$\boldsymbol{M_{a,b}}$ for a matrix M, submatrix with rows indexed by states in $\mathcal{E}_a$, and columns by states in $\mathcal{E}_b$.

$\boldsymbol{M_{\mathrm{r}}(x)}$ matrix of expected occupation time in $[0, x]$ during a regeneration interval for $\{Z, \varphi\}$.

$\overline{M}(x)$: matrix of expected occupation time for the process with change of regime.

$\boldsymbol{N_{\varepsilon}}$ number of crossings of level 0 for the fast-switching process.

$\boldsymbol{\mu = [\mu_i : i \in \mathcal{E}]}$ mean drift parameters of the MMBM.

$\boldsymbol{P_0}$ **and** $\boldsymbol{P_b}$ in Sect. 5.3: phase transition matrices upon change of regime.

$\boldsymbol{\pi = [\pi_i : i \in \mathcal{E}]}$ stationary probability vector of the phase.

$\boldsymbol{\Psi^{(a,b)}}$ **and** $\boldsymbol{\widehat{\Psi}^{(a,b)}}$ transition probability matrices for the phase process, upon hitting level $-a$ or b, respectively, after starting at 0.

$\boldsymbol{Q}$ generator of the Markov process $\{\varphi\}$.

Q_{ε} generator of the fast-switching approximation.

$\boldsymbol{R}$ **and** $\boldsymbol{\widehat{R}}$ generators of e^{Rx} and $e^{\widehat{R}x}$, respectively.

$(e^{Rx})_{ij}$: expected local time in (x, j) per unit of local time in $(0, i)$.

$(e^{\widehat{R}x})_{ij}$: expected local time in $(-x, j)$ per unit of local time in $(0, i)$.

$\boldsymbol{\rho(M)}$ spectrum of a matrix M.

$\boldsymbol{\sigma = [\sigma_i : i \in \mathcal{E}]}$ volatility parameters of the MMBM.

$\boldsymbol{U}$ **and** $\boldsymbol{\widehat{U}}$ generators of the ladder processes, for lower and higher levels, respectively.

$\boldsymbol{u}$ and $\widehat{\boldsymbol{u}}$: left eigenvectors for the eigenvalue closest to 0.

$\boldsymbol{v}$ and $\widehat{\boldsymbol{v}}$: right eigenvectors for the eigenvalue closest to 0.

vec(·) vectorisation operator for the columns of a matrix.

$\boldsymbol{W_x}$ **and** $\boldsymbol{\widehat{W}_x}$ defined in (3.7), factors to express exit probabilities from an interval.

W^*_x: scale matrix.

Miscellanea

0 column or row vector with all components equal to 0.

1 column vector with all components equal to 1.

a.s. almost surely, with probability one.

$\boldsymbol{\alpha_-((\mu, \sigma, \lambda)}$ **and** $\boldsymbol{\alpha_+(\mu, \sigma, \lambda)}$ in Lemma 2.6: parameters of $M(T)$ and $M(T) - X(T)$ of a Brownian motion (μ, σ), where T is exponential with parameter λ, also parameters of $X(T) - m(T)$ and $|m(T)|$.

$\boldsymbol{b}$ upper boundary for finite buffer.

$\mathbb{C}_<$, $\mathbb{C}_\leq$, $\mathbb{C}_\geq$, **and** $\mathbb{C}_>$ defined in Remark 2.5: subsets of complex numbers.

$\boldsymbol{C(t, h)}$ covariance between $X(t)$ and $X(t + h) - X(t)$.

$\boldsymbol{\frac{\partial}{\partial x} f(x)\big|_{x \downarrow 0}}$ short for $\lim_{x \overset{\geq}{\to} 0} \frac{\partial}{\partial x} f(x)$. See Eq. (4.43).

$\mathbf{E}_{x,i}[\cdot]$ short for $\mathrm{E}[\,\cdot\,|X(0)=x, \varphi(0)=i]$,
$\mathrm{E}_{x,\alpha}[\cdot]$ if $\varphi(0)$ has the distribution $\boldsymbol{\alpha}$.
$\mathrm{E}[\,\cdot\,;\varphi(t)]$ matrix of order m with elements

$$(\mathrm{E}[\,\cdot\,;\varphi(t)])_{ij} = \mathrm{E}[\,\cdot\,\mathbb{1}[\varphi(t)=j]|X(0)=0, \varphi(0)=i], \qquad i,j \in \mathcal{E},$$

$\mathrm{E}_{x,\varphi(0)}[\,\cdot\,;\varphi(t)]$ matrix with entries

$$(\mathrm{E}_{x,\varphi(0)}[\,\cdot\,;\varphi(t)])_{ij} = \mathrm{E}[\,\cdot\,\mathbb{1}[\varphi(t)=j]|X(0)=x, \varphi(0)=i].$$

$\boldsymbol{\eta}$ maximal eigenvalue of e^R, tail parameter of the stationary distribution.
i.i.d. independent identically distributed.
$\boldsymbol{\pi}^{\mathrm{T}}\boldsymbol{\mu}$ stationary drift of the MMBM.
$\boldsymbol{\mathcal{N}(\mu,\sigma^2)}$ Gaussian distribution with parameters μ and σ^2.
$\boldsymbol{\lambda_i : 1 \le i \le 2m}$ roots of $\det(\mathcal{F}(z))$, eigenvalues of U, $-\widehat{U}$, $-R$, and $\widehat{R}$.
$\boldsymbol{V_{a,b}(v,y)}$ see (2.17).
$\boldsymbol{\zeta}$ virtual stack associated with the functional iteration algorithm.

References

1. J. Abate, W. Whitt, Numerical inversion of Laplace transforms of probability distributions. ORSA J. Comput. **7**, 36–43 (1995)
2. V.M. Adukov, N.V. Adukova, G. Mishuris, An explicit Wiener-Hopf factorization algorithm for matrix polynomials and its exact realization within ExactMPF package. Proc. R. Soc. A **478**, 20210941 (2022). https://doi.org/10.1098/rspa.2021.0941
3. S. Ahn, Total shift during the first passages of Markov modulated Brownian motion with bilateral PH-type jumps: formulas driven by the minimal solution matrix of a Riccati equation. Stoch. Models **32**, 433–459 (2016)
4. S. Ahn, Time-dependent and stationary analyses of the two-sided reflected Markov modulated Brownian motion with bilateral PH-type jumps. J. Korean Stat. Soc., **46**, 45–69 (2017). https://doi.org/10.1016/j.jkss.2016.06.002
5. S. Ahn, V. Ramaswami, Fluid flow models and queues: a connection by stochastic coupling. Stoch. Models **19**, 325–348 (2003)
6. S. Ahn, V. Ramaswami, Bilateral phase type distributions. Stoch. Models **21**, 239–259 (2005)
7. S. Ahn, V. Ramaswami, Duality results for Markov-modulated fluid flow models. J. Appl. Probab. **48**, 309–318 (2011). https://doi.org/10.1239/jap/1318940473.
8. S. Ahn, V. Ramaswami, A quadratically convergent algorithm for first passage time distributions in the Markov-modulated Brownian motion. Stoch. Models **33**, 59–96 (2017). https://doi.org/10.1080/15326349.2016.1211018
9. O.F. Akbal, Regime-switching factor models and nowcasting with big data. Technical Report WP/24/190, International Monetary Fund Working Paper, 2024
10. A.S. Alfa, J. Xue, Q. Ye, Accurate computation of the smallest eigenvalue of a diagonally dominant M-matrix. Math. Comp. **71**, 217–236 (2002)
11. S.-D. Almousa, G. Horvàth, M. Telek, Parameter estimation of Markov modulated fluid arrival processes. Perf. Eval. **157–158**, 102316 (2022)

© The Author(s), under exclusive license to Springer Nature Switzerland AG 2026
G. Latouche, *Markov-Modulated Brownian Motion*, Springer Series in Operations Research and Financial Engineering, https://doi.org/10.1007/978-3-031-98576-8

12. D. Anick, D. Mitra, M.M. Sondhi, Stochastic theory of a data-handling system with multiple sources. Bell Syst. Tech. J. **61**, 1871–1894 (1982)
13. S. Asmussen, Stationary distributions for fluid flow models with or without Brownian noise. Commun. Stat. Stoch. Models **11**(1), 21–49 (1995)
14. S. Asmussen, *Applied Probability and Queues*, 2nd edn. (Springer, New York, 2003)
15. S. Asmussen, Lévy processes, phase-type distributions, and martingales. Stoch. Models **30**, 443–468 (2014). https://doi.org/10.1080/15326349.2014.958424
16. S. Asmussen, O. Kella, A multi-dimensional martingale for Markov additive processes and its applications. Adv. Appl. Probab. **32**, 376–393 (2000)
17. S. Asmussen, M. Pihlsgård, Loss rates for Lévy processes with two reflecting barriers. Math. Oper. Res. **32**, 308–321 (2007)
18. S. Asmussen, V. Ramaswami, Probabilistic interpretations of some duality results for the matrix paradigms in queueing theory. Stoch. Models **6**, 715–733 (1990)
19. S. Asmussen, O. Nerman, M. Olsson, Fitting phase-type distributions via the EM algorithm. Scand. J. Stat. **23**, 419–441 (1996)
20. S. Asmussen, F. Avram, M.R. Pistorius, Russian and American put options under exponential phase-type Lévy models. Stoch. Process. Appl. **109**, 79–111 (2004)
21. F. Avram, A.E. Kyprianou, M.R. Pistorius, Exit problems for spectrally negative Lévy processes and applications to (Canadized) Russian options. Ann. Appl. Probab. **14**, 215–238 (2004)
22. M.T. Barlow, L.C.G. Rogers, D. Williams, Wiener-Hop factorization for matrices, in *Séminaire de Probabilités XIV*. Lecture Notes in Mathematics, vol. 784 (Springer, Berlin, 1980), pp. 324–331
23. R. Bartels, G. Stewart, Solution of the matrix equation $AX + XB = C$. Comm. ACM **15**, 820–826 (1972)
24. N.G. Bean, M.M. O'Reilly, Performance measures of a multi-layer Markovian fluid model. Ann. Oper. Res. **160**, 99–120 (2008)
25. N.G. Bean, M.M. O'Reilly, Spatially-coherent uniformization of a stochastic fluid model to a quasi-birth-and-death process. Perf. Eval. **70**, 578–592 (2013)
26. N.G. Bean, M.M. O'Reilly, The stochastic fluid-fluid model: a stochastic fluid model driven by an uncountable-state process, which is a stochastic fluid model itself. Stoch. Proc. Appl. **124**, 1741–1772 (2014)
27. N.G. Bean, M.M. O'Reilly, P.G. Taylor, Algorithms for return probabilities for stochastic fluid flows. Stoch. Models **21**(1), 149–184 (2005)
28. N.G. Bean, M.M. O'Reilly, P.G. Taylor, Hitting probabilities and hitting times for stochastic fluid flows. Stoch. Proc. Appl. **115**, 1530–1556 (2005)
29. N.G. Bean, M.M. O'Reilly, J.E. Sargison, A stochastic fluid flow model of the operation and maintenance of power generation systems. IEEE Trans. Power Syst. **25**, 1361–1374 (2010)
30. N.G. Bean, G. Latouche, P.G. Taylor, Physical interpretations for quasi-birth-and-death process algorithms. Queueing Models Serv. Manag. **1**(2), 59–78 (2018)

31. A. Berman, R.J. Plemmons, *Nonnegative Matrices in the Mathematical Sciences*. Classics in Applied Mathematics, vol. 9 (SIAM, Philadelphia, 1994). Revised Reprint of the 1979 original
32. J. Bertoin, *Lévy Processes* (Cambridge University Press, Cambridge, 1996)
33. P. Billingsley, *Convergence of Probability Measures*, 2nd edn. (Wiley, New York, 1999)
34. D.A. Bini, B. Meini, On cyclic reduction applied to a class of Toeplitz-like matrices arising in queueing problems, in ed. by W.J. Stewart, *Computations with Markov Chains* (Kluwer Academic Publishers, Boston, 1995), pp. 21–38
35. D.A. Bini, G. Latouche, B. Meini, Solving matrix polynomial equations arising in queueing problems. Linear Algebra Appl. **340**, 225–244 (2002)
36. D.A. Bini, G. Latouche, B. Meini, *Numerical Methods for Structured Markov Chains*. Numerical Mathematics and Scientific Computation (Oxford University Press, Oxford, 2005)
37. D.A. Bini, B. Meini, S. Steffé, B.V. Houdt, Structured Markov chains solver: software tools, in *SMCtools '06: Proceeding from the 2006 workshop on Tools for solving structured Markov chains* (ACM, New York, 2006), pp. 14
38. D.A. Bini, B. Meini, V. Ramaswami, A probabilistic interpretation of cyclic reduction and its relationships with logarithmic reduction. Calcolo **45**, 207–216 (2008). https://doi.org/10.1007/s10092-008-0151-6
39. D.A. Bini, B. Meini, S. Steffè, SMCSolver (structured Markov chain solver) 2.1 (2009). http://bezout.dm.unipi.it/SMCSolver/
40. D.A. Bini, B. Iannazzo, B. Meini, *Numerical Solution of Algebraic Riccati Equations*. Number 9 in Fundamentals of Algorithms (SIAM, Philadelphia PA, 2012)
41. D.A. Bini, G. Latouche, B. Meini. Shift techniques for Quasi-Birth-and-Death processes: Canonical factorizations and matrix equations. Appl. Num. Math. **116**, 24–36 (2017). https://doi.org/10.1016/j.apnum.2016.09.001
42. M. Bražėnas, G. Horváth, M. Telek, Efficient implementations of the EM-algorithm for transient Markovian arrival processes, in ed. by S. Wittevrongel, T. Phung-Duc, *Analytical and Stochastic Modelling Techniques and Applications* (Springer International Publishing, Cham, 2016), pp. 107–122
43. L. Breiman, *Probability* (SIAM, Philadelphia, 1992)
44. L. Breuer, First passage times for Markov additive processes with positive jumps of Phase-Type. J. Appl. Probab. **45**, 778–799 (2008)
45. L. Breuer, Exit problems for reflected Markov-modulated Brownian motion. J. Appl. Probab. **49**, 697–709 (2012)
46. L. Breuer, Occupation times for Markov-modulated Brownian motion. J. Appl. Probab. **49**, 549–565 (2012)
47. L. Breuer, The resolvent and expected local times for Markov-modulated Brownian motion with phase dependent termination rates. J. Appl. Probab. **50**, 430–438 (2013)
48. S.L. Campbell, C.D. Meyer, *Generalized Inverses of Linear Transformations* (Dover Publications, New York, 1991). Republication

49. E. Çinlar, *Introduction to Stochastic Processes* (Prentice–Hall, Englewood Cliffs, 1975)
50. J. Chevallier, S. Goutte, On the estimation of regime-switching Lévy models. Stud. Nonlinear Dyn. Econ. **21**, 3–29 (2017)
51. K.L. Chung, *A Course in Probability Theory* (Harcourt, Brace and World, New York, 1968)
52. P. Coolen-Schrijner, E.A. van Doorn, The deviation matrix of a continuous-time Markov chain. Probab. Eng. Inf. Sci. **16**, 351–366 (2002)
53. A. da Silva Soares, G. Latouche, Further results on the similarity between fluid queues and QBDs, in ed. by G. Latouche, P. Taylor, *Matrix-Analytic Methods. Theory and Applications* (World Scientific, Singapore, 2002), pp. 89–106
54. A. da Silva Soares, G. Latouche, Matrix-analytic methods for fluid queues with finite buffers. Perf. Eval. **63**, 295–314 (2006)
55. B. D'Auria, J. Ivanovs, O. Kella, M. Mandjes, First passage of a Markov additive process and generalized Jordan chains. J. Appl. Probab. **47**(4), 899–1208 (2010)
56. B. D'Auria, J. Ivanovs, O. Kella, M. Mandjes, First passage process of a Markov additive process with applications to reflection problems. ArXiv e-prints (2010). https://doi.org/10.48550/arXiv.1006.2965
57. B. D'Auria, J. Ivanovs, O. Kella, M. Mandjes, Two-sided reflection of Markov-modulated Brownian motion. Stoch. Models **28**, 316–332 (2012). https://doi.org/10.1080/15326349.2012.672285.
58. G. Deelstra, G. Latouche, M. Simon, On barrier option pricing by Erlangization in a regime-switching model with jumps. J. Comput. Appl. Math. **371**, 112606 (2020). https://doi.org/10.1016/j.cam.2019.112606
59. E. Deiana, G. Latouche, M.-A. Remiche, Fluid flow model for energy-aware server performance evaluation. Methodol. Comput. Appl. Probab. **23**, 801–821 (2021). https://doi.org/10.1007/s11009-020-09784-z.
60. A.P. Dempster, N.M. Laird, D.B. Rubin, Maximum likelihood from incomplete data via the EM algorithm. J. R. Stat. Soc. Ser. B **39**, 1–22 (1977)
61. S. Dendievel, G. Latouche, Y. Liu, Y. Tang, Singularly perturbed Markov modulated fluid queue. SIAM J. Matrix Anal. Appl. **43**, 377–404 (2022). https://doi.org/10.1137/21M1395387
62. T. Duprey, B. Klaus, How to predict financial stress? an assessment of Markov switching models. *ERN: Forecasting Techniques (Topic)* (2017). https://api.semanticscholar.org/CorpusID:52242084
63. R.J. Elliott, L. Chan, T.K. Siu, Option pricing and Esscher transform under regime switching. Ann. Finance **1**, 423–432 (2005)
64. Y. Ephraim, W.J.J. Roberts, An EM algorithm for Markov modulated Markov processes. IEEE Trans. Signal Process. **57**, 463–470 (2009)
65. R.V. Evans, Geometric distribution in some two-dimensional queueing systems. Oper. Res. **15**, 830–846 (1967)
66. D. Freedman, *Brownian Motion and Diffusion*, 2nd edn. (Springer, New York, 1983)

67. F. Gantmacher, *The Theory of Matrices*, vol. 1 (AMS Chelsea Publishing, New York, 1959)
68. S. Goldstein, On diffusion by discontinuous movements, and on the telegraph equation. Quart. J. Mech. Appl. Math. **4**, 129–156 (1951)
69. G. Golub, S. Nash, C. Van Loan, A Hessenberg-Schur method for the problem $AX + XB = C$. IEEE Trans. Auto. Contr. **AC-24**, 909–913 (1979)
70. M. Govorun, G. Latouche, M.-A. Remiche, Stability for fluid queues: characteristic inequalities. Stoch. Models **29**(1), 64–88 (2013)
71. A. Graham, *Kronecker Products and Matrix Calculus: With Applications* (Wiley, New York, 1981)
72. W.K. Grassmann, M.J. Taksar, D.P. Heyman, Regenerative analysis and steady state distributions for Markov chains. Oper. Res. **33**, 1107–1116 (1985)
73. M. Gribaudo, D. Manini, B. Sericola, M. Telek, Second order fluid models with general boundary behaviour. Ann. Oper. Res. **160**, 69–82 (2008)
74. C.-H. Guo, On algebraic Riccati equations associated with M-matrices. Linear Algebra Appl. **439**, 2800–2814 (2013)
75. C.-H. Guo, N.J. Higham, Iterative solution of a nonsymmetric algebraic Riccati equation. SIAM J. Matrix Anal. Appl. **29**, 396–412 (2007)
76. C.-H. Guo, A.J. Laub, On the iterative solution of a class of nonsymmetric algebraic Riccati equations. SIAM J. Matrix Anal. Appl. **22**, 376–391 (2000)
77. X.-X. Guo, W.-W. Lin, S.-F. Xu, A structure-preserving doubling algorithm for nonsymmetric algebraic Riccati equation. Numer. Math. **103**, 393–412 (2006)
78. C.-H. Guo, B. Iannazzo, B. Meini, On the doubling algorithm for a (shifted) nonsymmetric algebraic Riccati equation. SIAM J. Matrix Anal. Appl. **29**, 1083–1100 (2007)
79. B. Hajek, Birth-and-death processes on the integers with phases and general boundaries. J. Appl. Probab. **19**, 488–499 (1982)
80. J.M. Harrison, *Brownian Motion and Stochastic Flow Systems*. Wiley Series on Probability and Mathematical Statistics (Wiley, New York, 1985)
81. J.M. Harrison, A.J. Lemoine, Sticky Brownian motion as the limit of storage processes. J. Appl. Probab. **18**, 216–226 (1981)
82. Q.-M. He, *Fundamentals of Matrix-Analytic Methods* (Springer New York, 2014). https://doi.org/10.1007/978-1-4614-7330-5
83. C. He, B. Meini, N.H. Rhee, A shifted cyclic reduction algorithm for Quasi-Birth-Death problems. SIAM J. Matrix Anal. Appl. **23**, 673–691 (2001/02)
84. N.J. Higham, *Functions of Matrices: Theory and Computation* (SIAM, Philadelphia, 2008)
85. G. Horváth, M. Telek, Matrix-analytic solution of infinite, finite and level-dependent second-order fluid models. Queueing Syst. **87**, 325–343 (2017)
86. I. Horváth, A. Mészáros, M. Telek, Numerical inverse Laplace transformation beyond the Abate-Whitt framework. J. Comput. Appl. Math. **418**, 114651 (2023). https://doi.org/10.1016/j.cam.2022.114651
87. J. Ivanovs, Markov-modulated Brownian motion with two reflecting barriers. J. Appl. Probab. **47**, 1034–1047 (2010)

88. J. Ivanovs, One-sided Markov additive processes and related exit problems. PhD Thesis, Universiteit van Amsterdam, 2011
89. J. Ivanovs, Potential measures of one-sided Markov additive processes with reflecting and terminating barriers. J. Appl. Probab. **51**(4), 1154–1170 (2014)
90. J. Ivanovs, Z. Palmowski, Occupation densities in solving exit problems for Markov additive processes and their reflections. Stoch. Process. Appl. **122**, 3342–3360 (2012)
91. J. Ivanovs, G. Latouche, P. Taylor, One-sided Markov additive processes with lattice and non-lattice increments. Stoch. Process. Appl. **190**, (2025). https://doi.org/10.1016/j.spa.2025.104771
92. M. Jamshidian, R.I. Jennrich, Acceleration of the EM algorithm by using quasi-Newton methods. J. R. Stat. Soc. Ser. B (Stat. Methodol.) **59**, 569–587 (1997)
93. Z. Jiang, M. Pistorius, On perpetual American put valuation and first-passage in a regime-switching model with jumps. Finance Stoch. **12**, 331–355 (2008). https://doi.org/10.1007/s00780-008-0065-9
94. M. Kac, A stochastic model related to the telegrapher's equation. Rocky Mountain J. Math. **4**, 497–510 (1974)
95. R.L. Karandikar, V. Kulkarni, Second-order fluid flow models: Reflected Brownian motion in a random environment. Oper. Res. **43**, 77–88 (1995)
96. P. Klusik, Z. Palmowski, A note on Wiener-Hopf factoriozation for Markov additive processes. J. Theor. Probab. **27**, 202–219 (2014)
97. L. Kruk, J. Lehoczky, K. Ramanan, S. Shreve, An explicit formula for the Skorokhod map on $[0, a]$. Ann. Probab. **35**, 1740–1768 (2007)
98. A.E. Kyprianou, *Introductory Lectures on Fluctuations of Lévy Processes with Applications*. Universitext (Springer, Berlin, 2006)
99. P. Lancaster, M. Tismenetsky, *The Theory of Matrices: With Applications* (Academic, New York, 1985)
100. G. Latouche, G.T. Nguyen, Fluid approach to two-sided Markov-modulated Brownian motion. Queueing Syst. **80**, 105–125 (2015). https://doi.org/10.1007/s11134-014-9432-8
101. G. Latouche, G.T. Nguyen, The morphing of fluid queues into Markov-modulated Brownian motion. Stoch. Syst. **5**, 62–86 (2015). https://doi.org/10.1214/13-SSY133
102. G. Latouche, G.T. Nguyen, Feedback control: Markov-modulated Brownian motion with instantaneous change of phase. Perf. Eval. **106**, 30–49 (2016). https://doi.org/10.1016/j.peva.2016.09.004
103. G. Latouche, G.T. Nguyen, Slowing time: Markov-modulated Brownian motion with a sticky boundary. Stoch. Models **33**, 297–321 (2017). https://doi.org/10.1080/15326349.2017.1284000
104. G. Latouche, G.T. Nguyen, Analysis of fluid flow models. Queueing Models Serv. Manag. **1**(2), 1–29 (2018)
105. G. Latouche, V. Ramaswami, A logarithmic reduction algorithm for quasi-birth-and-death processes. J. Appl. Probab. **30**, 650–674 (1993)

106. G. Latouche, V. Ramaswami, *Introduction to Matrix Analytic Methods in Stochastic Modeling*. ASA-SIAM Series on Statistics and Applied Probability (SIAM, Philadelphia PA, 1999)
107. G. Latouche, M. Simon, Markov-modulated Brownian motion with temporary change of regime at level zero. Methodo. Comput. Appl. Probab. **20**, 1199–1222 (2018). https://doi.org/10.1007/s11009-017-9602-3
108. G. Latouche, P. Taylor, First passage probability and occupation time for null recurrent skip-free Markov additive processes. *Work in progress* (2024)
109. G. Latouche, C. Pearce, P. Taylor, Invariant measures for quasi-birth-and-death processes. Stoch. Models **14**, 443–460 (1998)
110. R.M. Loynes, A continuous-time treatment of certain queues and infinite dams. J. Austral. Math. Soc. **2**, 484–498 (1962)
111. S. Mallat, *A Wavelet Tour of Signal Processing*, 2nd edn. (Academic, New York, 1999)
112. P. Mathews, Erlangized fluid methods for evaluating stochastic integrals. Master's Thesis, The University of Adelaide, Australia, 2014
113. B. Meini, New convergence results on functional iteration techniques for the numerical solution of M/G/1 type Markov chains. Numer. Math. **78**, 39–58 (1997)
114. Y. Meyer, *Wavelets and Operators. Advanced Mathematics* (Cambridge University Press, Cambridge, 1992)
115. M.F. Neuts, Markov chains with applications in queueing theory, which have a matrix-geometric invariant vector. Adv. Appl. Probab. **10**, 185–212 (1978)
116. M.F. Neuts, *Matrix-Geometric Solutions in Stochastic Models: An Algorithmic Approach* (The Johns Hopkins University Press, Baltimore, 1981)
117. M.F. Neuts, The caudal characteristic curve of queues. Adv. Appl. Probab. **18**, 221–254 (1986)
118. M.F. Neuts, *Structured Stochastic Matrices of M/G/1 Type and Their Applications* (Dekker, New York, 1989)
119. G.T. Nguyen, O. Peralta, Rate of strong convergence to Markov-modulated Brownian motion. J. Appl. Probab. **59**, 1–16 (2022). https://doi.org/10.1017/jpr.2021.30
120. G.T. Nguyen, F. Poloni, Componentwise accurate fluid queue computations using doubling algorithms. Numer. Math. **130**, 763–792 (2015)
121. G.T. Nguyen, F. Poloni, Componentwise accurate Brownian motion computations using cyclic reduction, 2016. https://doi.org/10.48550/arXiv.1605.01482
122. J.R. Norris, *Markov Chains*. Cambridge Series in Statistical and Probabilistic Mathematics (Cambridge University Press, Cambridge, 1997)
123. C.A. O'Cinneide, Entrywise perturbation theory and error analysis for Markov chains. Numer. Math. **65**, 109–120 (1993)
124. J.M. Ortega, W.C. Rheinboldt, *Iterative Solution of Nonlinear Equations in Several Variables*. Classics in Applied Mathematics (SIAM, Philadelphia, 2000)

125. A. Papapantoleon, An introduction to Lévy processes with applications in finance. Lecture notes, TU Vienna (2008). https://doi.org/10.48550/arXiv.0804.0482
126. M. Pinsky, Differential equations with a small parameter and the central limit theorem for functions defined on a finite Markov chain. Probab. Theory Related Fields **9**, 101–111 (1968)
127. M. Pistorius, On maxima and ladder processes for a dense class of Lévy processes. J. Appl. Prob. **43**, 208–220 (2006)
128. R.J. Plemmons, M-Matrix characterizations. I — nonsingular M-matrices. Linear Algebra Appl. **18**, 175–188 (1977)
129. V. Ramaswami, A duality theorem for the matrix paradigm in queueing theory. Stoch. Models **6**, 151–161 (1990)
130. V. Ramaswami, Matrix analytic methods for stochastic fluid flows, in ed. by D. Smith, P. Hey, *Teletraffic Engineering in a Competitive World (Proceedings of the 16th International Teletraffic Congress)* (Elsevier Science B.V., Edinburgh, 1999), pp. 1019–1030
131. V. Ramaswami, A fluid introduction to Brownian motion and stochastic integration, in ed. by G. Latouche, V. Ramaswami, J. Sethuraman, K. Sigman, M. Squillante, D. Yao, *Matrix-Analytic Methods in Stochastic Models*. Springer Proceedings in Mathematics & Statistics, vol. 27, chap. 10 (Springer Science, New York, 2013), pp. 209–225
132. S.I. Resnick, *Adventures in Stochastic Processes* (Birkhäuser Boston, Cambridge, 1992)
133. L.C.G. Rogers, Fluid models in queueing theory and Wiener-Hopf factorization of Markov chains. Ann. Appl. Probab. **4**, 390–413 (1994)
134. L.C.G. Rogers, Z. Shi, Computing the invariant law of a fluid model. J. Appl. Probab. **31**, 885–896 (1994)
135. K.-I. Sato, *Lévy Processes and Infinitely Divisible Distributions* (Cambridge University Press, Cambridge, 1999)
136. R.L. Schilling, L. Partzsch, *Brownian Motion: An Introduction to Stochastic Processes* (De Gruyter, Berlin, 2012). https://doi.org/10.1515/9783110278989
137. B. Sericola, *Markov Chains. Theory, Algorithms and Applications*. Applied Stochastic Methods Series (ISTE Ltd, London, 2013)
138. M. Simon, Markov-modulated Brownian motions perturbed by catastrophes. Stochastics **92**, 275–296 (2020). https://doi.org/10.1080/17442508.2019.1624753
139. D.A. van Dyk, Nesting EM algorithms for computational efficiency. Stat. Sinica **10**, 203–225 (2000)
140. S. Van Lierde, A. da Silva Soares, G. Latouche, Invariant measures for fluid queues. Stoch. Models **24**, 133–151 (2008)
141. R.S. Varga, *Matrix Iterative Analysis*. Springer Series in Computational Mathematics, vol. 27, Expanded edn. (Springer, Berlin, 2000)
142. V. Wallace, The Solution of Quasi Birth and Death Processes Arising from Multiple Access Computer Systems. PhD thesis, Systems Engineering Laboratory, University of Michigan, 1969

143. T. Watanabe, Approximation of uniform transport process on a finite interval to Brownian motion. Nagoya Math. J. **32**, 297–314 (1968)
144. R. Weron, *Modeling and Forecasting Electricity Loads and Prices: A Statistical Approach*. The Wiley Finance Series (Wiley, Chichester, 2006)
145. X. Zhou, Exit problems for spectrally negative Lévy processes reflected at either the supremum or the infimum. J. Appl. Prob. **44**, 1012–1030 (2007)

Index

© The Editor(s) (if applicable) and The Author(s), under exclusive license to Springer Nature Switzerland AG 2026
G. Latouche, *Markov-Modulated Brownian Motion*, Springer Series in Operations Research and Financial Engineering, https://doi.org/10.1007/978-3-031-98576-8

MIX
Papier aus verantwortungsvollen Quellen
Paper from responsible sources
FSC® C105338

If you have any concerns about our products,
you can contact us on
ProductSafety@springernature.com

In case Publisher is established outside the EU,
the EU authorized representative is:
Springer Nature Customer Service Center GmbH
Europaplatz 3, 69115 Heidelberg, Germany

Printed by Libri Plureos GmbH
in Hamburg, Germany